Student Solutions Manual
to Accompany

Technical Calculus

Fifth Edition

Dale Ewen
Parkland Community College

Joan S. Gary
Parkland Community College

James E. Trefzger
Parkland Community College

PEARSON
Prentice
Hall

Upper Saddle River, New Jersey
Columbus, Ohio

Editor in Chief: Stephen Helba
Senior Acquisitions Editor: Gary Bauer
Editorial Assistant: Natasha Holden
Media Development Editor: Michelle Churma
Production Editor: Louise N. Sette
Design Coordinator: Diane Ernsberger
Cover Designer: Eric Davis
Production Manager: Pat Tonneman
Marketing Manager: Leigh Ann Sims

Pearson Prentice Hall™ is a trademark of Pearson Education, Inc.
Pearson® is a registered trademark of Pearson plc
Prentice Hall® is a registered trademark of Pearson Education, Inc.

Pearson Education Ltd.
Pearson Education Singapore Pte. Ltd.
Pearson Education Canada, Ltd.
Pearson Education—Japan

Pearson Education Australia Pty. Limited
Pearson Education North Asia Ltd.
Pearson Educación de Mexico, S.A. de C.V.
Pearson Education Malaysia Pte. Ltd.

10 9 8 7 6 5 4
ISBN 0-13-118744-9

Contents

Solutions to Odd-Numbered Exercises

CHAPTER 1

Section 1.1

	Function	Domain	Range
1.	Yes	$\{2,3,9\}$	$\{2,4,7\}$
3.	No	$\{1,2,7\}$	$\{1,3,5\}$
5.	Yes	$\{-2,2,3,5\}$	$\{2\}$
7.	Yes	Real Numbers	Real Numbers
9.	Yes	Real Numbers	Real Numbers where $y \geq 1$

11. $y^2 = x + 2$

$y = \pm\sqrt{x+2}$ Real Numbers   Real Numbers

No   where $x \geq -2$

	Function	Domain	Range
13.	Yes	Real Numbers where $x \geq -3$	Real Numbers where $y \geq 0$
15.	Yes	Real Numbers where $x \geq 4$	Real Numbers where $y \geq 6$

17. $f(x) = 8x - 12$

(a) $f(4) = 8(4) - 12$ (b) $f(0) = 8(0) - 12$ (c) $f(-2) = 8(-2) - 12$

$\quad\ f(4) = 20$ $\qquad\qquad\quad f(0) = -12$ $\qquad\qquad\quad f(-2) = -28$

19. $g(x) = 10x + 15$

(a) $g(2) = 10(2) + 15$ (b) $g(0) = 10(0) + 15$ (c) $g(-4) = 10(-4) + 15$

$\quad\ g(2) = 35$ $\qquad\qquad\quad g(0) = 15$ $\qquad\qquad\quad g(-4) = -25$

21. $h(x) = 3x^2 + 4x$

(a) $h(5) = 3(5)^2 + 4(5)$ (b) $h(0) = 3(0)^2 + 4(0)$ (c) $h(-2) = 3(-2)^2 + 4(-2)$

$\quad\ h(5) = 3(25) + 20$ $\qquad\ h(0) = 0$ $\qquad\qquad\ h(-2) = 3(4) - 8$

$\quad\ h(5) = 95$ $\qquad\qquad\qquad\qquad\qquad\qquad\quad h(-2) = 4$

Section 1.1

23. $f(t) = \dfrac{5 - t^2}{2t}$

(a) $f(1) = \dfrac{5 - (1)^2}{2(1)}$

$f(1) = \dfrac{5 - 1}{2} = 2$

(b) $f(-3) = \dfrac{5 - (-3)^2}{2(-3)}$

$f(-3) = \dfrac{5 - 9}{-6}$

$f(-3) = \dfrac{-4}{-6} = \dfrac{2}{3}$

(c) 0 is not in the domain of $f(t)$.

25. $f(x) = 6x + 8$

(a) $f(a) = 6a + 8$

(b) $f(4a) = 6(4a) + 8$

$f(4a) = 24a + 8$

(c) $f(c^2) = 6c^2 + 8$

27. $h(x) = 4x^2 - 12x$

(a) $h(x + 2) = 4(x + 2)^2 - 12(x + 2)$

$h(x + 2) = 4(x^2 + 4x + 4) - 12x - 24$

$h(x + 2) = 4x^2 + 16x + 16 - 12x - 24$

$h(x + 2) = 4x^2 + 4x - 8$

(b) $h(x - 3) = 4(x - 3)^2 - 12(x - 3)$

$h(x - 3) = 4(x^2 - 6x + 9) - 12x + 36$

$h(x - 3) = 4x^2 - 24x + 36 - 12x + 36$

$h(x - 3) = 4x^2 - 36x + 72$

(c) $h(2x + 1) = 4(2x + 1)^2 - 12(2x + 1)$

$h(2x + 1) = 4(4x^2 + 4x + 1) - 24x - 12$

$h(2x + 1) = 16x^2 + 16x + 4 - 24x - 12$

$h(2x + 1) = 16x^2 - 8x - 8$

29. (a) $f(x) + g(x) = (3x - 1) + (x^2 - 6x + 1)$

$= x^2 - 3x$

(b) $f(x) - g(x) = (3x - 1) - (x^2 - 6x + 1)$

$= 3x - 1 - x^2 + 6x - 1$

$= -x^2 + 9x - 2$

(c) $[f(x)][g(x)] = (3x - 1)(x^2 - 6x + 1)$

$= 3x^3 - 18x^2 + 3x$

$-x^2 + 6x - 1$

$= 3x^3 - 19x^2 + 9x - 1$

(d) $f(x) = 3x - 1$

$f(x + h) = 3(x + h) - 1$

$f(x + h) = 3x + 3h - 1$

31. Domain is all real numbers x where $x \neq 2$.

33. Domain is all real numbers t where $t \neq 6$ or $t \neq -3$.

35. Domain is all real numbers where $x < 5$.

Section 1.1

Section 1.2

Graphs appear in the text answer section.

1. $y = 2x + 1$

x	y
0	1
1	3
2	5

3. $-2x - 3y = 6$

x	y
0	-2
3	-4
-3	0

5. $y = x^2 - 9$

x	y	x	y
-3	0	1	-8
-2	-5	2	-5
-1	-8	3	0
0	9	4	7

7. $y = x^2 - 5x + 4$

x	y	x	y
-1	10	3	-2
0	4	4	0
1	0	5	4
2	-2	6	10

9. $y = 2x^2 + 3x - 2$

x	y	x	y
-4	18	0	-2
-3	7	1	3
-2	0	2	12
-1	-3	3	25

11. $y = x^2 + 2x$

x	y	x	y
-4	8	0	0
-3	3	1	3
-2	0	2	8
-1	-1	3	15

13. $y = -2x^2 + 4x$

x	y	x	y
-2	-16	2	0
-1	-6	3	-6
0	0	4	-16
1	2	5	-30

15. $y = x^3 - x^2 - 10x + 8$

x	y	x	y
-4	-32	1	-2
-3	2	2	-8
-2	16	3	-4
-1	16	4	16
0	8	5	58

17. $y = x^3 + 2x^2 - 7x + 4$

x	y	x	y
-5	-36	0	4
-4	0	1	0
-3	16	2	6
-2	18	3	28
-1	12	4	72

19. $y = \sqrt{x + 4}$

x	y	x	y
-4	0	5	3
-3	1	12	4
0	2	21	5

21. $y = \sqrt{12 - 6x}$

x	y	x	y
-13	9.5	-4	6
-10	8.4	0	3.5
-7	7.3	2	0

Section 1.2

23. $y = x^2 - 9$

(a) To solve graphically, graph $y_1 = x^2 - 9$ and $y_2 = 0$. Find the values of x for which the graphs intersect.

$0 = x^2 - 9$

$9 = x^2$

$x = 3$ and -3

(b) Graph

$y_1 = x^2 - 9$

$y_2 = -5$

Find x where the graphs intersect.

$-5 = x^2 - 9$

$4 = x^2$

$x = 2$ and -2

(c) Graph

$y_1 = x^2 - 9$

$y_2 = 2$

Find x where the graphs intersect.

$2 = x^2 - 9$

$11 = x^2$

$x = \pm\sqrt{11}$

$x = 3.3$ and -3.3

25. $y = x^2 - 5x + 4$

(a) Graph

$y_1 = x^2 - 5x + 4$

$y_2 = 0$

Find the values of x for which the graphs intersect.

$0 = x^2 - 5x + 4$

$0 = (x - 4)(x - 1)$

$x = 4$ and 1

(b) Graph

$y_1 = x^2 - 5x + 4$

$y_2 = 2$

Find the values of x for which the graphs intersect.

$2 = x^2 - 5x + 4$

$0 = x^2 - 5x + 2$

$x = 4.5$ and 0.5

(c) Graph

$y_1 = x^2 - 5x + 4$

$y_2 = -4$

Find the values of x for which the graphs intersect. The graphs do not intersect.

No solution.

27. $y = 2x^2 + 3x - 2$

(a) Graph

$y_1 = 2x^2 + 3x - 2$

$y_2 = 0$

Find the values of x for which the graphs intersect.

$0 = 2x^2 + 3x - 2$

$0 = (2x - 1)(x + 2)$

$x = \dfrac{1}{2}$ and -2

(b) Graph

$y_1 = 2x^2 + 3x - 2$

$y_2 = 3$

Find the values of x for which the graphs intersect.

$3 = 2x^2 + 3x - 2$

$0 = 2x^2 + 3x - 5$

$0 = (2x + 5)(x - 1)$

$x = -\dfrac{5}{2}$ and 1

(c) Graph

$y_1 = 2x^2 + 3x - 2$

$y_2 = 5$

Find the values of x for which the graphs intersect.

$5 = 2x^2 + 3x - 2$

$0 = 2x^2 + 3x - 7$

By quadratic formula or graph

$x = 1.3$ and -2.8

Section 1.2

29.(a) $y_1 = x^2 + 2x$
$y_2 = 0$
Find the values of x for which the graphs intersect.
$0 = x^2 + 2x$
$0 = x(x + 2)$
$x = 0$ and -2

(b) $y_1 = x^2 + 2x$
$y_2 = 3$
Find the values of x for which the graphs intersect.
$3 = x^2 + 2x$
$0 = x^2 + 2x - 3$
$0 = (x + 3)(x - 1)$
$x = -3$ and 1

(c) $y_1 = x^2 + 2x$
$y_2 = 6$
Find the values of x for which the graphs intersect.
$6 = x^2 + 2x$
$0 = x^2 + 2x - 6$
By quadratic formula or graph
$x = 1.6$ and -3.6

31.(a) $y_1 = 2x^2 + 4x$
$y_2 = 0$
Find the values of x for which the graphs intersect.
$0 = -2x^2 + 4x$
$0 = -2x(x - 2)$
$x = 0$ and 2

(b) $y_1 = -2x^2 + 4x$
$y_2 = 5$
Find the values of x for which the graphs intersect.
These graphs do not intersect.
No solution.

(c) $y_1 = -2x^2 + 4x$
$y_2 = -4$
Find the values of x for which the graphs intersect.
$-4 = -2x^2 + 4x$
$2x^2 - 4x - 4 = 0$
By quadratic formula or graph
$x = 2.7$ and -0.7

(d) $y_1 = -2x^2 + 4x$
$y_2 = -1.5$
Find the values of x for which the graphs intersect.
$-1.5 = -2x^2 + 4x$
$2x^2 - 4x - \dfrac{3}{2} = 0$
By the quadratic formula or graph
$x = 2.3$ and -0.3

33.(a) $y_1 = x^3 - x^2 - 10x + 8$
$y_2 = 0$
Find the values of x for which the graphs intersect.
$0 = x^3 - x^2 - 10x + 8$
By graph
$x = -3.1$ and 0.8 and 3.3

(b) $y_1 = x^3 - x^2 - 10x + 8$
$y_2 = 2$
Find the values of x for which the graphs intersect.
$2 = x^3 - x^2 - 10x + 8$
$0 = x^3 - x^2 - 10x + 6$
By graph
$x = -3.0$ and 0.6 and 3.4

(c) $y_1 = x^3 - x^2 - 10x + 8$
$y_2 = -2$
Find the values of x for which the graphs intersect.
$-2 = x^3 - x^2 - 10x + 8$
$0 = x^3 - x^2 - 10x + 10$
$0 = x^2(x - 1) - 10(x - 1)$
$0 = (x - 1)(x^2 - 10)$
$x = 1.0$ and $\sqrt{10}$ and $-\sqrt{10}$
$x = 1.0$ and 3.2 and -3.2

35.(a) $y_1 = x^3 + 2x^2 - 7x + 4$
$y_2 = 0$
Find the values of x for which the graphs intersect.
$0 = x^3 + 2x^2 - 7x + 4$
$0 = (x + 4)(x - 1)^2$
or by graph
$x = -4$ and 1

Section 1.2

35.(b) $y_1 = x^3 + 2x^2 - 7x + 4$

$y_2 = 4$

Find the values of x for which the graphs intersect.

$4 = x^3 + 2x^2 - 7x + 4$

$0 = x^3 + 2x^2 - 7x$

$0 = x(x^2 + 2x - 7)$

By the quadratic formula or graph

$x = 0$ and 1.8 and -3.8

(c) $y_1 = x^3 + 2x^2 - 7x + 4$

$y_2 = 8$

Find the values of x for which the graphs intersect.

$8 = x^3 + 2x^2 - 7x + 4$

$0 = x^3 + 2x^2 - 7x - 4$

By graph

$x = -3.6$ and -0.5 and 2.1

37.(a) $y_1 = x^2 + 3x - 4$

$y_2 = 0$

Find the values of x for which the graphs intersect.

$0 = x^2 + 3x - 4$

$0 = (x + 4)(x - 1)$

$x = -4$ and 1

(b) $y_1 = x^2 + 3x - 4$

$y_2 = 6$

Find the values of x for which the graphs intersect.

$6 = x^2 + 3x - 4$

$0 = x^2 + 3x - 10$

$0 = (x + 5)(x - 2)$

$x = -5$ and 2

(c) $y_1 = x^2 + 3x - 4$

$y_2 = -2$

Find the values of x for which the graphs intersect.

$-2 = x^2 + 3x - 4$

$0 = x^2 + 3x - 2$

By the quadratic formula or graph

$x = -3.6$ and 0.6

39.(a) $y_1 = -\dfrac{1}{2}x^2 + 2$

$y_2 = 0$

Find the values of x for which the graphs intersect.

$0 = -\dfrac{1}{2}x^2 + 2$

$\dfrac{1}{2}x^2 = 2$

$x^2 = 4$

$x = 2$ and -2

(b) $y_1 = -\dfrac{1}{2}x^2 + 2$

$y_2 = 4$

Find the values of x for which the graphs intersect. These graphs do not intersect. No solution.

(c) $y_1 = -\dfrac{1}{2}x^2 + 2$

$y_2 = -4$

Find the values of x for which the graphs intersect.

$-4 = -\dfrac{1}{2}x^2 + 2$

$\dfrac{1}{2}x^2 = 6$

$x^2 = 12$

$x = \pm\sqrt{12}$

$x = 3.5$ and -3.5

Section 1.2

Solutions to Odd-Numbered Exercises

41.(a)
$y_1 = x^3 - 3x^2 + 1$
$y_2 = 0$
Find the values of
x where the graphs
intersect.
$0 = x^3 - 3x^2 + 1$
By graph
$x = -0.5$ and 0.7
and 2.9

(b)
$y_1 = x^3 - 3x^2 + 1$
$y_2 = -2$
Find the values of
x where the graphs
intersect.
$-2 = x^3 - 3x^2 + 1$
$0 = x^3 - 3x^2 + 3$
By graph
$x = -0.9$ and 1.3
and 2.5

(c)
$y_1 = x^3 - 3x^2 + 1$
$y_2 = -0.5$
Find the values of
x where the graphs
intersect.
$-0.5 = x^3 - 3x^2 + 1$
$0 = x^3 - 3x^2 + 1.5$
By graph
$x = 0.6$ and 0.8
and 2.8

43. $r = 10t^2 + 20$

(a) $r_1 = 10t^2 + 20$
$r_2 = 90\Omega$
Find the values of
t for which the
graphs intersect.
$90 = 10t^2 + 20$
$70 = 10t^2$
$7 = t^2$
$t = \sqrt{7}$
By graph or
calculator
$t = 2.6$ ms

(b) $r_1 = 10t^2 + 20$
$r_2 = 180\Omega$
Find the values of
t for which the
graphs intersect.
$180 = 10t^2 + 20$
$160 = 10t^2$
$16 = t^2$
$t = 4$ ms

(c) $r_1 = 10t^2 + 20$
$r_2 = 320$
Find the values of
t for which the
graphs intersect.
$320 = 10t^2 + 20$
$300 = 10t^2$
$30 = t^2$
$t = \sqrt{30}$
By graph or
calculator
$t = 5.5$ ms

45. $w = 5t^2 + 6t$

(a) $w_1 = 5t^2 + 6t$
$w_2 = 2$
Find the values of
t for which the
graphs intersect.
$2 = 5t^2 + 6t$
$0 = 5t^2 + 6t - 2$
By graph or the
quadratic formula
$t = 0.27$ ms

(b) $w_1 = 5t^2 + 6t$
$w_2 = 4$
Find the values of t
for which the graphs
intersect.
$4 = 5t^2 + 6t$
$0 = 5t^2 + 6t - 4$
By graph or the
quadratic formula
$t = 0.48$ ms

(c) $w_1 = 5t^2 + 6t$
$w_2 = 10$
Find the values of
t for which the
graphs intersect.
$10 = 5t^2 + 6t$
$0 = 5t^2 + 6t - 10$
By graph or the
quadratic formula
$t = 0.94$ ms

Section 1.2

47. $i = t^3 - 15$

   (a) $i_1 = t^3 - 15$

          $i_2 = 5A$

          Find the values of t for which the graphs intersect.

          $5 = t^3 - 15$

          $20 = t^3$

          $t = \sqrt[3]{20}$

          By graph or calculator

          $t = 2.7$ s

   (b) $i_1 = t^3 - 15$

          $t_2 = 15A$

          Find the values of t for which the graphs intersect.

          $15 = t^3 - 15$

          $30 = t^3$

          $t = \sqrt[3]{30}$

          By graph or calculator

          $t = 3.1$ s

49. See Figure 1.15.

    Let x = horizontal length

    $-y$ = vertical length

   (a) $\sin 36° = \dfrac{x}{2}$

          $x = 1.18$

          $\cos 36° = \dfrac{-y}{2}$

          $y = -1.62$

          $A(1.18, -1.62)$

   (b) $\sin 36° = \dfrac{x}{4}$

          $x = 2.35$

          $\cos 36° = \dfrac{-y}{4}$

          $y = -3.24$

          $B(2.35, -3.24)$

   (c) $\sin 36° = \dfrac{x}{6}$

          $x = 3.53$

          $\cos 36° = \dfrac{-y}{6}$

          $y = -4.85$

          $C(3.53, -4.85)$

Section 1.3

1. $m = \dfrac{y_2 - y_1}{x_2 - x_1}$

    $m = \dfrac{2 - 1}{4 - 3}$

    $m = \dfrac{1}{1} = 1$

3. $m = \dfrac{y_2 - y_1}{x_2 - x_1}$

    $m = \dfrac{-5 - 3}{4 - 2}$

    $m = \dfrac{-8}{2} = -4$

5. $m = \dfrac{y_2 - y_1}{x_2 - x_1}$

    $m = \dfrac{2 - 2}{-3 - 6}$

    $m = \dfrac{0}{-9} = 0$

7. $m = \dfrac{y_2 - y_1}{x_2 - x_1}$

    $m = \dfrac{7 - 2}{5 - (-3)}$

    $m = \dfrac{5}{8}$

9. $(2, -1)$, $m = 2$

    Points on the line go up 2 for every 1 (unit) moved to the right, so another point is (3, 1).

11. $(-3, -2)$, $m = \dfrac{1}{2}$

    Points on the line go up 1 for every 2 (units) moved to the right, so another point is $(-1, -1)$.

Sections 1.2 – 1.3

13. $(4,0)$, $m = -2$

Points on the line drop 2 for every 1 (unit) moved to the right, so another point is $(5, -2)$.

15. $(0,-3)$, $m = -\frac{3}{4}$

Points on the line drop 3 for every 4 (unit) moved to the right, so another point is $(4, -6)$.

17. $(-2, 8)$, $m = -3$

$y = mx + b$
where $x = -2$
and $y = 8$.
$8 = -3(-2) + b$
$8 = 6 + b$
$2 = b$
$y = -3x + 2$
$3x + y - 2 = 0$

19. $(-3, -4)$, $m = \frac{1}{2}$

$y = mx + b$
where $x = -3$
and $y = -4$.
$-4 = \frac{1}{2}(-3) + b$
$-4 = \frac{-3}{2} + b$
$-\frac{5}{2} = b$
$y = \frac{1}{2}x - \frac{5}{2}$
$2y = x - 5$
$0 = x - 2y - 5$

21. $(-2, 7)$ and $(1, 4)$

$m = \frac{y_2 - y_1}{x_2 - x_1} = \frac{7 - 4}{-2 - 1}$
$m = \frac{3}{-3} = -1$

Use either point for $y = mx + b$.
Use $x = -2$ and $y = 7$.
$7 = -1(-2) + b$
$7 = 2 + b$
$5 = b$
$y = -1x + 5$
$x + y - 5 = 0$

23. $(6, -8)$ and $(-4, -3)$

$m = \frac{y_2 - y_1}{x_2 - x_1}$
$m = \frac{-8 - (-3)}{6 - (-4)}$
$m = \frac{-5}{10} = \frac{-1}{2}$

Use either point for $y = mx + b$.
Use $x = 6$ and $y = -8$.
$-8 = \frac{-1}{2}(6) + b$
$-8 = -3 + b$
$-5 = b$
$y = -\frac{1}{2}x - 5$
$2y = -x - 10$
$x + 2y + 10 = 0$

25. $m = -5$ and y-intercept is -2.

$y = mx + b$
$y = -5x - 2$

27. $m = 2$ and y-intercept is 7.

$y = mx + b$
$y = 2x + 7$

29. Horizontal line and 5 units above x-axis.

$y = b$ becomes
$y = 5$

31. Vertical line through $(-2, 0)$.

$x = a$ becomes
$x = -2$

33. Horizontal line through $(2, -3)$.

$y = b$ becomes
$y = -3$

35. Vertical line through $(-7, 9)$.

$x = a$ becomes
$x = -7$

Section 1.3

37.

$x + 4y = 12$

$4y = -x + 12$

$y = -\dfrac{1}{4}x + 3$

$m = -\dfrac{1}{4}$, y-intercept is 3.

39.

$4x - 2y + 14 = 0$

$-2y = -4x - 14$

$y = 2x + 7$

$m = 2$, y-intercept is 7.

41.

$y = 6$

$y = 0x + 6$

$m = 0$, y-intercept is 6.

43. $y = 3x - 2$

x	y
-1	-5
0	-2
1	1
2	4

45.

$5x - 2y + 4 = 0$

$-2y = -5x - 4$

$y = \dfrac{5}{2}x + 2$

x	y
-2	-3
0	2
2	7
4	12

47. $x = 7$

Vertical line.

x	y
7	-1
7	0
7	1
7	2

See p. 724

49. $y = -3$

Horizontal line.

x	y
-1	-3
0	-3
1	-3
2	-3

51.

$6x + 8y = 24$

$8y = -6x + 24$

$y = -\dfrac{3}{4}x + 3$

x	y
-4	6
0	3
4	0
8	-3

53.

$x - 3y = -12$

$-3y = -x - 12$

$y = \dfrac{1}{3}x + 4$

x	y
-6	2
-3	3
0	4
3	5

55. $(-15.0, 43.0)$ and $(55.0, 43.2)$

$m = \dfrac{y_2 - y_1}{x_2 - x_1}$

$m = \dfrac{43.0 - 43.2}{-15.0 - 55.0}$

$m = \dfrac{-0.2}{-70} = \dfrac{1}{350}$

Section 1.3

Solutions to Odd-Numbered Exercises

Section 1.4

1.
$$x + 3y - 7 = 0$$
$$3y = -x + 7$$
$$y = -\frac{1}{3}x + \frac{7}{3}$$
and
$$-3x + y + 2 = 0$$
$$y = 3x - 2$$
perpendicular,
product of
slopes is -1.

3.
$$-x + 4y + 7 = 0$$
$$4y = x - 7$$
$$y = \frac{1}{4}x - \frac{7}{4}$$
and
$$x + 4y - 5 = 0$$
$$4y = -x + 5$$
$$y = -\frac{1}{4}x + \frac{5}{4}$$
neither

5.
$$y - 5x + 13 = 0$$
$$y = 5x - 13$$
and
$$y - 5x + 9 = 0$$
$$y = 5x - 9$$
parallel,
slopes are
equal.

7.
Parallel to
$$-2x + y + 13 = 0$$
$$y = 2x - 13$$
Desired slope
is 2.
For $(-1, 5)$, $x = -1$
and $y = 5$.
$$y = mx + b$$
$$5 = 2(-1) + b$$
$$5 = -2 + b$$
$$7 = b$$
$$y = 2x + 7$$
$$0 = 2x - y + 7$$

9.
Perpendicular
to $5y = x$.
$$y = \frac{1}{5}x$$
$$y = 2x - 13$$
Desired slope
is -5.
For $(-7, 4)$, $x = -7$
and $y = 4$.
$$y = mx + b$$
$$4 = -5(-7) + b$$
$$4 = 35 + b$$
$$-31 = b$$
$$y = -5x - 31$$
$$5x + y + 31 = 0$$

11.
Parallel to
$$3x - 4y = 12$$
$$-4y = -3x + 12$$
$$y = \frac{3}{4}x - 3$$
Desired slope
is $\frac{3}{4}$. Origin is
$(0, 0)$, so $x = 0$
and $y = 0$
$$y = mx + b$$
$$0 = \frac{3}{4}(0) + b$$
$$0 = b$$
$$y = \frac{3}{4}x$$
$$4y = 3x$$
$$0 = 3x - 4y$$

13.
Perpendicular to
$$4x + 6y = 9$$
$$6y = -4x + 9$$
$$y = -\frac{2}{3}x + \frac{3}{2}$$
Desired slope
is $\frac{3}{2}$.
x-intercept 6 is
the point $(6, 0)$,
so $x = 6$ and $y = 0$.
$$y = mx + b$$
$$0 = \frac{3}{2}(6) + b$$
$$0 = 9 + b$$
$$b = -9$$
$$y = \frac{3}{2}x - 9$$
$$2y = 3x - 18$$
$$18 = 3x - 2y$$

15.
Parallel to
$y = 2$ means a
horizontal line.
y-intercept 8
is the point $(0, 8)$.
$$y = b$$
$$y = 8$$

17.
Parallel to
$x = -4$ means a
vertical line.
x-intercept 7
is the point $(7, 0)$.
$$x = a$$
$$x = 7$$

Section 1.4

19.(a) Find the slopes of the line segments connecting the points to see if opposite sides have the same slopes.

$$m = \frac{y_2 - y_1}{x_2 - x_1}$$

For $\overline{AD}$:

$$m = \frac{7-3}{5-(-2)} = \frac{4}{7}$$

For $\overline{BC}$:

$$m = \frac{6-2}{9-2} = \frac{4}{7}$$

Thus $\overline{AD}$ and $\overline{BC}$ are parallel.

For $\overline{AB}$:

$$m = \frac{3-2}{-2-2}$$

$$m = \frac{1}{-4}$$

For $\overline{DC}$:

$$m = \frac{7-6}{5-9}$$

$$m = \frac{1}{-4}$$

Thus $\overline{AB}$ and $\overline{DC}$ are parallel.

Yes, it is a parallelogram.

19.(b) Use the slopes already calculated.

For $\overline{AD}$, $m = \frac{4}{7}$.

For $\overline{AB}$, $m = -\frac{1}{4}$.

These sides are adjacent, but the product of their slopes is not -1. The figure is not a rectangle.

Section 1.5

1.
$$d = \sqrt{(x_2 - x_1)^2 + (y_2 - y_1)^2}$$
$(4,-7)$ and $(-5,5)$
$$d = \sqrt{(-5-4)^2 + (5-(-7))^2}$$
$$d = \sqrt{(-9)^2 + (12)^2}$$
$$d = \sqrt{81+144} = \sqrt{225}$$
$$d = 15$$

3.
$$d = \sqrt{(x_2 - x_1)^2 + (y_2 - y_1)^2}$$
$(3,-2)$ and $(10,-2)$
$$d = \sqrt{(10-3)^2 + (-2-(-2))^2}$$
$$d = \sqrt{7^2 + 0^2}$$
$$d = 7$$

5.
$$d = \sqrt{(x_2 - x_1)^2 + (y_2 - y_1)^2}$$
$(5,-2)$ and $(1,2)$
$$d = \sqrt{(5-1)^2 + (-2-2)^2}$$
$$d = \sqrt{4^2 + (-4)^2} = \sqrt{32}$$
$$d = \sqrt{16}\sqrt{2} = 4\sqrt{2}$$

7.
$$d = \sqrt{(x_2 - x_1)^2 + (y_2 - y_1)^2}$$
$(3,-5)$ and $(3,2)$
$$d = \sqrt{(3-3)^2 + (-5-2)^2}$$
$$d = \sqrt{0^2 + (-7)^2}$$
$$d = \sqrt{49} = 7$$

9.
$$x_m = \frac{x_1 + x_2}{2}, \ y_m = \frac{y_1 + y_2}{2}$$
$(2,3)$ and $(5,7)$
$$x_m = \frac{2+5}{2} = \frac{7}{2} = 3.5$$
$$y_m = \frac{3+7}{2} = 5$$
$(3.5, 5)$

11.
$$x_m = \frac{x_1 + x_2}{2}, \ y_m = \frac{y_1 + y_2}{2}$$
$(3,-2)$ and $(0,0)$
$$x_m = \frac{3+0}{2} = 1.5$$
$$y_m = \frac{-2+0}{2} = -1$$
$(1.5, -1)$

13.
$$x_m = \frac{x_1 + x_2}{2}, \ y_m = \frac{y_1 + y_2}{2}$$
$(11,4)$ and $(-11,-9)$
$$x_m = \frac{11+(-11)}{2} = \frac{0}{2} = 0$$
$$y_m = \frac{4+(-9)}{2} = \frac{-5}{2} = -2.5$$
$(0, -2.5)$

Sections $1.4 - 1.5$

15. $A(2,8)$, $B(10,2)$, $C(10,8)$ **a.** $P = 10 + 6 + 8 = 24$

First, we find the
lengths of the sides.

b.   Yes, it is a right triangle because

$$d = \sqrt{(x_2 - x_1)^2 + (y_2 - y_1)^2}$$

$$AB = \sqrt{(2 - 10)^2 + (8 - 2)^2}$$

$$AB = \sqrt{(-8)^2 + 6^2}$$

$$AB = \sqrt{64 + 36} = \sqrt{100}$$

$$AB = 10$$

$$BC = \sqrt{(10 - 10)^2 + (2 - 8)^2}$$

$$BC = \sqrt{0^2 + (-6)^2} = \sqrt{36}$$

$$BC = 6$$

$$AC = \sqrt{(2 - 10)^2 + (8 - 8)^2}$$

$$AC = \sqrt{(-8)^2 + 0^2} = \sqrt{64}$$

$$AC = 8$$

   $AB^2 = BC^2 + AC^2$

     i.e. $10^2 = 6^2 + 8^2$

           $100 = 36 + 64$

c.   No, it is not isosceles; none of the sides are equal in length.

d. $A = \dfrac{1}{2} bh$

     $A = \dfrac{1}{2}(6)(8)$

     $A = 24$

17. $A(-3,6)$, $B(5,0)$, $C(4,9)$ **a.** $P = 10 + \sqrt{82} + \sqrt{58} = 26.7$

First find the lengths
of the sides.

b.   No, it is not a right triangle because

$$d = \sqrt{(x_2 - x_1)^2 + (y_2 - y_1)^2}$$

$$AB = \sqrt{(-3 - 5)^2 (6 - 0)^2}$$

$$AB = \sqrt{(-8)^2 + 6^2}$$

$$AB = \sqrt{64 + 36} = \sqrt{100}$$

$$AB = 10$$

$$BC = \sqrt{(5 - 4)^2 + (0 - 9)^2}$$

$$BC = \sqrt{1^2 + (-9)^2}$$

$$BC = \sqrt{82}$$

$$AC = \sqrt{(-3 - 4)^2 + (6 - 9)^2}$$

$$AC = \sqrt{(-7)^2 + (-3)^2}$$

$$AC = \sqrt{49 + 9} = \sqrt{58}$$

   $10^2 \neq \left(\sqrt{82}\right)^2 + \left(\sqrt{58}\right)^2$

   $100 \neq 82 + 58$

   $100 \neq 140$

c.   No, it is not isosceles; none of the sides are equal in length.

Section 1.5

19. $A(7,-1)$, $B(9,1)$, $C(-3,5)$

midpoint of BC:

$$x_m = \frac{x_1 + x_2}{2}, \; y_m = \frac{y_1 + y_2}{2}$$

$$x_m = \frac{9 + (-3)}{2} = \frac{6}{2} = 3$$

$$y_m = \frac{1+5}{2} = \frac{6}{2} = 3$$

midpoint of BC is $(3,3)$

$$d = \sqrt{(x_2 - x_1)^2 + (y_2 - y_1)^2}$$

$$d = \sqrt{(7-3)^2 + (-1-3)^2}$$

$$d = \sqrt{4^2 + (-4)^2}$$

$$d = \sqrt{16 + 16} = \sqrt{32}$$

$$d = \sqrt{16} = 4\sqrt{2}$$

21. Parallel to

$$3x - 6y = 10$$

$$-6y = -3x + 10$$

$$y = \frac{1}{2}x - \frac{5}{3}$$

Desired slope is $\frac{1}{2}$.

midpoint of $A(4,2)$ and $B(8,-6)$:

$$x_m = \frac{x_1 + x_2}{2} = \frac{4+8}{2} = \frac{12}{2} = 6$$

$$y_m = \frac{y_1 + y_2}{2} = \frac{2 + (-6)}{2} = \frac{-4}{2} = -2$$

$(6,-2)$.

Use $y = mx + b$; from

the midpoint, $x = 6$ and $y = 2$.

$$-2 = \frac{1}{2}(6) + b$$

$$-2 = 3 + b$$

$$-5 = b$$

$$y = \frac{1}{2}x - 5$$

$$2y = x - 10$$

$$10 = x - 2y$$

23. $A(-8,12)$ and $(6,10)$ midpoint of AB:

$$x_m = \frac{x_1 + x_2}{2} = \frac{-8+6}{2}$$

$$x_m = -1$$

$$y_m = \frac{y_1 + y_2}{2} = \frac{12 + 10}{2}$$

$$y_m = 11$$

midpoint $AB = (-1,11)$

Perpendicular to

$$4x + 8y = 16$$

$$8y = -4x + 16$$

$$y = -\frac{1}{2}x + 2$$

Desired slope is 2.

$y = mx + b$; from the

midpoint, $x = -1$, $y = 11$.

$$11 = 2(-1) + b$$

$$11 = -2 + b$$

$$13 = b$$

$$y = 2x + 13$$

$$2x - y = -13$$

25. $A(-2,2)$, $B(1,3)$, $C(2,0)$, $D(-1,-1)$.

First find the lengths of the sides of $ABCD$.

$$d = \sqrt{(x_2 - x_1)^2 + (y_2 - y_1)^2}$$

$$AB = \sqrt{(-2-1)^2 + (2-3)^2} = \sqrt{(-3)^2 + (-1)^2}$$

$$AB = \sqrt{9+1} = \sqrt{10}$$

$$BC = \sqrt{(1-2)^2 + (3-0)^2} = \sqrt{(-1)^2 + (3)^2}$$

$$BC = \sqrt{1+9} = \sqrt{10}$$

$$CD = \sqrt{(2-(-1))^2 + (0-(-1))^2}$$

$$CD = \sqrt{(3)^2 + (1)^2}$$

$$CD = \sqrt{9+1} = \sqrt{10}$$

$$DA = \sqrt{(-1-(-2))^2 + (-1-2)^2}$$

$$DA = \sqrt{(1)^2 + (-3)^2}$$

$$DA = \sqrt{1+9} = \sqrt{10}$$

Slope of $AB = \dfrac{y_2 - y_1}{x_2 - x_1}$

$$m = \frac{2-3}{-2-1} = \frac{-1}{-3} = \frac{1}{3}$$

Slope of $BC = \dfrac{3-0}{1-2} = -3$

The sides are equal and adjacent sides are perpendicular, thus ABCD is a square, a special rectangle.

Section 1.5

27. $A(-12,8)$, $B(3,2)$, $C(5,7)$, $D(-5,11)$.

Find slopes to show opposite sides are parallel and two adjacent sides are perpendiculuar.

$$m_{AB} = \frac{y_2 - y_1}{x_2 - x_1} = \frac{8-2}{-12-3}$$

$$m_{AB} = \frac{6}{-15} = \frac{-2}{5}$$

$$m_{BC} = \frac{2-7}{3-5} = \frac{-5}{-2}$$

$$m_{BC} = \frac{5}{2}$$

$$m_{CD} = \frac{7-11}{5-(-5)} = \frac{-4}{10}$$

$$m_{CD} = \frac{-2}{5}$$

$$m_{DA} = \frac{11-8}{-5-(-12)} = \frac{3}{7}$$

Since $m_{AB}(m_{BC}) = -1$,

AB is perpendicular to BC.

Since $m_{AB} = m_{CD}$,

AB and CD are parallel.

Thus, $ABCD$ is a trapezoid with one right angle.

Section 1.6

1.

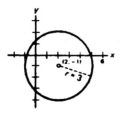

3.

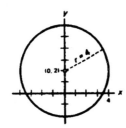

5. $(x-1)^2 + (y+1)^2 = 16$

7. $(x+2)^2 + (y+4)^2 = 34$

9. $x^2 + y^2 = 36$

11. $C(0, 0)$; $r = 4$

13. $C(-3, 4)$; $r = 4$

15. $x^2 - 8x + y^2 + 12y = 8$
    $x^2 - 8x + 16 + y^2 + 12y + 36 = 8 + 16 + 36$
    $(x-4)^2 + (y+6)^2 = 60$
    $C(4, -6)$; $r = 2\sqrt{15}$

17. $x^2 - 12x + y^2 - 2y = 12$
    $x^2 - 12x + 36 + y^2 - 2y + 1 = 12 + 36 + 1$
    $(x-6)^2 + (y-1)^2 = 49$; $C(6, 1)$; $r = 7$

Sections 1.5 – 1.6

19. $x^2 + 7x + y^2 + 3y = 9$
   $x^2 + 7x + 49/4 + y^2 + 3y + 9/4 = 9 + 49/4 + 9/4$
   $(x + 7/2)^2 + (y + 3/2)^2 = 47/2;$ $C(-7/2, -3/2); r = \sqrt{94}/2$

21. $\sqrt{(0-1)^2 + (y-4)^2} = \sqrt{(0+3)^2 + (y-2)^2}$
   $1 + y^2 - 8y + 16 = 9 + y^2 - 4y + 4$
   $4 = 4y$ or $y = 1$
   $C(0, 1); r = \sqrt{(1-0)^2 + (4-1)^2} = \sqrt{10}$
   $x^2 + (y - 1)^2 = 10$ or $x^2 + y^2 - 2y - 9 = 0$

23. $x^2 + y^2 + Dx + Ey + F = 0$; substitute $(3, 1)$, $(0, 0)$ and $(8, 4)$
   $3D + E + F = -10$
   $F = 0$
   $8D + 4E + F = -80$

   Solve simultaneously: $D = 10,$
   $E = -40, F = 0$
   $x^2 + y^2 + 10x - 40y = 0$
   $(x + 5)^2 + (y - 20)^2 = 425$
   $C(-5, 20); r = 5\sqrt{17}$

Section 1.7

1.
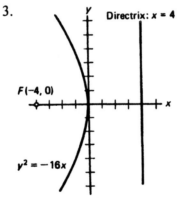

3.
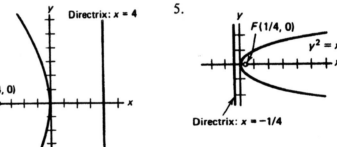

5.

7.
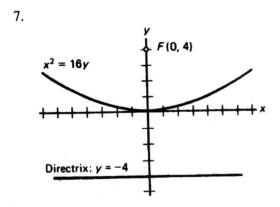

9.
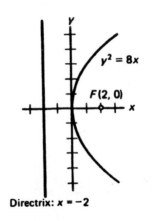

Sections 1.6 – 1.7

Solutions to Odd-Numbered Exercises

11. $y^2 = 8x$ 13. $y^2 = -32x$ 15. $x^2 = 24y$ 17. $y^2 = -16x$

19. $\sqrt{(x+1)^2 + (y-3)^2} = \sqrt{(x-3)^2 + (y-y)^2}$;

     $y^2 - 6y + 8x + 1 = 0$

21. $x^2 = 4py$

     $200^2 = 4p(-16); p = -625; x^2 = -2500y$

     at 50 m, $50^2 = -625y_1$; $y_1 = -1$ so 15 m high

     at 150 m, $150^2 = -625y_2$; $y_2 = -9$ m so 7 m high

23. $x^2 = 32y$

25. $y = 2x^2 + 7x - 15$ 27. $y = -2x^2 + 4x + 16$

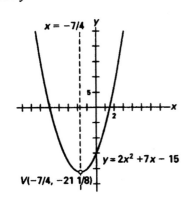

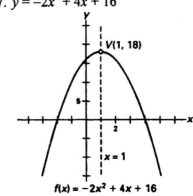

29. a)   Maximum occurs at $x = -\dfrac{b}{2a} = 512$; $h(512) = 1024$ meters.

     b)   horizontal and vertical directions from 0 to 1024 meters.

31. $2\ell + 2w = 240$; $A = \ell w = \ell(120 - \ell)$, $A = -\ell^2 + 120\ell$

     maximum at $\ell = \dfrac{-b}{2a} = 60$, $A = 60(60) = 3600$ sq. m.

Section 1.8

1.   3.

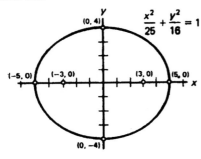

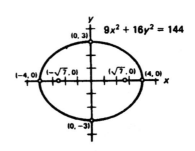

5.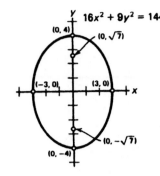

7.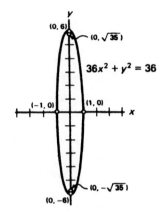

9. $\dfrac{x^2}{16} + \dfrac{y^2}{12} = 1$
$3x^2 + 4y^2 = 48$

11. $\dfrac{x^2}{45} + \dfrac{y^2}{81} = 1$
$9x^2 + 5y^2 = 405$

13. $\dfrac{x^2}{36} + \dfrac{y^2}{25} = 1$
$25x^2 + 36y^2 = 900$

15. $\dfrac{x^2}{39} + \dfrac{y^2}{64} = 1$
$64x^2 + 39y^2 = 2496$

17. $\dfrac{x^2}{5600^2} + \dfrac{y^2}{5000^2} = 1$; $625x^2 + 784y^2 = 1.96 \times 10^{10}$

Section 1.9

1.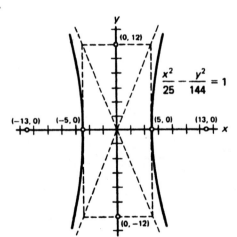

3.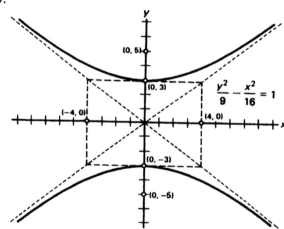

Sections 1.8 – 1.9

5.

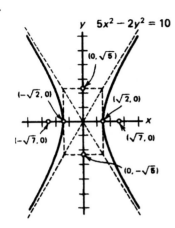

7.

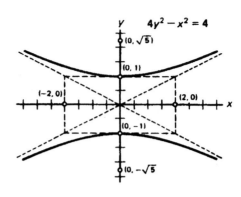

9. $\dfrac{x^2}{16} - \dfrac{y^2}{20} = 1$
$5x^2 - 4y^2 = 80$

11. $\dfrac{y^2}{36} - \dfrac{x^2}{28} = 1$
$7y^2 - 9x^2 = 252$

13. $\dfrac{x^2}{9} - \dfrac{y^2}{25} = 1$
$25x^2 - 9y^2 = 225$

15. $\dfrac{x^2}{25} - \dfrac{y^2}{11} = 1$
$11x^2 - 25y^2 = 275$

17.

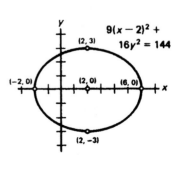

Section 1.10

1. $\dfrac{(x-1)^2}{16} + \dfrac{(y+1)^2}{12} = 1$
3. $\dfrac{(y-1)^2}{36} - \dfrac{(x-1)^2}{28} = 1$
5. $(y+1)^2 = 8(x-3)$

7. parabola; $V(2, -3)$
9. hyperbola; $C(-2, 0)$
11. ellipse; $C(2, 0)$

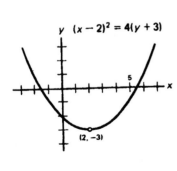

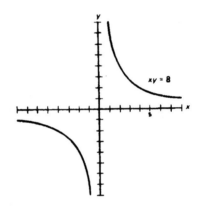

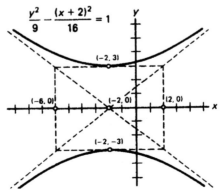

Sections 1.9 – 1.10

13. ellipse; $C(3, 1)$

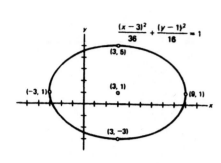

$$\frac{(x-3)^2}{36} + \frac{(y-1)^2}{16} = 1$$

15. parabola; $V(1, -3)$

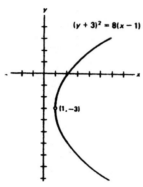

$(y+3)^2 = 8(x-1)$

17. hyperbola; $C(-1, -1)$

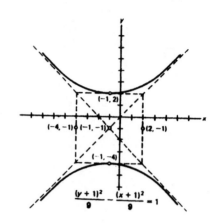

$$\frac{(y+1)^2}{9} - \frac{(x+1)^2}{9} = 1$$

19. parabola; $V(2, -1)$

$(x-2)^2 = -2(y+1)$

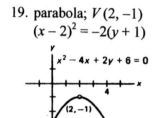

$x^2 - 4x + 2y + 6 = 0$

21. ellipse; $C(-2, 1)$

$$\frac{(x+2)^2}{16} + \frac{(y-1)^2}{4} = 1$$

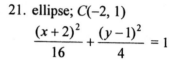

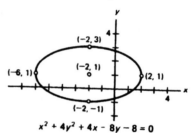

$x^2 + 4y^2 + 4x - 8y - 8 = 0$

23. hyperbola; $C(1, 1)$

$$\frac{(x-1)^2}{1} - \frac{(y-1)^2}{4} = 0$$

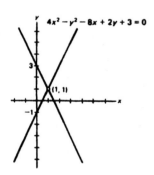

$4x^2 - y^2 - 8x + 2y + 3 = 0$

25. hyperbola; $C(-3, 3)$

$$\frac{(y-3)^2}{4} - \frac{(x+3)^2}{25} = 1$$

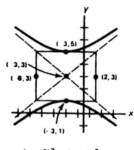

$$\frac{(y-3)^2}{4} - \frac{(x+3)^2}{25} = 1$$

Section 1.10

27. parabola; $V(-8, -2)$

$$(x + 8)^2 = -12(y + 2)$$

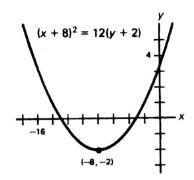

29. ellipse; $C(-6, -2)$

$$\frac{(x + 6)^2}{16} + \frac{(y + 2)^2}{64} = 1$$

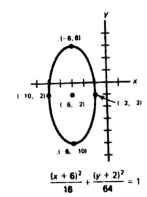

Section 1.11

1. A and C both positive and not equal, thus ellipse

3. $A = 0$ and C not zero, thus parabola

5. $A > 0$, $B < 0$, thus hyperbola

7. $A = C$, thus circle

9. $A = C$, thus circle

11. A and C both positive and not equal, thus ellipse

13. $A > 0$, $B < 0$, thus hyperbola

15. $A = 0$ and C not zero, thus parabola

Section 1.12

1. $x^2 = 3y$ and $y = 2x - 3$; substitute, $x^2 = 3(2x - 3)$,
    $x^2 - 6x + 9 = 0$, $x = 3$, $(3, 3)$.

3. $x^2 + 4x + y^2 - 8 = 0$ and $x^2 + y^2 = 4$; subtract, $4x - 4 = 0$,
    $x = 1$, $(1, \sqrt{3})$, $(1, -\sqrt{3})$.

5. $y^2 - x^2 = 12$ and $x^2 = 4y$; substitute, $y^2 - 4y - 12 = 0$,
    $(y - 6)(y + 2) = 0$, $y = 6$, y cannot be negative, $(2\sqrt{6}, 6)$, $(-2\sqrt{6}, 6)$.

7. $x^2 + y^2 = 4$ and $x^2 - y^2 = 4$; add, $2x^2 = 8$, $x \pm 2$, $(2, 0)$, $(-2, 0)$.

9. $x^2 = 6y$ and $y = 6$; substitute, $x^2 = 6(6)$, $x = \pm 6$, $(6, 6)$, $(-6, 6)$.

Sections $1.10 - 1.12$

11. $y^2 = 4x + 12$ and $y^2 = -4x - 4$; substitute, $4x + 12 = -4x - 4$, $8x = -16$, $x = -2$, $(-2, 2)$, $(-2, -2)$.

13. $x^2 + y^2 = 36$ and $y = x^2$; substitute, $y + y^2 = 36$, $y = \dfrac{-1 \pm \sqrt{145}}{2}$,

$y = \dfrac{-1 + \sqrt{145}}{2} \approx 5.5$, y cannot be negative. Approximate points $(2.3, 5.5)$, $(-2.3, 5.5)$.

15. $x^2 - y^2 = 9$ and $x^2 + 9y^2 = 169$; subtract, $10y^2 = 160$, $y^2 = 16$, $y = \pm 4$, $(5, 4)$, $(5, -4)$ $(-5, 4)$, $(-5, -4)$.

17. $x^2 + y^2 = 17$ and $xy = 4$; substitute, $\dfrac{16}{2} + y^2 = 17$,

$y^4 - 17y^2 + 16 = 0$, $(y^2 - 16)(y^2 - 1) = 0$, $y = \pm 4$ or ± 1.
$(1, 4)$, $(-1, -4)$, $(4, 1)$, $(-4, -1)$.

Section 1.13

1.
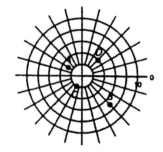

3.
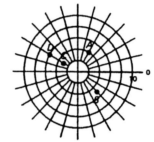

5. $(-3, 240°)$, $(-3, -120°)$, $(3, -300°)$

7. $(5, 135°)$, $(-5, -45°)$, $(5, -225°)$

9. $(-4, -315°)$, $(-4, 45°)$, $(4, 225°)$

11. $(-3, 7\pi/6)$, $\left(-3, \dfrac{-5\pi}{6}\right)$, $\left(3, \dfrac{-11\pi}{6}\right)$

13. $(9, 5\pi/3)$, $(9, -\pi/3)$, $(-9, -4\pi/3)$

15. $(4, -3\pi/4)$, $(-4, \pi/4)$, $(4, 5\pi/4)$

17. $r = 10 \sin \theta$

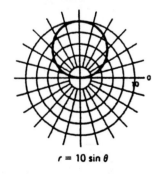

$r = 10 \sin \theta$

19. $r = 4 + 4 \cos \theta$

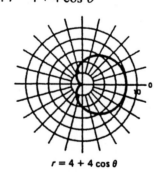

$r = 4 + 4 \cos \theta$

Sections 1.12 − 1.13

21. $r \cos \theta = 4$

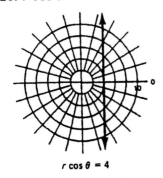

$r \cos \theta = 4$

23. $r = -10 \cos \theta$

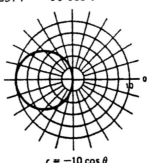

$r = -10 \cos \theta$

25. $r = 4 - 4 \sin \theta$

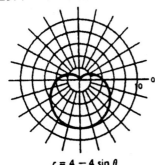

$r = 4 - 4 \sin \theta$

27. $r = \theta$

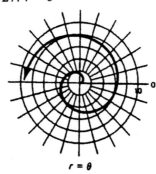

$r = \theta$

Note: A sample of the different kinds of exercises is given here. The other exercises are done in a similar manner.

29. $x = r \cos \theta, x = 3 \cos 30°, x = \dfrac{3\sqrt{3}}{2}$; $y = r \sin \theta, y = 3 \sin 30°, y = 3\left(\dfrac{1}{2}\right) = \dfrac{3}{2}$. $\left(\dfrac{3\sqrt{3}}{2}, \dfrac{3}{2}\right)$

31. $x = r \cos \theta, x = 2 \cos \dfrac{\pi}{3}, x = 2\left(\dfrac{1}{2}\right) = 1$; $y = r \sin \theta, y = 2 \sin \dfrac{\pi}{3}, y = 2\left(\dfrac{\sqrt{3}}{2}\right) = \sqrt{3}$. $(1, \sqrt{3})$

33. $(2\sqrt{3}, -2)$ 35. $(0, 6)$ 37. $(2.5, -4.33)$ 39. $(1.4, 1.4)$

41. $x^2 + y^2 = r^2, 5^2 + 5^2 = r^2, 50 = r^2, r = 5\sqrt{2} \approx 7.1$;

$\theta = \arctan\dfrac{y}{x}, \theta = \arctan\dfrac{5}{5}, \theta = 45°; (7.1, 45°)$.

43. $(4, 90°)$ 45. $(4, 240°)$ 47. $\left(4\sqrt{2}, \dfrac{3\pi}{4}\right)$

49. $x^2 + y^2 = r^2, \left(\sqrt{-6}^2\right) + (\sqrt{2})^2 = r^2, 6 + 2 = r^2, r = 2\sqrt{2}$;

$\theta = \arctan\dfrac{y}{x}, \theta = \arctan\dfrac{\sqrt{2}}{-\sqrt{6}}, \theta = \arctan\left(-\dfrac{1}{\sqrt{3}}\right), \theta = \dfrac{5\pi}{6}, \left(2\sqrt{2}, \dfrac{5\pi}{6}\right)$

Section 1.13

51. $\left(4, \dfrac{3\pi}{2}\right)$ 53. $r\,\cos\theta = 3$ 55. $r = 6$

57. $x^2 + y^2 + 2x + 5y = 0$, $r^2 + 2r\cos\theta + 5r\sin\theta = 0$, $r + 2\cos\theta + 5\sin\theta = 0$

59. $4r\cos\theta - 3r\sin\theta = 12$, $r = \dfrac{12}{(4\cos\theta - 3\sin\theta)}$.

61. $9(r\cos\theta)^2 + 4(r\sin\theta)^2 = 36$, $9r^2\cos^2\theta + 4r^2\sin^2\theta = 36$,

$r^2(9(1 - \sin^2\theta) + 4\sin^2\theta) = 36$, $r^2(9 - 5\sin^2\theta) = 36$, $r^2 = \dfrac{36}{(9 - 5\sin^2\theta)}$.

63. $(r\cos\theta)^3 = 4(r\sin\theta)^2$, $r^3\cos^3\theta = 4r^2\sin^2\theta$, $r = \dfrac{4\sin^2\theta}{\cos^3\theta}$, $r = 4\sec\theta\tan^2\theta$.

65. $y = -3$ 67. $x^2 + y^2 = 25$ 69. $\tan\theta = \tan\dfrac{\pi}{4}$, $\dfrac{y}{x} = 1$, $y = x$

71. $r^2 = 5r\cos\theta$. $x^2 + y^2 = 5x$

73. $r^2 = 6r\cos\left(\theta + \dfrac{\pi}{3}\right)$, $r^2 = 6r\left[\cos\theta\cos\dfrac{\pi}{3} - \sin\theta\sin\dfrac{\pi}{3}\right]$,

$r^2 = 6r\left[\dfrac{1}{2}\cos\theta - \dfrac{\sqrt{3}}{2}\sin\theta\right]$, $x^2 + y^2 = 3x - 3\sqrt{3}y$

75. $r\sin\theta\sin\theta = 3\cos\theta$, $y = \dfrac{3\cos\theta}{\sin\theta}$, $y = 3\cot\theta$, $y = 3\left(\dfrac{x}{y}\right)$, $y^2 = 3x$

77. $r^2(2\sin\theta\cos\theta) = 2$, $2r\sin\theta\,r\cos\theta = 2$, $xy = 1$

79. $r^4 = 2r\sin\theta\,r\cos\theta$, $\left(x_2^2 + y^2\right)^2 = 2xy$, $x^4 + 2x^2y^2 + y^4 - 2xy = 0$

81. $r^2 = \tan^2\theta$, $x^2 + y^2 = \left(\dfrac{y}{x}\right)^2$, $x^2(x^2 + y^2) = y^2$

83. $r + r\sin\theta = 3$, $\sqrt{x^2 + y^2} = 3 - y$, $x^2 + y^2 = (3 - y)^2$,

$x^2 + y^2 = 9 - 6y + y^2$, $x^2 + 6y - 9 = 0$

85. $r = 4\sin(2\theta + \theta)$, $r = 4[\sin 2\theta\cos\theta + \cos 2\theta\sin\theta]$

$r = 4[2\sin\theta\cos\theta\cos\theta + \sin\theta(\cos^2\theta - \sin^2\theta)]$,

$r = 4[2\sin\theta\cos^2\theta + \sin\theta\cos^2\theta - \sin^3\theta]$,

$r = 4[3\sin\theta\cos^2\theta - \sin^3\theta]$

$r^4 = 12r\sin\theta\,r^2\cos^2\theta - 4r^3\sin^3\theta$

$(x^2 + y^2)^2 = 12yx^2 - 4y^3$

Section 1.13

87. $r^2 = 2r + 4r\sin\theta$, $x^2 + y^2 = 2\left(\pm\sqrt{x^2 + y^2 + 4y}\right)$

$(x^2 + y^2 - 4y)^2 = 2\left(\pm\sqrt{x^2 + y^2}\right)^2$

$x^4 + 2x^2y^2 - 8x^2y - 8y^3 + 16y^2 + y^4 = 4(x^2 + y^2)$
$x^4 + 2x^2y^2 - 8x^2y - 8y^3 + 12y^2 + y^4 - 4x^2 = 0$

89. $\sqrt{13}$

91. $P_1(r_1\cos\theta_1, r_1\sin\theta_1)$, $P_2(r_2\cos\theta_2, r_2\sin\theta_2)$;
$d^2 = (r_1\cos\theta_1 - r_2\cos\theta_2)^2 + (r_1\sin\theta_1 - r_2\sin\theta)^2$
$d^2 = r_1^2\cos^2\theta_1 - 2r_1\cos\theta_1\, r_2\cos_2 + r_2^2\cos^2\theta_2 + r_1^2\sin^2\theta_1 - 2r_1\sin\theta_1\, r_2\sin\theta_2 + r_2^2\sin^2\theta_2$
$d^2 = r_1^2\left(\sin^2\theta_1 + \cos^2\theta_1\right) + r_2^2\left(\sin^2\theta_2 + \cos^2\theta_2\right) - 2r_1r_2(\cos\theta_1\cos\theta_2 + \sin\theta_1\sin\theta_2)$
$d^2 = r_1^2 + r_2^2 - 2r_1r_2\cos(\theta_1 - \theta_2)$
$d = \sqrt{r_1^2 + r_2^2 - 2r_1r_2\cos(\theta_1 - \theta_2)}$

Section 1.14

1.

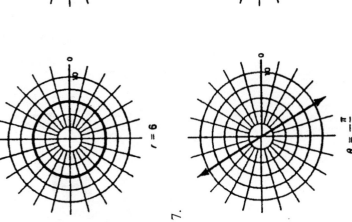

$r = 6$

3.

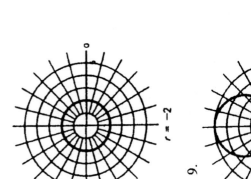

$r = -2$

5.

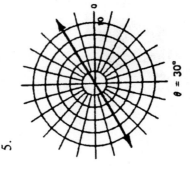

$\theta = 30°$

7.

$\theta = -\dfrac{\pi}{3}$

9.

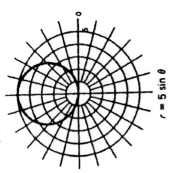

$r = 5\sin\theta$

11.

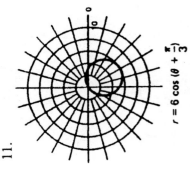

$r = 6\cos\left(\theta + \dfrac{\pi}{3}\right)$

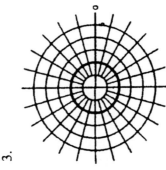

Sections 1.13 − 1.14

Chapter 1

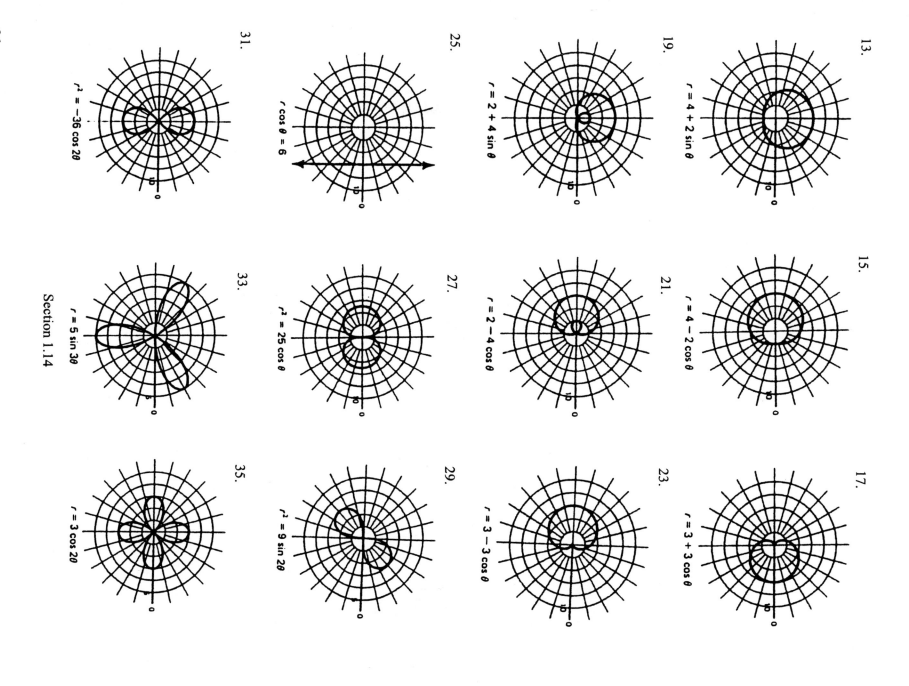

13. $r = 4 + 2\sin\theta$

15. $r = 4 - 2\cos\theta$

17. $r = 3 + 3\cos\theta$

19. $r = 2 + 4\sin\theta$

21. $r = 2 - 4\cos\theta$

23. $r = 3 - 3\cos\theta$

25. $r\cos\theta = 6$

27. $r^2 = 25\cos\theta$

29. $r^2 = 9\sin 2\theta$

31. $r^2 = -36\cos 2\theta$

33. $r = 5\sin 3\theta$

35. $r = 3\cos 2\theta$

Section 1.14

Solution to Odd-Numbered Exercises

37.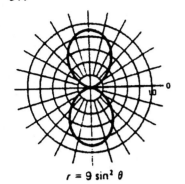

$r = 9 \sin^2 \theta$

39.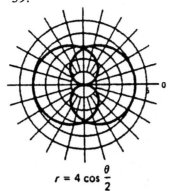

$r = 4 \cos \dfrac{\theta}{2}$

41.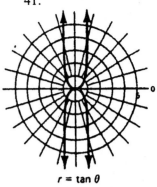

$r = \tan \theta$

Chapter 1 Review

	Function	Domain	Range
1.	Yes	$\{2,3,4,5\}$	$\{3,4,5,6\}$
2.	No	$\{2,4,6\}$	$\{1,3,4,6\}$
3.	Yes	Real Numbers	Real Numbers
4.	Yes	Real Numbers	Real Numbers where $y \geq -5$

5.
$$x = y^2 + 4$$
$$x - 4 = y^2$$
$$\pm\sqrt{x-4} = y$$
No

Domain: Real Numbers where $x \geq 4$

Range: Real Numbers

6.
$$y = \sqrt{4 - 8x}$$
Yes

Domain: Real Numbers where $x \leq \dfrac{1}{2}$

Range: Real Numbers where $y \geq 0$

7. $f(x) = 5x + 14$

(a) $f(2) = 5(2) + 14 = 24$

(b) $f(0) = 5(0) + 14 = 14$

(c) $f(-4) = 5(-4) + 14 = -6$

8. $g(t) = 3t^2 + 5t - 12$

(a) $g(2) = 3(2)^2 + 5(2) - 12$
$g(2) = 3(4) + 10 - 12$
$g(2) = 12 + 10 - 12$
$g(2) = 10$

(b) $g(0) = 3(0)^2 + 5(0) - 12$
$g(0) = 0 + 0 - 12$
$g(0) = -12$

(c) $g(-5) = 3(-5)^2 + 5(-5) - 12$
$g(-5) = 3(25) - 25 - 12$
$g(-5) = 75 - 25 - 12$
$g(-5) = 38$

Section 1.14 – Chapter 1 Review

9. $h(x) = \dfrac{4x^2 - 3x}{2\sqrt{x-1}}$

(a) $h(2) = \dfrac{4(2)^2 - 3(2)}{2\sqrt{2-1}}$

$h(2) = \dfrac{4(4) - 6}{2\sqrt{1}}$

$h(2) = \dfrac{16 - 6}{2}$

$h(2) = 5$

(b) $h(5) = \dfrac{4(5)^2 - 3(5)}{2\sqrt{5-1}}$

$h(5) = \dfrac{4(25) - 15}{2\sqrt{4}}$

$h(5) = \dfrac{100 - 15}{2(2)}$

$h(5) = \dfrac{85}{4}$

(c) $h(-15)$

-15 is not in the domain of $h(x)$.

10. $g(x) = x^2 - 6x + 4$

(a) $g(a) = a^2 - 6a + 4$

(b) $g(2x) = (2x)^2 - 6(2x) + 4$

$g(2x) = 4x^2 - 12x + 4$

(c) $g(z-2) = (z-2)^2 - 6(z-2) + 4$

$g(z-2) = z^2 - 4z + 4 - 6z + 12 + 4$

$g(z-2) = z^2 - 10z + 20$

11. $y = 4x - 5$

x	y
−1	−9
0	−5
1	−1
2	3

12. $y = x^2 + 4$

x	y	x	y
−3	13	1	5
−2	8	2	8
−1	5	3	13
0	4	4	20

13. $y = x^2 + 2x - 8$

x	y	x	y
−4	0	0	−8
−3	−5	1	−5
−2	−8	2	0
−1	−9	3	7

14. $y = 2x^2 + x - 6$

x	y	x	y
−3	9	1	−3
−2	0	2	4
−1	−5	3	15
0	−6	4	30

15. $y = -x^2 - x + 4$

x	y	x	y
−4	−8	0	4
−3	−2	1	2
−2	2	2	−2
−1	4	3	−8

16. $y = \sqrt{2x}$

x	y	x	y
0	0	4	2.8
1	1.4	5	3.2
2	2	6	3.5
3	2.4	7	3.7
		8	4

17. $y = \sqrt{-2 - 4x}$

x	y	x	y
−1	−1.4	−9	5.8
−3	3.2	−11	6.5
−5	4.2	−13	7.1
−7	5.1	−15	7.6

18. $y = x^3 - 6x$

x	y	x	y
−3	−9	1	−5
−2	4	2	−4
−1	5	3	−9
0	0	4	40

Chapter 1 Review

19. $y = x^2 + 4$

    (a) $y_1 = x^2 + 4$

        $y_2 = 5$

        Find values of x
for which the
graphs intersect.

        $5 = x^2 + 4$

        $1 = x^2$

        $x = 1$ and -1

    (b) $y_1 = x^2 + 4$

        $y_2 = 7$

        Find values of x
for which the
graphs intersect.

        $7 = x^2 + 4$

        $3 = x^2$

        $x = \pm\sqrt{3}$

        $x = 1.7$ and -1.7

    (c) $y_1 = x^2 + 4$

        $y_2 = 2$

        Find values of x
for which the
graphs intersect.

        These graphs do

        not intersect.

        no solution

20.   (a) $y_1 = x^2 + 2x - 8$

        $y_2 = 0$

        Find values of x
for which the
graphs intersect.

        $0 = x^2 + 2x - 8$

        $0 = (x + 4)(x - 2)$

        $x = -4$ and 2

    (b) $y_1 = x^2 + 2x - 8$

        $y_2 = -2$

        Find values of x
for which the
graphs intersect.

        $-2 = x^2 + 2x - 8$

        $0 = x^2 + 2x - 6$

        By graph or
quadratic formula,

        $x = 1.6$ and -3.6

    (c) $y_1 = x^2 + 2x - 8$

        $y_2 = 3$

        Find values of x
for which the
graphs intersect.

        $3 = x^2 + 2x - 8$

        $0 = x^2 + 2x - 11$

        By graph or
quadratic formula,

        $x = 2.5$ and -4.5

21.   (a) $y_1 = -x^2 - x + 4$

        $y_2 = 2$

        Find values of x
for which the
graphs intersect.

        $2 = -x^2 - x + 4$

        $x^2 + x - 2 = 0$

        $(x + 2)(x - 1) = 0$

        $x = -2$ and 1

    (b) $y_1 = -x^2 - x + 4$

        $y_2 = 0$

        Find values of x
for which the
graphs intersect.

        $0 = -x^2 - x + 4$

        By graph or
quadratic formula,

        $x = -2.6$ and 1.6

    (c) $y_1 = -x^2 - x + 4$

        $y_2 = -2$

        Find values of x
for which the
graphs intersect.

        $-2 = -x^2 - x + 4$

        $x^2 + x - 6 = 0$

        $(x + 3)(x - 2) = 0$

        $x = -3$ and 2

22.   (a) $y_1 = x^3 - 6x$

        $y_2 = 0$

        Find values of x
for which the
graphs intersect.

        $0 = x^3 - 6x$

        $0 = x(x^2 - 6)$

        $x = 0$ and $\pm\sqrt{6}$

        $x = 0, 2.4$ and -2.4

    (b) $y_1 = x^3 - 6x$

        $y_2 = 2$

        Find values of x
for which the
graphs intersect.

        $2 = x^3 - 6x$

        $0 = x^3 - 6x - 2$

        By graph,

        $x = -2.3, -0.3$, and 2.6

    (c) $y_1 = x^3 - 6x$

        $y_2 = -3$

        Find values of x
for which the
graphs intersect.

        $-3 = x^3 - 6x$

        $0 = x^3 - 6x + 3$

        By graph

        $x = -2.7, 0.5$, and 2.1

Chapter 1 Review

23. (a) $i_1 = 2t^2$

$i_2 = 2$

Find values of t for which the graphs intersect.

$2 = 2t^2$

$1 = t^2$

$t = 1$

(b) $i_1 = 2t^2$

$i_2 = 6$

Find values of t for which the graphs intersect.

$6 = 2t^2$

$3 = T^2$

$T = 1.7$

(c) $i_1 = 2t^2$

$i_2 = 8$

Find values of t for which the graphs intersect.

$8 = 2t^2$

$4 = t^2$

$t = 2$

24. (a) $V_1 = 4t^3 + t$

$V_2 = 40$

Find values of t for which the graphs intersect.

$0 = 4t^3 + t$

$0 = 4t^3 + t - 40$

By graph,

$t = 2.1$

(b) $V_1 = 4t^3 + t$

$V_2 = 60$

Find values of t for which the graphs intersect.

$60 = 4t^3 + t$

$0 = 4t^3 + t - 60$

By graph,

$t = 2.4$

25. $(3, -4)$ and $(-6, -2)$

$m = \dfrac{y_2 - y_1}{x_2 - x_1} = \dfrac{-4 - (-2)}{3 - (-6)}$

$m = \dfrac{-2}{9}$

26. $(3, -4)$ and $(-6, -2)$

$d = \sqrt{(x_2 - x_1)^2 + (y_2 - y_1)^2}$

$d = \sqrt{(3 - (-6))^2 + (-4 - (-2))^2}$

$d = \sqrt{9^2 + (-2)^2} = \sqrt{81 + 4}$

$d = \sqrt{85}$

27. $(3, -4)$ and $(-6, -2)$

$x_m = \dfrac{x_1 + x_2}{2}$

$x_m = \dfrac{3 + (-6)}{2} = \dfrac{-3}{2}$

$y_m = \dfrac{y_1 + y_2}{2}$

$y_m = \dfrac{-4 + (-2)}{2}$

$y_m = -3$

$(-1.5, -3)$

28. $(4, 7)$ and $(6, -4)$

$m = \dfrac{y_2 - y_1}{x_2 - x_1}$

$m = \dfrac{7 - (-4)}{4 - 6} = \dfrac{11}{-2}$

Use either point;

for $x = 4$, $y = 7$

$y = mx + b$

$7 = -\dfrac{11}{2}(4) + b$

$7 = -22b$

$29 = b$

$y = -\dfrac{11}{2}x + 29$

$2y = -11x + 58$

$11 + 2y - 58 = 0$

Chapter 1 Review

29.

$$m = \frac{2}{3} \text{ and } (-3,1)$$
$$y = mx + b$$
$$x = -3, y = 1$$
$$1 = -3\left(\frac{2}{3}\right) + b$$
$$1 = -2 + b$$
$$3 = b$$
$$y = \frac{2}{3}x + 3$$
$$3y = 2x + 9$$
$$0 = 2x - 3y + 9$$

30.

$$m = -\frac{1}{3} \text{ and}$$
y-intercept is $(0,-3)$.
$$y = mx + b$$
$$y = -\frac{1}{3}x - 3$$
$$3y = -x - 9$$
$$x + 3y + 9 = 0$$

31. Parallel to the y-axis is a vertical line, $x = a$. Three units left of the y-axis means $x = -3$

32.

$$3x - 2y - 6 = 0$$
$$-2y = -3x + 6$$
$$y = \frac{3}{2}x - 3$$
$$m = \frac{3}{2} \text{ and}$$
y-intercept is -3.

33.

$$3x - 4y = 12$$
$$-4y = -3x + 12$$
$$y = \frac{3}{4}x - 3$$

x	y
-4	-6
0	-3
4	0
8	3

34.

$$x - 2y + 3 = 0$$
$$-2y = -x - 3$$
$$y = \frac{1}{2}x + \frac{3}{2}$$
and $8x + 4y - 9 = 0$
$$4y = -8x + 9$$
$$y = -2x + \frac{9}{4}$$
Lines are perpendicular because $m_1 m_2 = -1$

35.

$$2x - 3y + 4 = 0$$
$$-3y = -2x - 4$$
$$y = \frac{2}{3}x + \frac{4}{3}$$
and $-8x + 12y = 16$
$$12y = 8x + 16$$
$$y = \frac{2}{3}x + \frac{4}{3}$$
Parallel and the same line since the slopes and y-intercepts are equal.

36.

$$3x - 2y + 5 = 0$$
$$-2y = -3x - 5$$
$$y = \frac{3}{2}x + \frac{5}{2}$$
and $2x - 3y + 9 = 0$
$$-3y = -2x - 9$$
$$y = \frac{2}{3}x + 3$$
neither

37. $x = 2$ and $y = -3$
A vertical line and a horizontal line are perpendicular.

38. $x = 4$ and $x = 7$
Two vertical lines are parallel.

Chapter 1 Review

39. Parallel to
$$2x - y + 4 = 0$$
$$-y = -2x - 4$$
$$y = 2x + 4$$
Desired slope is 2.
For the point,
$$x = 5, y = 2$$
$$y = mx + b$$
$$2 = 2(5) + b$$
$$2 = 10 + b$$
$$-8 = b$$
$$y = 2x - 8$$
$$0 = 2x - y - 8$$

40. Perpendicular to
$$3x + 5y - 6 = 0$$
$$5y = -3x + 6$$
$$y = -\frac{3}{5}x + \frac{6}{5}$$
Desired slope is $\frac{5}{3}$.
For the point,
$$x = -4, y = 0$$
$$y = mx + b$$
$$0 = \frac{5}{3}(-4) + b$$
$$0 = \frac{-20}{3} + b$$
$$\frac{20}{3} = b$$
$$y = \frac{5}{3}x + \frac{20}{3}$$
$$3y = 5x + 20$$
$$0 = 5x - 3y + 20$$

41. $(x - 5)^2 + (y + 7)^2 = 36$
   $x^2 + y^2 - 10x + 14y + 38 = 0$

42. $x^2 - 8x + y^2 + 6y = 24$
   $x^2 - 8x + 16 + y^2 + 6y + 9 = 24 + 16 + 9$
   $(x - 4)^2 + (y + 3)^2 = 49$, thus $C(4, -3); r = 7$

43. $F(0, 3/2)$; directrix $y = -3/2$

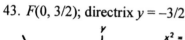
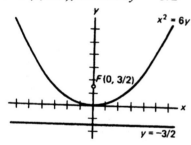

44. $y^2 = -16x$

45. $(y - 3)^2 = 8(x - 2)$
   $y^2 - 6y - 8x + 25 = 0$

46. $\dfrac{x^2}{49} + \dfrac{y^2}{4} = 1; V(\pm 7, 0); F\left(\pm 3\sqrt{5}, 0\right)$

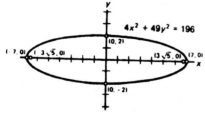

47. $\dfrac{x^2}{4} + \dfrac{y^2}{16} = 1$
   $4x^2 + 16y^2 = 16$

Chapter 1 Review

32

48. $\dfrac{x^2}{36} - \dfrac{y^2}{16} = 1$; $V(\pm 6, 0)$; $F\left(\pm 2\sqrt{13}, 0\right)$

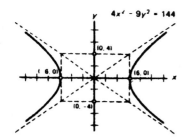

49. $\dfrac{y^2}{25} - \dfrac{x^2}{16} = 1$

$16y^2 - 25x^2 = 400$

50. $\dfrac{(x-3)^2}{9} + \dfrac{(y+4)^2}{25} = 1$

51. $\dfrac{(x+7)^2}{81} - \dfrac{(y-4)^2}{9} = 1$

52. $16x^2 - 64x - 4y^2 - 24y = -12$

$16(x^2 - 4x) - 4(y^2 + 6y) = -12$

$16(x^2 - 4x + 4) - 4(y^2 + 6y + 9)$
$\quad = -12 - 36 + 64$

$16(x-2)^2 - 4(y+3)^2 = 16$

$\dfrac{(x-2)^2}{1} - \dfrac{(y+3)^2}{4} = 1$

hyperbola; $C(2, -3)$

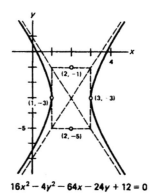

53. $y^2 + 4y + x = 0$

$\qquad\qquad x = 2y$

$y^2 + 4y + 2y = 0$
$\qquad y^2 + 6y = 0$
$\qquad y(y + 6) = 0$

$y = 0$ or -6

$(0, 0), (-12, -6)$

54. $3x^2 - 4y^2 = 36$

$\quad 5x^2 - 8y^2 = 56$

$-6x^2 + 8y^2 = -72$
$\quad 5x^2 - 8y^2 = 56$

$-x^2 = -16$

$x = 4$ or -4

$\left(4\sqrt{3}\right), \left(4, -\sqrt{3}\right), \left(-4, \sqrt{3}\right), \left(-4, -\sqrt{3}\right)$

55.

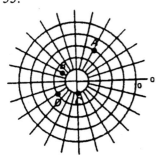

56.

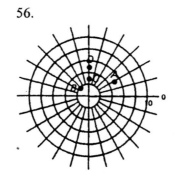

Chapter 1 Review

57. (5, –225°)
   (–5, –45°)
   (–5, 315°)

58. (2, –11π/6), (–2, –5π/6), (2, π/6)

59. a) (–2.6, –1.5)   b) (–1, –1.7)   c) (–4.3, 2.5)   d) (0, 6)

60. a) (4.2, 135°)   b) (6, 270°)   c) (2, 120°)

61. a) (5, π)   b) (12, 5π/6)   c) ($\sqrt{2}$, 7π/4)

62. $r = 7$ 63. $r \sin^2\theta = 9 \cos \theta$ 64. $r = 8/(5 \cos \theta + 2 \sin \theta)$

65. $r^2 = 12/(1 – 5 \sin^2\theta)$ 66. $r = 6 \csc \theta \cot^2 \theta$ 67. $r = \cos \theta \cot \theta$

68. $x = 12$ 69. $x^2 + y^2 = 81$ 70. $y = -\sqrt{3}x$ 71. $x^2 + y^2 – 8x = 0$

72. $y^2 = 5x$ 73. $xy = 4$ 74. $x^4 + 2x^2y^2 + y^4 + 4y^2 – 4x^2 = 0$

75. $y = 1$ 76. $x^4 + y^4 + 2x^2y^2 – 2x^2y – 2y^3 – x^2 = 0$ 77. $x^2 = 4(y + 1)$

78.

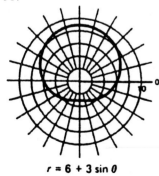

$r = 7$

79.

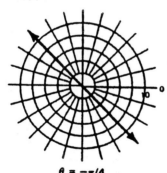

$\theta = -\pi/4$

80.

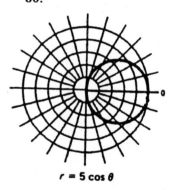

$r = 5 \cos \theta$

81.

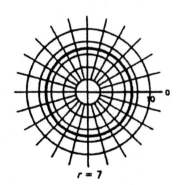

$r = 6 + 3 \sin \theta$

82.

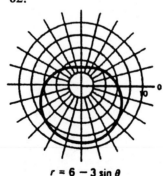

$r = 6 – 3 \sin \theta$

83.

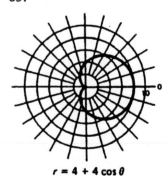

$r = 4 + 4 \cos \theta$

Chapter 1 Review

84.

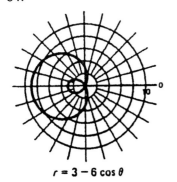

$r = 3 - 6 \cos \theta$

85.

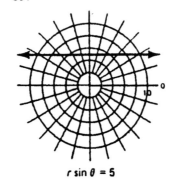

$r \sin \theta = 5$

86.

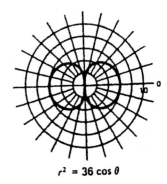

$r^2 = 36 \cos \theta$

87.

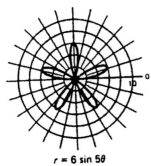

$r = 6 \sin 5\theta$

88.

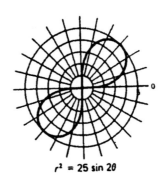

$r^2 = 25 \sin 2\theta$

89.

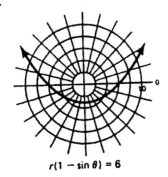

$r(1 - \sin \theta) = 6$

Chapter 1 Review

CHAPTER 2

Section 2.1

1. $f(1) = 2(1)^2 + 7 = 9$

3. $h(-2) = 3(-2)^3 - 2(-2) + 4 = -16$

5. $f(-2) = \dfrac{2^2 - 3}{2 + 5} = \dfrac{1}{7}$

7. $f(-5) = \sqrt{(-5)^2 + 3} = 2\sqrt{7}$

9. $f(h + 3) = 3(h + 3) - 2 = 3h + 7$

11. $f(2 + \Delta t) = 3(2 + \Delta t)^2 + 2(2 + \Delta t) - 5 = 11 + 14\Delta t + 3(\Delta t)^2$

13. a) $\Delta s = f(t + \Delta t) - f(t) = [3(t + \Delta t) - 4] - [3t - 4] = 3\Delta t$

    b) $\bar{v} = \dfrac{\Delta s}{\Delta t} = \dfrac{3\Delta t}{\Delta t} = 3$

15. a) $\Delta s = [2(t + \Delta t)^2 + 5] - [2t^2 + 5] = 4t\,\Delta t + 2(\Delta t)^2$

    b) $\bar{v} = \dfrac{4t\Delta t + 2(\Delta t)^2}{\Delta t} = 4t + 2\Delta t$

17. a) $\Delta s = [(t + \Delta t)^2 - 2(t + \Delta t) + 8] - [t^2 - 2t + 8] = 2t\,\Delta t + (\Delta t)^2 - 2\Delta t$

    b) $\bar{v} = \dfrac{2t\Delta t + (\Delta t)^2 - 2\Delta t}{\Delta t} = 2t + \Delta t - 2$

19. $\bar{v} = \dfrac{\Delta s}{\Delta t} = \dfrac{f(7) - f(3)}{4} = \dfrac{[5(7)^2 + 6] - [5(3)^2 + 2 + 4]}{4} = \dfrac{251 - 51}{4} = 50 \text{ m/s}$

    $t = 3,\ \Delta t = 4$

21. $\bar{v} = \dfrac{\Delta s}{\Delta t} = \dfrac{f(4) - f(2)}{2} = \dfrac{[3(4)^2 - 4 + 4] - [3(2)^2 - 2 + 4]}{2} = \dfrac{48 - 14}{2} = 17 \text{ m/s}$

    $t = 2,\ \Delta t = 2$

23. $\bar{v} = \dfrac{\Delta s}{\Delta t} = \dfrac{0.5 \text{ m} - 0.2 \text{ m}}{0.5 \times 10^{-6} \text{ s}} = 6 \times 10^5 \text{ m/s}$

25. $i_{av} = C\dfrac{\Delta V}{\Delta t} = 10\ \mu\text{F}\left[\dfrac{(5^2 + 3(5) + 160) - (2^2 + 3(2) + 160)}{3s}\right] = 100\ \mu\text{A}$

27. $\bar{v} = \dfrac{f(2 + \Delta t) - f(t)}{\Delta t} = \dfrac{[3(2 + \Delta t)^2 - 6(2 + \Delta t) + 1] - [3(2)^2 - 6(2) + 1]}{\Delta t}$

    $= \dfrac{12\Delta t + 3(\Delta t)^2 - 6\Delta t}{\Delta t} = 6 + 3\Delta t,\ \text{when } \Delta t \to 0,\ v = 6 \text{ m/s}$

Section 2.1

29. $\bar{v} = \dfrac{f(1 + \Delta t) - f(1)}{\Delta t} = \dfrac{[5(1 + \Delta t)^2 - 7] - [5(1)^2 - 7]}{\Delta t} = \dfrac{10\Delta t + 5(\Delta t)^2}{\Delta t}$

$= 10 + 5\Delta t$, when $\Delta t \to 0$, $v = 10$ m/s

31. $\bar{v} = \dfrac{f(3 + \Delta t) - f(3)}{\Delta t} = \dfrac{\dfrac{1}{2(3 + \Delta t)} - \dfrac{1}{2(3)}}{\Delta t} = \dfrac{1}{\Delta t} \cdot \dfrac{-2\,\Delta t}{6(6 + \Delta t)} = \dfrac{-2}{6(6 + \Delta t)}$,

when $\Delta t \to 0$, $v = -\dfrac{1}{18}$ m/s

33. $\bar{v} = \dfrac{f(4 + \Delta t) - f(4)}{\Delta t} = \dfrac{\dfrac{1}{(4 + \Delta t) - 2} - \dfrac{1}{4 - 2}}{\Delta t} = \dfrac{1}{\Delta t} \cdot \dfrac{-\Delta t}{2(2 + \Delta t)} = \dfrac{-1}{2(2 + \Delta t)}$,

when $\Delta t \to 0$, $v = -\dfrac{1}{4}$ m/s

35. $\bar{v} = \dfrac{\Delta s}{\Delta t} = \dfrac{16(2 + \Delta t)^2 - 16(2)}{\Delta t} = \dfrac{64\Delta t + 16(\Delta t)^2}{\Delta t} = 64 + 16\Delta t$, when $\Delta t \to 0$,

$v = 64$ ft/s

Section 2.2

1. $\displaystyle\lim_{x \to 2} \dfrac{x^2 - 4}{x - 2} = \lim_{x \to 2}(x + 2) = 4$

3. $\displaystyle\lim_{x \to \infty} \dfrac{3x + 2}{x} = \lim_{x \to \infty}(3 + 2/x) = 3$

5. $\displaystyle\lim_{x \to 0} \dfrac{\sin x}{x} = 1$

7. $\displaystyle\lim_{x \to 2}(x^2 - 5x) = 2^2 - 5(2) = -6$

9. $\displaystyle\lim_{x \to -1}(2x^3 + 5x^2 - 2) = 2(-1)^3 + 5(-1)^2 - 2 = 1$

11. $\displaystyle\lim_{x \to 1} \dfrac{x^2 - 1}{x - 1} = \lim_{x \to 1}(x + 1) = 2$

13. $\displaystyle\lim_{x \to -3/2} \dfrac{4x^2 - 9}{2x + 3} = \lim_{x \to -3/2}(2x - 3) = -6$

15. $\displaystyle\lim_{x \to -1} \sqrt{2x + 3} = 1$

17. $\displaystyle\lim_{x \to 6} \sqrt{4 - x} = $ no limit (does not exist)

19. $\displaystyle\lim_{x \to \infty} \dfrac{1}{2x} = 0$

21. $\displaystyle\lim_{x \to \infty} \dfrac{3x^2 - 5x + 2}{4x^2 + 8x - 11} = \lim_{x \to \infty} \dfrac{3 - \dfrac{5}{x} + \dfrac{2}{x^2}}{4 + \dfrac{8}{x} - \dfrac{11}{x^2}} = \dfrac{3}{4}$

Sections 2.1 – 2.2

23. $v = \lim\limits_{\Delta t \to \infty} \dfrac{f(2 + \Delta t) - f(2)}{\Delta t} = \lim\limits_{\Delta t \to \infty} \dfrac{[4(2 + \Delta t)^2 - 3(2 + \Delta t)] - [4(2)^2 - 3(2)]}{\Delta t}$

$= \lim\limits_{\Delta t \to 0} \dfrac{13\Delta t + 4(\Delta t)^2}{\Delta t} = \lim\limits_{\Delta t \to 0}(13 + 4\Delta t) = 13$

25. $v = \lim\limits_{\Delta t \to 0} \dfrac{f(4 + \Delta t) - f(4)}{\Delta t} = \lim\limits_{\Delta t \to 0} \dfrac{[(4 + \Delta t)^2 + 3(4 + \Delta t) - 10] - [4^2 + 2]}{\Delta t}$

$= \lim\limits_{\Delta t \to 0} \dfrac{11\Delta t + (\Delta t)^2}{\Delta t} = \lim\limits_{\Delta t \to 0}(11 + \Delta t) = 11$

27. $\lim\limits_{x \to 2} x^2 + \lim\limits_{x \to 2} x = 4 + 2 = 6$

29. $\lim\limits_{x \to 1} 4x^2 + \lim\limits_{x \to 1} 100x + \lim\limits_{x \to 1}(-2) = 102$

31. $\lim\limits_{x \to 1}(x + 3) \cdot \lim\limits_{x \to 1}(x - 4) = 4(-3) = -12$

33. $\lim\limits_{x \to -2}(x^2 + 3x + 1) \cdot \lim\limits_{x \to -2}(x^4 - 2x^2 + 3) = (-1)(11) = -11$

35. $\dfrac{\lim\limits_{x \to 2}(x^2 + 3x + 2)}{\lim\limits_{x \to 2}(x^2 + 1)} = \dfrac{12}{5}$

37. $\lim\limits_{x \to -7} \dfrac{x^2 - 49}{x + 7} = \lim\limits_{x \to -7}(x - 7) = -14$

39. $\lim\limits_{x \to 5/2} \dfrac{4x^2 - 25}{2x - 5} = \lim\limits_{x \to 5/2}(2x + 5) = 10$

41. $\dfrac{\lim\limits_{x \to 3}(x^2 + 3x + 1)\lim\limits_{x \to 3}(x + 5)}{\lim\limits_{x \to 3}(x - 2)} = \dfrac{(19)(8)}{1} = 152$

43. $\lim\limits_{h \to 0} \dfrac{(x + h)^2 - x^2}{h} = \lim\limits_{h \to 0} \dfrac{2xh + h^2}{h} = \lim\limits_{h \to 0}(2x + h) = 2x$

45. $\lim\limits_{h \to 0} \dfrac{\dfrac{1}{x + h} - \dfrac{1}{x}}{h} = \lim\limits_{h \to 0} \dfrac{\dfrac{x - (x + h)}{x(x + h)}}{h} = \lim\limits_{h \to 0} \dfrac{-1}{x(x + h)} = -\dfrac{1}{x^2}$

47. $\lim\limits_{x \to a} \dfrac{\left(\sqrt{x} - \sqrt{a}\right)\left(\sqrt{x} + \sqrt{a}\right)}{(x - a)\left(\sqrt{x} + \sqrt{a}\right)} = \lim\limits_{x \to a} \dfrac{x - a}{(x - a)\left(\sqrt{x} + \sqrt{a}\right)} = \lim\limits_{x \to a} \dfrac{1}{\sqrt{x} + \sqrt{a}} = \dfrac{1}{2\sqrt{a}}$

49. Does not exist   51. b   53. Does not exist   55. b

57. no   59. no   61. no   63. no

65. Does not exist   67. 0   69. b   71. Does not exist

Section 2.2

Section 2.3

1. $m_{\tan} = \lim_{\Delta x \to 0} \dfrac{(3 + \Delta x)^2 - 3^2}{\Delta x} = \lim_{\Delta x \to 0} \dfrac{6\Delta x + (\Delta x)^2}{\Delta x} = \lim_{\Delta x \to 0}(6 + \Delta x) = 6$

3. $m_{\tan} = \lim_{\Delta x \to 0} \dfrac{[3(-1 + \Delta x)^2 - 4] - [3(-1)^2 - 4]}{\Delta x} = \lim_{\Delta x \to 0} \dfrac{-6(\Delta x) + 3(\Delta x)^2}{\Delta x} = \lim_{\Delta x \to 0}(-6 + 3\Delta x) = -6$

5. $m_{\tan} = \lim_{\Delta x \to 0} \dfrac{[2(2 + \Delta x)^2 + (2 + \Delta x) - 3] - [2(2)^2 + 2 - 3)]}{\Delta x} = \lim_{\Delta x \to 0} \dfrac{9\Delta x + 2(\Delta x)^2}{\Delta x} = \lim_{\Delta x \to 0}(9 + 2\Delta x) = 9$

7. $m_{\tan} = \lim_{\Delta x \to 0} \dfrac{[-4(1 + \Delta x)^2 + 3(1 + \Delta x) - 2] - [-4(1)^2 + 3(1) - 2]}{\Delta x}$

   $= \lim_{\Delta x \to 0} \dfrac{-5\Delta x - 4(\Delta x)^2}{\Delta x} = \lim_{\Delta x \to 0}(-5 - 4\Delta x) = -5$

9. $m_{\tan} = \lim_{\Delta x \to 0} \dfrac{(2 + \Delta x)^3 - 2^3}{\Delta x} = \lim_{\Delta x \to 0} \dfrac{12\Delta x + 6(\Delta x)^2 + (\Delta x)^3}{\Delta x} = \lim_{\Delta x \to 0}[12 + 6\Delta x + (\Delta x)^2] = 12$

11. $m_{\tan} = -4$, $(-2, 4)$
    $y - 4 = -4(x + 2)$
    $4x + y + 4 = 0$

13. $m_{\tan} = -8$, $(-2, 5)$
    $y - 5 = -8(x + 2)$
    $8x + y + 11 = 0$

15. $m_{\tan} = -13$, $(-1, 10)$
    $y - 10 = -13(x + 1)$
    $13x + y + 3 = 0$

17. $m_{\tan} = -13$; $x = 3$ then $y = -10$; $y + 10 = -13(x - 3)$ or $13x + y - 29 = 0$

19. $m_{\tan} = 4$, $x = 1$ then $y = 1$; $y - 1 = 4(x - 1)$; $4x - y - 3 = 0$

21. $m_{\tan} = \lim_{\Delta x \to 0} \dfrac{(x + \Delta x)^2 - x^2}{\Delta x} = \lim_{\Delta x \to 0} \dfrac{2x(\Delta x) + (\Delta x)^2}{\Delta x} = \lim_{\Delta x \to 0}(2x + \Delta x) = 2x$
    $m_{\tan} = -1/3$; thus $2x = -1/3$ so $x = -1/6$ and $(-1/6, 1/36)$

23. $m_{\tan} = \lim_{\Delta x \to 0} \dfrac{[(x + \Delta x)^3 + (x + \Delta x)] - [x^3 + x]}{\Delta x} = \lim_{\Delta x \to 0} \dfrac{3x^2(\Delta x) + 3x(\Delta x)^2 + (\Delta x)^3 + \Delta x}{\Delta x}$

    $= \lim_{\Delta x \to 0}[3x^2 + 3x(\Delta x) + (\Delta x)^2 + 1] = 3x^2 + 1$
    $m_{\tan} = 4$ thus $3x^2 + 1 = 4$ so $x = \pm 1$; $(1, 2)$ and $(-1, -2)$

Section 2.3

Section 2.4

Note: Samples of the different kinds of exercises are given here. The other exercises are done in a similar manner.

1. $\dfrac{dy}{dx} = \lim\limits_{\Delta x \to 0} \dfrac{f(x + \Delta x) - f(x)}{\Delta x} = \lim\limits_{\Delta x \to 0} \dfrac{[3(x + \Delta x) + 4] - [3x + 4]}{\Delta x} = \lim\limits_{\Delta x \to 0} \dfrac{3\Delta x}{\Delta x} = \lim\limits_{\Delta x \to 0} 3 = 3$

3. -2 5. $6x$ 7. $2x - 2$

9. $\dfrac{dy}{dx} = \lim\limits_{\Delta x \to 0} \dfrac{f(x + \Delta x) - f(x)}{\Delta x} = \lim\limits_{\Delta x \to 0} \dfrac{[3(x + \Delta x)^2 - 4(x + \Delta x) + 1] - [3x^2 - 4x + 1]}{\Delta x}$

$\qquad = \lim\limits_{\Delta x \to 0} \dfrac{3x^2 + 6x\Delta x + 3(\Delta x)^2 - 4x - 4\Delta x + 1 - 3x^2 + 4x - 1}{\Delta x}$

$\qquad = \lim\limits_{\Delta x \to 0} \dfrac{6x(\Delta x) + 3(\Delta x)^2 - 4\Delta x}{\Delta x} = \lim\limits_{\Delta x \to 0} [6x + 3\Delta x - 4] = 6x - 4$

11. $-12x$ 13. $3x^2 + 4$ 15. $-1/x^2$

17. $\dfrac{dy}{dx} = \lim\limits_{\Delta x \to 0} \dfrac{f(x + \Delta x) - f(x)}{\Delta x} = \lim\limits_{\Delta x \to 0} \dfrac{\dfrac{2}{(x + \Delta x) - 3} - \dfrac{2}{x - 3}}{\Delta x} = \lim\limits_{\Delta x \to 0} \dfrac{2(x - 3) - 2(x + \Delta x - 3)}{(\Delta x)(x - 3)(x + \Delta x - 3)}$

$\qquad = \lim\limits_{\Delta x \to 0} \dfrac{-2\Delta x}{(\Delta x)(x - 3)(x + \Delta x - 3)} = \lim\limits_{\Delta x \to 0} \dfrac{-2}{(x - 3)(x + \Delta x - 3)} = \dfrac{-2}{(x - 3)^2}$

19. $-2/x^3$ 21. $2x/(4 - x^2)^2$

23. $\dfrac{dy}{dx} = \lim\limits_{\Delta x \to 0} \dfrac{f(x + \Delta x) - f(x)}{\Delta x} = \lim\limits_{\Delta x \to 0} \dfrac{\sqrt{(x + \Delta x) + 1} - \sqrt{x + 1}}{\Delta x}$

$\qquad = \lim\limits_{\Delta x \to 0} \left[\dfrac{\sqrt{(x + \Delta x) + 1} - \sqrt{x + 1}}{\Delta x} \cdot \dfrac{\sqrt{(x + \Delta x) + 1} + \sqrt{x + 1}}{\sqrt{x + \Delta x + 1} + \sqrt{x + 1}} \right]$

$\qquad = \lim\limits_{\Delta x \to 0} \dfrac{[(x + \Delta x) + 1] - [x + 1]}{\Delta x \left[\sqrt{(x + \Delta x) + 1} + \sqrt{x + 1} \right]} = \lim\limits_{\Delta x \to 0} \dfrac{\Delta x}{\Delta x \sqrt{x + \Delta x + 1} + \sqrt{x + 1}}$

$\qquad = \lim\limits_{\Delta x \to 0} \dfrac{1}{\sqrt{x + \Delta x + 1} + \sqrt{x + 1}} = \dfrac{1}{2\sqrt{x + 1}}$

25. $-1/\sqrt{1 - 2x}$

Section 2.4

27. $\dfrac{dy}{dx} = \lim\limits_{\Delta x \to 0} \dfrac{f(x+\Delta x)-f(x)}{\Delta x} = \lim\limits_{\Delta x \to 0} \dfrac{\dfrac{1}{\sqrt{x+\Delta x -1}} - \dfrac{1}{\sqrt{x-1}}}{\Delta x}$

$\qquad = \lim\limits_{\Delta x \to 0} \dfrac{\sqrt{x-1}-\sqrt{x+\Delta x -1}}{\Delta x \sqrt{x+\Delta x -1}\sqrt{x-1}} \cdot \dfrac{\sqrt{x-1}+\sqrt{x+\Delta x -1}}{\sqrt{x-1}+\sqrt{x+\Delta x -1}}$

$\qquad = \lim\limits_{\Delta x \to 0} \dfrac{(x-1)-(x+\Delta x -1)}{\Delta x \sqrt{x+\Delta x -1}\sqrt{x-1}\,(\sqrt{x-1}+\sqrt{x+\Delta x -1})}$

$\qquad = \lim\limits_{\Delta x \to 0} \dfrac{-\Delta x}{\Delta x \sqrt{x+\Delta x -1}\sqrt{x-1}\,(\sqrt{x-1}+\sqrt{x+\Delta x -1})}$

$\qquad = \lim\limits_{\Delta x \to 0} \dfrac{-1}{\sqrt{x+\Delta x -1}\sqrt{x-1}\,(\sqrt{x-1}+\sqrt{x+\Delta x -1})}$

$\qquad = \dfrac{-1}{2(x-1)\sqrt{x-1}} = \dfrac{-1}{2(x-1)^{3/2}}$

29. 6   31. 3/2   33. $y = x - 4$ 35. $x - 4y = 3$

37. (4, 1), (2, –1)   39. (–3, 1)

Section 2.5

1. 0   3. $5x^4$ 5. 4   7. –3   9. $10x$ 11. $2x - 3$

13. $8x - 3$ 15. $-16x$ 17. $9x^2 + 4x - 6$ 19. $20x^4 - 6x^2 + 1$

21. $20x^7 - 6x^4 + 30x^3 - 3x^2$ 23. $4\sqrt{7}x^3 - 3\sqrt{5}x^2 - \sqrt{3}$

25. $f'(x) = 6x + 2$
$\quad f'(-1) = 6(-1) + 2 = -4$

27. $f'(x) = 6x^2 - 12x + 2$
$\quad f'(-3) = 6(-3)^2 - 12(-3) + 2 = 92$

29. $f'(x) = 20x^4 + 6x$
$\quad f'(1) = 20(1)^4 + 6(1) = 26$

31. $f'(x) = 20x^3 + 24x^2 + 2$
$\quad f'(0) = 2$

33. $f'(x) = -16x - 15x^2 + 30x^5$
$\quad f'(3) = -16(3) - 15(3)^2 + 30(3)^5 = 7107$

35. $m_{\tan} = 3x^2 + 8x - 1\Big|_{x=-2} = -5;\ (-2, 12)\quad y - 12 = -5(x + 2)$
$\qquad\qquad\qquad\qquad\qquad\qquad\qquad 5x + y - 2 = 0$

Sections 2.4 – 2.5

37. $p = 30i^2$; $\dfrac{dp}{dt} = 60i \Big|_{i=2} = 120$ W/A

39. $\dfrac{dV}{dr} = i \Big|_{i=0.4} = 0.4$ V/Ω

41. $\dfrac{3}{2} x^{1/2}$

43. $-4x^{-5}$

45. $120x^{19}$

47. $-112x^{-9}$

49. $-\dfrac{5}{3} x^{-4/3}$

51. $\dfrac{dV}{dr} = i \Big|_{i=0.5} = 0.5$ V/Ω

Section 2.6

1. $y' = x^2(2) + (2x + 1)(2x) = 6x^2 + 2x$
   or $y = 2x^3 + x^2$, thus $y' = 6x^2 + 2x$

3. $y' = 2x(8x + 3) + 2(4x^2 + 3x - 5)$ or $y = 8x^3 + 6x^2 - 10x$
   $= 24x^2 + 12x - 10$ $y' = 24x^2 + 12x - 10$

5. $y' = 5(2x + 3) + 2(5x - 4)$ or $y = 10x^2 + 7x - 12$
   $= 20x + 7$ $y' = 20x + 7$

7. $y' = (4x + 7)(2x) + (x^2 - 1)(4)$ or $y = 4x^3 + 7x^2 - 4x - 7$
   $= 12x^2 + 14x - 4$ $y' = 12x^2 + 14x - 4$

9. $y' = (x^2 + 3x + 4)(3x^2 - 4) + (x^3 - 4x)(2x + 3)$
   $= 5x^4 + 12x^3 - 24x - 16$

11. $y' = (x^4 - 3x^2 - x)(6x^2 - 4) + (2x^3 - 4x)(4x^3 - 6x - 1)$
    $= 14x^6 - 50x^4 - 8x^3 + 36x^2 + 8x$

13. $y' = \dfrac{(2x + 5)((1)) - x(2)}{(2x + 5)^2}$
    $= \dfrac{5}{(2x + 5)^2}$

15. $y' = \dfrac{(x^2 + x)(0) - (1)(2x + 1)}{(x^2 + x)^2}$
    $= \dfrac{-2x - 1}{(x^2 + x)^2}$

17. $y' = \dfrac{3(2x + 4) - 2(3x - 1)}{(2x + 4)^2}$
    $= \dfrac{14}{(2x + 4)^2} = \dfrac{7}{2(x + 2)^2}$

19. $y' = \dfrac{(2x + 1)(2x) - x^2(2)}{(2x + 1)^2}$
    $= \dfrac{2x^2 + 2x}{(2x + 1)^2}$

21. $y' = \dfrac{(x^2 + x + 1)(1) - (x - 1)(2x + 1)}{(x^2 + x + 1)^2} = \dfrac{-x^2 + 2x + 2}{(x^2 + x + 1)^2}$

Sections 2.5 – 2.6

23. $y' = \dfrac{(3x^3 - 4x^2)(8x) - (4x^2 + 9)(9x^2 - 8x)}{(3x^3 - 4x^2)^2} = \dfrac{-12x^4 - 81x^2 + 72x}{(3x^3 - 4x^2)^2} = \dfrac{-12x^3 - 81x + 72}{x^3(3x - 4)^2}$

25. $f'(x) = (x^2 - 4x + 3)(3x^2 - 5) + (x^3 - 5x)(2x - 4) = 5x^4 - 16x^3 - 6x^2 + 40x - 15$

   $f'(2) = 5(2)^4 - 16(2)^3 - 6(2)^2 + 40(2) - 15 = -7$

27. $f'(x) = \dfrac{(x + 2)(3) - (3x - 4)(1)}{(x + 2)^2} = \dfrac{10}{(x + 2)^2} \; ; f'(-1) = 10$

29. $y' = \dfrac{(2 - 5x)(1) - (x - 3)(-5)}{(2 - 5x)^2} = \dfrac{-13}{(2 - 5x)^2} \; ; m_{\tan} = f'(2) = -\dfrac{13}{64}$

31. $y' = \dfrac{(x - 2) - (x + 3)}{(x - 2)^2} = \dfrac{-5}{(x - 2)^2} \; ; m_{\tan} = f'(3) = -5$ at $(3, 6)$

   then $y - 6 = -5(x - 3)$ or $5x + y - 21 = 0$

33. $\left. \dfrac{dV}{dt} \right|_{t = 3} = i\dfrac{dr}{dt} + r\dfrac{di}{dt} = (6 + 0.02t^3)(-0.05) + (20 - 0.05t)(0.6t^2) \left. \right|_{t = 3}$

   $= 10.4$ V/s

Section 2.7

1. $y' = 40(4x + 3)^{39}(4) = 160(4x + 3)^{39}$

3. $y' = 5(3x^2 - 7x + 4)^4(6x - 7) = 5(6x - 7)(3x^2 - 7x + 4)^4$

5. $y' = -4(x^3 + 3)^{-5}(3x^2) = \dfrac{-12x^2}{(x^3 + 3)^5}$

7. $y' = \dfrac{1}{2}(5x^2 - 7x + 2)^{-1/2}(10x - 7) = \dfrac{10x - 7}{2\sqrt{5x^2 - 7x + 2}}$

9. $y' = \dfrac{2}{3}(8x^3 + 3x)^{-1/3}(24x^2 + 3) = \dfrac{16x^2 + 2}{\sqrt[3]{8x^3 + 3x}}$

11. $y' = -\dfrac{3}{4}(2x + 3)^{-7/4}(2) = \dfrac{-3}{2(2x + 3)^{7/4}}$

13. $y' = 3x\left[4(4x + 5)^3(4)\right] + (4x + 5)^4(3)$

   $= 48x(4x + 5)^3 + 3(4x + 5)^4$

   $= 3(4x + 5)^3[16x + (4x + 5)]$

   $= 3(4x + 5)^3[20x + 5]$

   $= 15(4x + 5)^3(4x + 1)$

Sections 2.6 – 2.7

15. $y' = x^3\left[3(x^3 - x)^2(3x^2 - 1)\right] + (x^3 - x)^3(3x^2)$

$= 3x^3(3x^2 - 1)(x^3 - x)^2 + 3x^2(x^3 - x)^3$

$= 3x^2(x^3 - x)^2\left[x(3x^2 - 1) + (x^3 - x)\right]$

$= 3x^2(x^3 - x)^2\left[3x^2 - x + x^3 - x\right]$

$= 3x^2(x^3 - x)^2(4x^3 - 2x)$

$= 6x^2(x[x^2 - 1])^2(x[2x^2 - 1])$

$= 6x^5(x^2 - 1)^2(2x^2 - 1)$

17. $y' = (2x + 1)^2\left[2(x^2 + 1)(2x)\right] + (x^2 + 1)^2[2(2x + 1)(2)]$

$= 4x(2x + 1)^2(x^2 + 1) + 4(2x + 1)(x^2 + 1)^2$

$= 4(2x + 1)(x^2 + 1)\left[x(2x + 1) + (x^2 + 1)\right]$

$= 4(2x + 1)(x^2 + 1)\left[2x^2 + x + x^2 + 1\right]$

$= 4(2x + 1)(x^2 + 1)(3x^2 + x + 1)$

19. $y' = (x^2 + 1)\left(\dfrac{1}{2}\right)(9x^2 - 2x)^{-1/2}(18x - 2) + (9x^2 - 2x)^{1/2}(2x)$

$= (9x^2 - 2x)^{-1/2}[(9x - 1)(x^2 + 1) + 2x(9x^2 - 2x)] = \dfrac{27x^3 - 5x^2 + 9x - 1}{\sqrt{9x^2 - 2x}}$

21. $y' = (3x + 4)^{3/4}(8x) + (4x^2 + 8)\left(\dfrac{3}{4}\right)(3x + 4)^{-1/4}(3)$

$= (3x + 4)^{-1/4}[(3x + 4)(8x) + 9(x^2 + 2)] = \dfrac{33x^2 + 32x + 18}{(3x + 4)^{1/4}}$

23. $y' = -4x^{-5} - 4(2x + 1)^3(2)$

$= \dfrac{-4}{x^5} - 8(2x + 1)^3$

25. $y' = \dfrac{(3x - 1)^2(10x) - 5x^2(2)(3x - 1)(3)}{(3x - 1)^4}$

$= \dfrac{10x(3x - 1)[(3x - 1) - 3x]}{(3x - 1)^4} = \dfrac{10x[-1]}{(3x - 1)^3} = \dfrac{-10x}{(3x - 1)^3}$

27. $y' = \dfrac{(4x^2 - 3x)(4)(x^3 + 2)^3(3x^2) - (x^3 + 2)^4(8x - 3)}{(4x^2 - 3x)^2}$

$= \dfrac{(x^3 + 2)^3[12x^2(4x^2 - 3x) - (x^3 + 2)(8x - 3)]}{(4x^2 - 3x)^2} = \dfrac{(x^3 + 2)^3(40x^4 - 33x^3 - 16x + 6)}{(4x^2 - 3x)^2}$

Section 2.7

29. $y' = \dfrac{(2x-1)^3(5)(3x+2)^4(3) - (3x+2)^5(3)(2x-1)^2(2)}{(2x-1)^6}$

$= \dfrac{3(2x-1)^2(3x+2)^4[5(2x-1) - 2(3x+2)]}{(2x-1)^6} = \dfrac{3(3x+2)^4(4x-9)}{(2x-1)^4}$

31. $y' = \dfrac{(4x+3)^{1/2}\left(\dfrac{2}{3}\right)(3x-1)^{-1/3}(3) - (3x-1)^{2/3}\left(\dfrac{1}{2}\right)(4x+3)^{-1/2}(4)}{\left(\sqrt{4x+3}\right)^2}$

$= \dfrac{2(4x+3)^{-1/2}(3x-1)^{-1/3}[(4x+3) - (3x-1)]}{4x+3} = \dfrac{2(x+4)}{(3x-1)^{1/3}(4x+3)^{3/2}}$

33. $y' = 4\left(\dfrac{1+x}{1-x}\right)^3\left[\dfrac{(1-x)(1) - (1+x)(-1)}{(1-x)^2}\right] = \dfrac{4(1+x)^3}{(1-x)^3}\left[\dfrac{1-x+1+x}{(1-x)^2}\right]$

$= \dfrac{4(1+x)^3}{(1-x)^3}\left[\dfrac{2}{(1-x)^2}\right] = \dfrac{8(1+x)^3}{(1-x)^5}$

35. $s' = \dfrac{(t^2-1)^{1/2}(1) - (t+1)\left(\dfrac{1}{2}\right)(t^2-1)^{-1/2}(2t)}{\left(\sqrt{t^2-1}\right)^2} = \dfrac{(t^2-1)^{-1/2}[(t^2-1) - t(t+1)]}{t^2-1}$

$= \dfrac{-1-t}{(t^2-1)^{3/2}}$ thus $s'(3) = \dfrac{-1-3}{8^{3/2}} = 0.177$ m/s

Section 2.8
Note: To save space, we will let $y' = dy/dx$.

1. $4 + 3y' = 0$
$\quad y' = -4/3$

3. $2x - 2yy' = 0$
$\quad y' = x/y$

5. $2x + 2yy' + 4y' = 0$
$\quad (2y+4)y' = -2x$
$\quad y' = \dfrac{-2x}{2y+4} = \dfrac{-x}{y+2}$

7. $6x - 3y^2y' - 3xy' - 3y = 0$
$\quad (y^2+x)y' = 2x - y$
$\quad y' = \dfrac{2x-y}{y^2+x}$

9. $4y^3y' - 2xyy' - y^2 + 2x = 0$
$\quad (4y^3 - 2xy)y' = y^2 - 2x$
$\quad y' = \dfrac{y^2 - 2x}{4y^3 - 2xy}$

11. $4y^3 - 4yx^2y' - 4y^2x + 6x = 0$
$\quad (2y^3 - 2yx^2)y' = 2xy^2 - 3x$
$\quad y' = \dfrac{2xy^2 - 3x}{2y^3 - 2x^2y}$

13. $2(y^2+2)(2yy') = 3(x^3-4x)^2(3x^2-4)$
$\quad y' = \dfrac{3(x^3-4x)^2(3x^2-4)}{4y(y^2+2)}$

Sections 2.7 – 2.8

15. $3(x + y)^2(1 + y') = 2(x - y + 4)(1 - y')$

$\quad (3x^2 + 6xy + 3y^2)(1 + y') = 2(x - y + 4)(1 - y')$

$\quad (3x^2 + 6xy + 3y^2)y' + 2(x - y + 4)y' = 2(x - y + 4) - (3x^3 + 6xy + 3y^2)$

$\quad y' = \dfrac{2x - 2y + 8 - 3x^2 - 6xy - 3y^2}{3x^2 + 6xy + 3y^2 + 2x - 2y + 8}$

17. $\dfrac{(x - y)(1 + y') - (x + y)(1 - y')}{(x - y)^2} = 2yy'$

$\quad \dfrac{(x - y) + (x - y)y' - (x + y) + (x + y)y'}{(x - y)^2} = 2yy'$

$\quad \dfrac{-2y + 2xy'}{(x - y)^2} = 2yy'$

$\quad -2y + 2xy' = (x - y)^2 2yy'$

$\quad xy' - y(x - y)^2 y' = y$

$\quad [x - y(x - y)^2]y' = y$

$\quad y' = \dfrac{y}{x - y(x - y)^2}$

19. $8x + 10yy' = 0$

$\quad m_{\tan} = y' = -\dfrac{4x}{5y}\Bigg|_{\substack{x = 2 \\ y = -2}} = \dfrac{4}{5}$

21. $2x + 2yy' - 6 - 2y' = 0$

$\quad (y - 1)y' = 3 - x$

$\quad m_{\tan} = y' = \dfrac{3 - x}{y - 1}\Bigg|_{\substack{x = 2 \\ y = 4}} = \dfrac{1}{3}$

23. $m_{\tan} = y' = 6x + 4\Big|_{x = 0} = 4;$

$\quad (0, 9);\ y - 9 = 4(x - 0)$

$\quad 4x - y + 9 = 0$

25. $2yy' + 3xy' + 3y = 0$

$\quad (2y + 3x)y' = -3y$

$\quad m_{\tan} = y' = \dfrac{-3y}{2y + 3x}\Bigg|_{\substack{x = 1 \\ y = -4}} = -\dfrac{12}{5}$

$\quad y + 4 = (-12/5)(x - 1)$

$\quad 12x + 5y + 8 = 0$

Section 2.8

27. a) $y = \dfrac{-x^2}{2} + \dfrac{7}{2}$, $y' = -x$, $y' = -(1)$, $y' = -1$

   b) $2x + 2y' = 0$, $y' = -x$, $y' = -1$

29. a) $y = \sqrt{x-2}$, $y' = \dfrac{1}{2}(x-2)^{-1/2}$, $y' = \dfrac{1}{2\sqrt{11-2}}$,

      $y' = \dfrac{1}{2(3)}$, $y' = \dfrac{1}{6}$

   b) $2yy' = 1$, $y' = \dfrac{1}{2y}$, $y' = \dfrac{1}{2(3)}$, $y' = \dfrac{1}{6}$

Section 2.10

1. $y' = 5x^4 + 6x$; $y'' = 20x^3 + 6$; $y''' = 60x^2$; $y^{(4)} = 120x$

3. $y' = 25x^4 + 6x^2 - 8$; $y' = 100x^3 + 12x$; $y''' = 300x^2 + 12$; $y^{(4)} = 600x$

5. $y' = -\dfrac{1}{x^2}$; $y'' = \dfrac{2}{x^3}$; $y''' = -\dfrac{6}{x^4}$

7. $y' = 9(3x - 5)^2$; $y'' = 54(3x - 5)$; $y''' = 162$

9. $y' = \dfrac{1}{2}(3x + 2)^{-1/2}(3)$; $y'' = (3/2)(-1/2)(3x + 2)^{-3/2}(3) = (-9/4)(3x + 2)^{-3/2}$

    $y''' = (-9/4)(-3/2)(3x + 2)^{-5/2}(3) = (81/8)(3x + 2)^{-5/2}$

    $y^{(4)} = (81/8)(-5/2)(3x + 2)^{-7/2}(3) = (-1215/16)(3x + 2)^{-7/2} = \dfrac{-1215}{16(3x + 2)^{7/2}}$

11. $y = (x^2 + 1)^{-1}$; $\quad y' = -1(x^2 + 1)^{-2}(2x) = -2x(x^2 + 1)^{-2}$

                $y'' = -2x[-2(x^2 + 1)^{-3}(2x)] - 2(x^2 + 1)^{-2}$

                   $= -2(x^2 + 1)^{-3}[-4x^2 + (x^2 + 1)]$

                   $= \dfrac{6x^2 - 2}{(x^2 + 1)^3}$

13. $y' = \dfrac{(x - 1) - (x + 1)}{(x - 1)^2} = \dfrac{-2}{(x - 1)^2} = -2(x - 1)^{-2}$; $\quad y'' = 4(x - 1)^{-3} = \dfrac{4}{(x - 1)^3}$

15. $2x + 2yy' = 0$; $\quad y' = -x/y$; $\quad y'' = \dfrac{-y + xy'}{y^2} = \dfrac{-y + x(-x/y)}{y^2} = \dfrac{y^2 + x^2}{y^3} = \dfrac{-1}{y^3}$

Sections 2.8 – 2.10

17. $2x - xy' - y + 2yy' = 0$

$$y' = \frac{y - 2x}{2y - x} = \frac{2x - y}{x - 2y}$$

$$y'' = \frac{(x - 2y)(2 - y') - (2x - y)(1 - 2y')}{(x - 2y)^2}$$

$$y'' = \frac{(x - 2y)\left\{2 - \dfrac{2x - y}{x - 2y}\right\} - (2x - y)\left\{1 - 2\left\{\dfrac{2x - y}{x - 2y}\right\}\right\}}{(x - 2y)^2} \cdot \frac{(x - 2y)}{(x - 2y)}$$

$$= \frac{2(x - 2y)^2 - (2x - y)(x - 2y) - (2x - y)[(x - 2y) - 2(2x - y)]}{(x - 2y)^3}$$

$$= \frac{2(x - 2y)^2 - (2x - y)(x - 2y) - (2x - y)(x - 2y) + 2(2x - y)^2}{(x - 2y)^3}$$

$$= \frac{6x^2 - 6xy + 6y^2}{(x - 2y)^3} = \frac{6(x^2 - xy + y^2)}{(x - 2y)^3} = \frac{6}{(x - 2y)^3} \quad \text{(since } x^2 - xy + y^2 = 1\text{)}$$

19. $\dfrac{1}{2}x^{-1/2} + \dfrac{1}{2}y^{-1/2}y' = 0; \quad y' = -\dfrac{y^{1/2}}{x^{1/2}}$

$$y'' = \frac{-\dfrac{1}{2}x^{1/2}y^{-1/2}y' + \dfrac{1}{2}y^{1/2}x^{-1/2}}{(x^{1/2})^2} = \frac{-x^{1/2}y^{-1/2}\left\{\dfrac{-y^{1/2}}{x^{1/2}}\right\} + \dfrac{y^{1/2}}{x^{1/2}}}{2x}$$

$$= \frac{1 + \dfrac{y^{1/2}}{x^{1/2}}}{2x} = \frac{x^{1/2} + y^{1/2}}{2x^{3/2}} = \frac{1}{2x^{3/2}}$$

21. $-\dfrac{1}{x^2} + \dfrac{1}{y^2} \cdot y' = 0$

$$y' = \frac{1/x^2}{1/y^2} = \frac{y^2}{x^2}$$

$$y'' = \frac{x^2(2yy') - y^2(2x)}{(x^2)^2} = \frac{2x^2y[y^2/x^2] - 2xy^2}{x^4} = \frac{2y^3 - 2xy^2}{x^4}$$

23. $2(1 + y)y' = 1$

$$y' = \frac{1}{2(1 + y)}$$

$$y'' = -\frac{1}{2}(1 + y)^{-2}y' = -\frac{1}{2} \cdot \frac{1}{(1 + y)^2} \cdot \frac{1}{2(1 + y)} = -\frac{1}{4(1 + y)^3}$$

Section 2.10

Solutions to Odd-Numbered Exercises

25. $v = \dfrac{ds}{dt} = 2t^3 - 18t^2 - 8t$

$a = \dfrac{dv}{dt} = 6t^2 - 36t - 8$

27. $v = \dfrac{1}{2}(6t-4)^{-1/2}(6) = 3(6t-2)^{-1/2}$

$a = (-3/2)(6t-4)^{-3/2}(6) = \dfrac{-9}{(6t-4)^{3/2}}$

29. $2x - xy' - y + 2yy' = 0$

$y' = \dfrac{y - 2x}{2y - x}$

For $(1, 3)$, $m_{\tan} = y' = 1/5$

$y - 3 = \dfrac{1}{5}(x - 1)$

$x - 5y + 14 = 0$

If $x = 1$, then $1 - y + y^2 = 7$

$y^2 - y - 6 = 0$ or $y = 3, -2$

For $(1, -2)$ $m_{\tan} = y' = 4/5$

$y + 2 = \dfrac{4}{5}(x - 1)$

$4x - 5y - 14 = 0$

31. $m_{\tan} = y' = \dfrac{-b^2 x}{a^2 y}$

Equation of tangent to ellipse at (x_o, y_o) is given by

$y - y_o = \dfrac{-b^2 x}{a^2 y}(x - x_o)$

$a^2 y^2 - a^2 yy_o = -b^2 x^2 + b^2 xx_o$

$b^2 xx_o + a^2 yy_o = a^2 y^2 + b^2 x^2$

but $a^2 y^2 + b^2 x^2 = a^2 b^2$ so

$b^2 xx_o + a^2 yy_o = a^2 b^2$

$\dfrac{x_o x}{a^2} + \dfrac{y_o y}{b^2} = 1$

Chapter 2 Review

1. a) $\Delta s = f(t + \Delta t) - f(t) = [3(t + \Delta t)^2 + 4] - [3t^2 + 4]$
$= 3t^2 + 6t\Delta t + 3(\Delta t)^2 + 4 - 3t^2 - 4 = 6t\Delta t + 3(\Delta t)^2$

   b) $\bar{v} = \dfrac{\Delta s}{\Delta t} = \dfrac{6t\Delta t + 3(\Delta t)^2}{\Delta t} = 6t + 3\Delta t$

2. a) $\Delta s = f(t + \Delta t) - f(t) = [5(t + \Delta t)^2 - 6] - [5t^2 - 6]$
$= 5t^2 + 10t\Delta t + 5(\Delta t)^2 - 6 - 5t^2 + 6 = 10t\Delta t + 5(\Delta t)^2$

   b) $\bar{v} = \dfrac{\Delta s}{\Delta t} = \dfrac{10t\Delta t + 5(\Delta t)^2}{\Delta t} = 10t + 5\Delta t$

3. a) $\Delta s = f(t + \Delta t) - f(t) = [(t + \Delta t)^2 - 3(t + \Delta t) + 5] - [t^2 - 3t + 5]$
$= t^2 + 2t\Delta t + (\Delta t)^2 - 3t - 3\Delta t + 5 - t^2 + 3t - 5 = 2t\Delta t + (\Delta t)^2 - 3\Delta t$

   b) $\bar{v} = \dfrac{\Delta s}{\Delta t} = \dfrac{2t\Delta t + (\Delta t)^2 - 3\Delta t}{\Delta t} = 2t + \Delta t - 3$

4. a) $\Delta s = f(t + \Delta t) - f(t) = [3(t + \Delta t)^2 - 6(t + \Delta t) + 8] - [3t^2 - 6t + 8]$
$= 3t^2 + 6t\Delta t + 3(\Delta t)^2 - 6t - 6\Delta t + 8 - 3t^2 + 6t - 8$
$= 6t\Delta t + 3(\Delta t)^2 - 6\Delta t$

   b) $\bar{v} = \dfrac{\Delta s}{\Delta t} = \dfrac{6t\Delta t + 3(\Delta t)^2 - 6\Delta t}{\Delta t} = 6t + 3\Delta t - 6$

Section 2.10 – Chapter 2 Review

5. $\bar{v} = \dfrac{\Delta s}{\Delta t} = \dfrac{f(t + \Delta t) - f(t)}{\Delta t} = \dfrac{f(2 + 5) - f(2)}{5} = \dfrac{140 - 5}{5} = 27$ m/s

6. $\bar{v} = \dfrac{f(3) - f(1)}{2} = 20$ m/s   7. $\bar{v} = \dfrac{f(5) - f(2)}{3} = 10$ m/s

8. $\bar{v} = \dfrac{f(7) - f(3)}{4} = 33$ m/s

9. $v = \lim\limits_{\Delta t \to 0} \dfrac{f(t + \Delta t) - f(t)}{\Delta t} = \lim\limits_{\Delta t \to 0} \dfrac{f(2 + \Delta t) - f(2)}{\Delta t}$

$= \lim\limits_{\Delta t \to 0} \dfrac{[3(2 + \Delta t)^2 - 7] - [3(2)^2 - 7]}{\Delta t} = \lim\limits_{\Delta t \to 0} \dfrac{12\Delta t + 3(\Delta t)^2}{\Delta t}$

$= \lim\limits_{\Delta t \to 0} (12 + 3\Delta t) = 12$ m/s

10. $v = \lim\limits_{\Delta t \to 0} \dfrac{f(1 + \Delta t) - f(1)}{\Delta t} = \lim\limits_{\Delta t \to 0} \dfrac{[5(1 + \Delta t)^2 - 3] - [5(1)^2 - 3]}{\Delta t}$

$= \lim\limits_{\Delta t \to 0} \dfrac{10\Delta t + 5(\Delta t)^2}{\Delta t} = \lim\limits_{\Delta t \to 0} (10 + 5\Delta t) = 10$ m/s

11. $v = \lim\limits_{\Delta t \to 0} \dfrac{f(2 + \Delta t) - f(2)}{\Delta t}$

$= \lim\limits_{\Delta t \to 0} \dfrac{[2(2 + \Delta t)^2 - 4(2 + \Delta t) + 7] - [2(2)^2 - 4(2) + 7]}{\Delta t}$

$= \lim\limits_{\Delta t \to 0} \dfrac{4\Delta t + 2(\Delta t)^2}{\Delta t} = \lim\limits_{\Delta t \to 0} (4 + 2\Delta t) = 4$ m/s

12. $v = \lim\limits_{\Delta t \to 0} \dfrac{f(3 + \Delta t) - f(3)}{\Delta t}$

$= \lim\limits_{\Delta t \to 0} \dfrac{[4(3 + \Delta t)^2 - 7(3 + \Delta t) + 2] - [4(3)^2 - 7(3) + 2]}{\Delta t}$

$= \lim\limits_{\Delta t \to 0} \dfrac{17\Delta t + 4(\Delta t)^2}{\Delta t} = \lim\limits_{\Delta t \to 0} (17 + 4\Delta t) = 17$ m/s

13. $\lim\limits_{x \to 3} (2x^2 - 5x + 1) = 4$ 14. $\lim\limits_{x \to -2} (x^2 + 4x - 7) = -11$

15. $\lim\limits_{x \to -2} (x - 2) = -4$ 16. $\lim\limits_{x \to 5} (5 + x) = 10$

17. No limit ($\sqrt{-8}$ is not real.)   18. $\lim\limits_{x \to -3} \sqrt{6 + x^2} = \sqrt{15}$

19. $\lim\limits_{x \to 2} \dfrac{5x^2 + 2}{3x^2 - 2x + 1} = \dfrac{22}{9}$ 20. $\lim\limits_{x \to -3} \dfrac{2x^2 - 4x + 7}{x^3 - x} = -\dfrac{37}{24}$

Chapter 2 Review

21. $\lim\limits_{x\to-3} (x^2 - 4x + 3)(2x^2 + 5x + 4) = 24(7) = 168$

22. $\lim\limits_{x\to2} (x^3 + x - 2)(x^3 + x^2 + x) = (8)(14) = 112$

23. $\lim\limits_{h\to0} \dfrac{2 - (2 + h)}{2(2 + h)h} = \lim\limits_{h\to0} \dfrac{-h}{2(2 + h)h} = \lim\limits_{h\to0} \dfrac{-1}{2(2 + h)} = -\dfrac{1}{4}$

24. $\lim\limits_{h\to0} \dfrac{\sqrt{1 + h} - 1}{h} \cdot \dfrac{\sqrt{1 + h} + 1}{\sqrt{1 + h} + 1} = \lim\limits_{h\to0} \dfrac{h}{h(\sqrt{1 + h} + 1)} = \lim\limits_{h\to0} \dfrac{1}{\sqrt{1 + h} + 1} = \dfrac{1}{2}$

25. $\lim\limits_{x\to\infty} \dfrac{(5x^2 - 2x + 3)/(x^2)}{(2x^2 - 4)/(x^2)} = \lim\limits_{x\to\infty} \dfrac{5 - \dfrac{2}{x} + \dfrac{3}{x^2}}{2 - \dfrac{4}{x^2}} = \dfrac{5}{2}$

26. $\lim\limits_{x\to\infty} \dfrac{(7x^3 - 4x + 2)/(x^3)}{(10x^3 - x^2 + 5)/(x^3)} = \lim\limits_{x\to\infty} \dfrac{7 - \dfrac{4}{x^2} + \dfrac{2}{x^3}}{10 - \dfrac{1}{x} + \dfrac{5}{x^3}} = \dfrac{7}{10}$

27. Does not exist   28. c   29. c   30. Does not exist   31. No   32. Yes

33. $m_{\tan} = y' = 6x - 4\Big|_{x = 1} = -10$

tangent line at $(-1, 12)$
$y - 12 = -10(x + 1)$
$10x + y - 2 = 0$

34. $m_{\tan} = y' = 2x - 5\Big|_{x = 2} = -1$

tangent line at $(2, -18)$
$y + 18 = -1(x - 2)$
$x + y + 16 = 0$

35. $m_{\tan} = y' = 4x + 2\Big|_{x = 3} = 14$

tangent line at $(3, 31)$
$y - 31 = 14(x - 3)$
$14x - y - 11 = 0$

Chapter 2 Review

36. $m_{\tan} = y' = 8x - 8 \Big|_{x=2} = -24$

tangent line at $(-2, 35)$
$y - 35 = -24(x + 2)$
$24x + y + 13 = 0$

37. $v = \dfrac{ds}{dt} = -\dfrac{6}{t^3}\Big|_{t=5} = -\dfrac{6}{125}$ cm/s or -0.048 cm/s

38. $v = \dfrac{ds}{dt} = -32t\Big|_{t=2} = -64$ ft/s

39. $y' = 20x^3 - 9x^2 + 4x + 5$

40. $y' = 100x^{99} + 400x^4$

41. $y' = (x^3 + 4)(3x^2 - 1) + (x^3 - x + 1)(3x^2)$
$= 6x^5 - 4x^3 + 15x^2 - 4$

42. $y' = (3x^2 - 5)(5x^4 + 2x - 4) + (x^5 + x^2 - 4x)(6x)$
$= 21x^6 - 25x^4 + 12x^3 - 36x^2 - 10x + 20$

43. $y' = \dfrac{(3x - 4)(2x) - (x^2 + 1)(3)}{(3x - 4)^2} = \dfrac{3x^2 - 8x - 3}{(3x - 4)^2}$

44. $y' = \dfrac{(3x^4 + 2)(2 - 2x) - (2x - x^2)(12x^3)}{(3x^4 + 2)^2} = \dfrac{6x^5 - 18x^4 - 4x + 4}{(3x^4 + 2)^2}$

45. $y' = 5(3x^2 - 8)^4(6x) = 30x(3x^2 - 8)^4$

46. $y' = \dfrac{3}{4}(x^4 + 2x^3 + 7)^{-1/4}(4x^3 + 6x^2)$

47. $y' = -4(3x + 5)^{-5}(3) = \dfrac{-12}{(3x + 5)^5}$

48. $y' = \dfrac{(x + 3)^2(1/2)(7x^2 - 5)^{-1/2}(14x) - (7x^2 - 5)^{1/2}(2)(x + 3)}{[(x + 3)^2]^2}$

$= \dfrac{(x + 3)(7x^2 - 5)^{-1/2}[7x(x + 3) - 2(7x^2 - 5)]}{(x + 3)^4} = \dfrac{-7x^2 + 21x + 10}{(7x^2 - 5)^{1/2}(x + 3)^3}$

Chapter 2 Review

52

49. $y' = \dfrac{(x+5)\left[(x\left(\dfrac{1}{2}\right)(2-3x)^{-1/2}(-3) + (2-3x)^{1/2} - x(2-3x)^{1/2}\right]}{(x+5)^2}$

$= \dfrac{\dfrac{1}{2}(2-3x)^{-1/2}\{(x+5)[-3x+2(2-3x)] - 2x(2-3x)\}}{(x+5)^2} = \dfrac{-3x^2 - 45x + 20}{2(x+5)^2\sqrt{2-3x}}$

50. $2x - 4y^3 - 12xy^2y' + 2yy' = 0$

$(2y - 12xy^2)y' = 4y^3 - 2x$

$y' = \dfrac{2y^3 - x}{y - 6xy^2}$

51. $4y^3y' - 2yy' = 2xy' + 2y$

$(4y^3 - 2y - 2x)y' = 2y$

$y' = \dfrac{y}{2y^3 - y - x}$

52. $3(y^2 + 1)^2(2yy') = 8x$

$y' = \dfrac{4x}{3y(y^2 + 1)^2}$

53. $4(y + 2)^3y' = 3(2x^3 - 3)^2(6x)^2$

$y' = \dfrac{9x^2(2x^3 - 3)^2}{2(y + 2)^3}$

54. $f'(x) = (x^2 + 2)^3 + (x + 1)(3)(x^2 + 2)^2(2x)$

$= (x^2 + 2)^3 + 6x(x + 1)(x^2 + 2)^2$

$= (x^2 + 2)^2(7x^2 + 6x + 2)$

$f'(-2) = (36)(18) = 648$

55. $f'(x) = \dfrac{(x + 3)(2x) - (x^2 - 8)(1)}{(x + 3)^2}$

$= \dfrac{x^2 + 6x + 8}{(x + 3)^2} \qquad f'(-4) = 0$

56. $f'(x) = \dfrac{(x - 2)(2x + 3) - (x^2 + 3x - 2)(1)}{(x - 2)^2}$

$= \dfrac{x^2 - 4x - 4}{(x - 2)^2} \qquad f'(3) = -7$

Chapter 2 Review

57. $f'(x) = \dfrac{(x+5)(1/2)(3x^2-1)^{-1/2}(6x)-(3x^2-1)^{1/2}(1)}{(x+5)^2}$

$\quad = \dfrac{(3x^2-1)^{-1/2}[3x(x+5)-(3x^2-1)]}{(x+5)^2} = \dfrac{15x+1}{(x+5)^2\sqrt{3x-1}}$

$f'(1) = \dfrac{4}{9\sqrt{2}}$ or $\dfrac{2\sqrt{2}}{9}$

58. $m_{\tan} = 6x + 1\Big|_{x=-2} = -11$

tangent line at $(-2, 8)$ is

$y - 8 = -11(x + 2)$

$11x + y + 14 = 0$

59. $m_{\tan} = y' = 3x^2 - 1\Big|_{x=-3} = 26$

tangent line at $(-3, -16)$ is

$y + 16 = 26(x + 3)$

$26x - y + 62 = 0$

60. $y' = \dfrac{(x-3)^{1/2}(2x)-(x^2-2)(1/2)(x-3)^{-1/2}}{[(x-3)^{1/2}]^2}$

$\quad = \dfrac{(1/2)(x-3)^{-1/2}[4x(x-3)-(x^2-2)]}{x-3} = \dfrac{3x^2-12x+2}{2(x-3)^{3/2}}$

$m_{\tan} = y'\Big|_{x=4} = 1$; tangent line at $(4, 14)$ is $y - 14 = 1(x - 4)$ or $x - y + 10 = 0$

61. $y' = \dfrac{(x-2)(1/2)(x^2+7)^{-1/2}(2x)-(x^2+7)^{1/2}(1)}{(x-2)^2}$

$\quad = \dfrac{(x^2+7)^{-1/2}[x(x-2)-(x^2+7)]}{(x-2)^2} = \dfrac{-2x-7}{\sqrt{x^2+7}(x-2)^2}$

$m_{\tan} = y'\Big|_{x=3} = -\dfrac{13}{4}$; tangent line at $(3, 4)$ is $y - 4 = -\dfrac{13}{4}(x-3)$

$13x + 4y - 55 = 0$

62. $v = s'(t) = 6t - 8\Big|_{t=2} = 4$ m/s

63. $v = s'(t) = 3t^2 - 18t\ \Big|_{t=-2} = 48$ m/s

64. $v = s'(t) = \dfrac{(t+1)^{1/2}(2t)-(t^2-5)(1/2)(t+1)^{-1/2}}{[(t+1)^{1/2}]^2}$

$\quad = \dfrac{(1/2)(t+1)^{-1/2}[(t+1)(4t)-(t^2-5)]}{t+1} = \dfrac{3t^2+4t+5}{2(t+1)^{3/2}}\Big|_{t=3} = 2.75$ m/s

Chapter 2 Review

65. $v = s'(t) = \dfrac{(t+5)(1/2)(t^2-3)^{-1/2}(2t) - (t^2-3)^{1/2}(1)}{(t+5)^2}$

$= \dfrac{(t^2-3)^{-1/2}[t(t+5) - (t^2-3)]}{(t+5)^2} = \dfrac{5t+3}{(t^2-3)^{1/2}(t+5)^2}\bigg|_{t=4}$

$= \dfrac{23}{81\sqrt{13}} \approx 0.0788$ m/s

66. $m_{\tan} = \dfrac{(x+3)(2x) - (x^2-6)(1)}{(x+3)^2} = \dfrac{x^2+6x+6}{(x+3)^2}\bigg|_{x=-4} = -2$

tangent line at $(-4, -10)$ is $y + 10 = -2(x+4)$ or $2x + y + 18 = 0$

67. $i = \dfrac{dq}{dt} = 3000t^2 + 50\bigg|_{t=0.01} = 50.3$ A

68. $\dfrac{dc}{dT} = 0.5 + 0.000012T$

69. $\dfrac{df}{dC} = -\dfrac{1}{2} \cdot \dfrac{1}{2\pi}(LC)^{-3/2}(L) = \dfrac{-L}{4\pi\sqrt{(LC)^3}} = \dfrac{-L}{4\pi LC\sqrt{LC}} = \dfrac{-1}{4\pi C\sqrt{LC}}$

70. $y' = 24x^5 - 32x^3 + 27x^2 - 6; \quad y'' = 120x^4 - 96x^2 + 54x;$
$y''' = 480x^3 - 192x + 54; \quad y^{(4)} = 1440x^2 - 192$

71. $y' = (1/2)(2x-3)^{-1/2}(2) = (2x-3)^{-1/2}$

$y'' = (-1/2)(2x-3)^{-3/2}(2) = \dfrac{-1}{(2x-3)^{3/2}}$

72. $y' = -3(2x^2+1)^{-2}(4x) = \dfrac{-12x}{(2x^2+1)^2}$

$y'' = \dfrac{(2x^2+1)^2(-12) - (-12x)(2)(2x^2+1)(4x)}{(2x^2+1)^4} = \dfrac{-12(2x^2+1)[(2x^2+1) - 8x^2]}{(2x^2+1)^4} = \dfrac{72x^2-12}{(2x^2+1)^3}$

73. $2yy' + 2xy' + 2y = 0$

$y' = -\dfrac{y}{y+x}$

$y'' = \dfrac{(y+x)(-y') - (-y)(y'+1)}{(y+x)^2} = \dfrac{(y+x)\left\{\dfrac{y}{y+x}\right\} + y\left\{\dfrac{-y}{y+x}+1\right\}}{(y+x)^2} \cdot \dfrac{y+x}{y+x}$

$= \dfrac{y(y+x) + y[-y+(y+x)]}{(y+x)^3} = \dfrac{y^2 + yx - y^2 + y^2 + xy}{(y+x)^3} = \dfrac{y^2 + 2xy}{(x+y)^3} = \dfrac{4}{(x+y)^3}$

Chapter 2 Review

74. $x^{-1/2} - y^{-1/2} = 1$

$(-1/2)x^{-3/2} + (1/2)y^{-3/2}y' = 0$

$y' = \dfrac{x^{-3/2}}{y^{-3/2}} = \dfrac{y^{3/2}}{x^{3/2}}$

$y'' = \dfrac{x^{3/2}(3/2)y^{1/2}y' - y^{3/2}(3/2)x^{1/2}}{(x^{3/2})^2} = \dfrac{3\left[x^{3/2}y^{1/2}(y^{3/2}/x^{3/2}) - y^{3/2}x^{1/2}\right]}{2x^3} = \dfrac{3(y^2 - y^{3/2}x^{1/2})}{2x^3}$

75. $v = \dfrac{ds}{dt} = (1/4)(2t+3)^{-3/4}(2) = (1/2)(2t+3)^{-3/4}$

$a = \dfrac{dv}{dt} = (-3/8)(2t+3)^{-7/4}(2) = (-3/4)(2t+3)^{-7/4}$

Chapter 2 Review

Section 3.1

KEY: I = intercepts; A = asymptotes; S = symmetry; D = domain;
      a = curve above the *x*-axis; b = curve below the *x*-axis

In Exercises 1-12, the domain is the set of real numbers; there are no asymptotes and no symmetry, except number 9, which has symmetry with respect to the *y*-axis.

1. I: $x = 0, x = -1, x = 4$
         $y = 0$

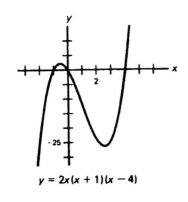

$$y = 2x(x + 1)(x - 4)$$

3. I: $x = -5, 1, 3; y = 15$

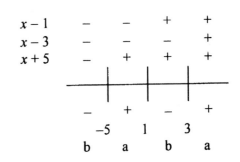

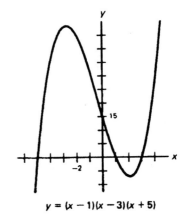

$$y = (x - 1)(x - 3)(x + 5)$$

5. $y = x(x+5)(x-3)$
   I: $x = -5, 0, 3$; $y = 0$

x	−	−	+	+
$x+5$	−	+	+	+
$x-3$	−	−	−	+

	−	+	−	+
		−5	0	3
	b	a	b	a

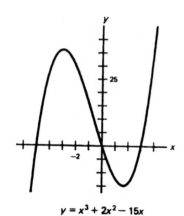

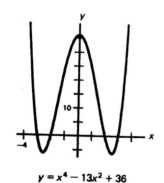

$y = x^3 + 2x^2 - 15x$

7. I: $x = 0, -1, 3/2$; $y = 0$

x^2	+	+	+	+
$x+1$	−	+	+	+
$3-2x$	+	+	+	−

	−	+	+	−
		−1	0	3/2
	b	a	a	b

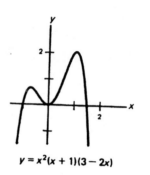

$y = x^2(x+1)(3-2x)$

9. $y = (x+2)(x+3)(x-2)(x-3)$
   I: $x = -2, -3, 2, 3$; $y = 36$

$x+2$	−	−	+	+	+
$x+3$	−	+	+	+	+
$x-2$	−	−	−	+	+
$x-3$	−	−	−	−	+

	+	−	+	−	+
		−3	−2	2	3
	a	b	a	b	a

$y = x^4 - 13x^2 + 36$

11. I: $x = -4, 0, 2$; $y = 0$

x^2	+	+	+	+
$(x-2)^2$	+	+	+	+
$(x+4)^2$	+	+	+	+

	+	+	+	+
		−4	0	2
	a	a	a	a

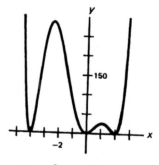

$y = x^2(x-2)^2(x+4)^2$

Section 3.1

13. I: $y = 3$

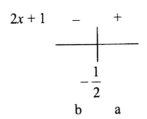

$2x + 1$ $-$ $+$

$-\dfrac{1}{2}$

b   a

A: $x = \dfrac{-1}{2}, y = 0$

D: $x \neq \dfrac{-1}{2}$

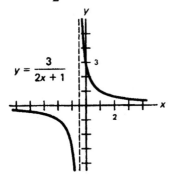

$y = \dfrac{3}{2x + 1}$

15. I: $x = 0, y = 0$

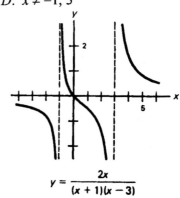

$2x$	$-$	$-$	$+$	$+$
$x + 1$	$-$	$+$	$+$	$+$
$x - 3$	$-$	$-$	$-$	$+$
	$-$	$+$	$-$	$+$

-1 0 3

b   a   b   a

A: $x = -1, x = 3, y = 0$

D: $x \neq -1, 3$

$y = \dfrac{2x}{(x + 1)(x - 3)}$

17. I: $y = 3/4$; graph always above the x-axis;

A: $y = 0$;

S: y-axis; $D = \{\text{reals}\}$

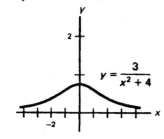

$y = \dfrac{3}{x^2 + 4}$

19. I: $x = 0, y = 0$

$4x$ $-$ $+$ $+$

$x - 2$ $-$ $-$ $+$

     $+$ $-$ $+$

0 2

a   b   a

A: $x = 2, y = 4$; D: $x \neq 2$

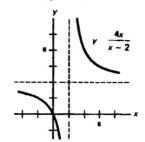

$y = \dfrac{4x}{x - 2}$

Section 3.1

21. I: $x = 0, y = 0$

$3x^2$	+	+	+	+
$x - 2$	−	−	−	+
$x + 2$	−	+	+	+

+		−		−		+
	−2		0		2	
	a		b		b		a

A: $x = 2, x = -2, y = 3$
S: y-axis; D: $x \neq \pm 2$

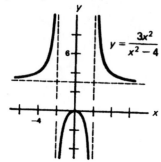

23. I: $x = \pm\sqrt{6}, y = -3$

$x + \sqrt{6}$	−	+	+	+
$x - \sqrt{6}$	−	−	−	+
$x + 2$	−	−	+	+

+		+		+		+
	$-\sqrt{6}$		−2		$\sqrt{6}$	
	b		a		b		a

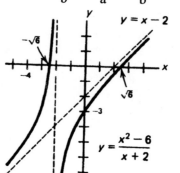

25. I: $x = 3/2, -1; y = 3/4$

$2x - 3$	−	−	+	+
$x + 1$	−	+	+	+
$x - 4$	−	−	−	+

−		+		−		+
	−1		3/2		4	
	b		a		b		a

A: $x = 4, y = 2x + 7$;
D: $x \neq 4$

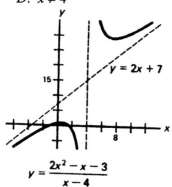

27. I: $x = -4, y = 2$
Graph always above x-axis.
D: $x \geq -4$

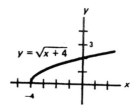

Section 3.1

29. I: $x = 0, y = 0$

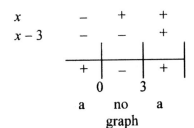

a   no   a

graph

A: $x = 3, y = 1$
D: $x \leq 0, x > 3$

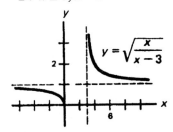

33. $y = \pm\sqrt{x/(x+4)}$ I: $x = 0, y = 0$

no

graph

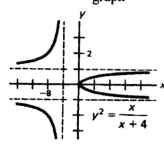

31. I: $x = -9, y = \pm 3$
   S: x-axis
   D: $x \geq 9$

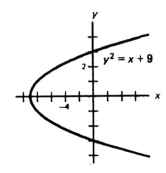

35. $y = \pm\sqrt{x^2/(x^2+4)}$

I: $x = 0, y = 0$
S: x-axis, y-axis, origin
A: $y = \pm 1$;   D: {reals}

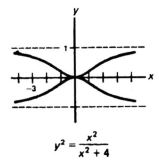

Section 3.1

Section 3.2

KEY: d = decreasing, i = increasing, and same as in Section 3.1.

1. $f'(x) = 2x + 6 = 0$; $x = -3$

$2x + 6$ $-$ $+$

$$\begin{array}{c} d \quad | \quad i \\ 0 \quad 3 \\ -3 \end{array}$$

$f(-3) = -25$; $(-3, -25)$ min
I: $x = -8, 2$; $y = -16$

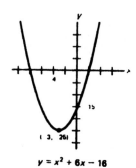

$y = x^2 + 6x - 16$

3. $f'(x) = -4 - 6x = 0$; $x = -2/3$

$-4 - 6x$ $+$ $-$

$$\begin{array}{c} | \\ -2/3 \\ i \qquad d \end{array}$$

$f(-2/3) = 5 \ 1/3$; $(-2/3, 16/3)$ max
I: $y = 4$; $x = -2, 2/3$

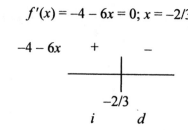

$y = 4 - 4x - 3x^2$

5. $f'(x) = 3x^2 + 6x = 3x(x + 2) = 0$

$3x$ $-$ $-$ $+$
$x + 2$ $-$ $+$ $+$

$$\begin{array}{c} + \quad | \quad - \quad | \quad + \\ -2 \qquad 0 \\ i \qquad d \qquad i \end{array}$$

max $(-2, 8)$; min: $(0, 4)$
I: $y = 4$

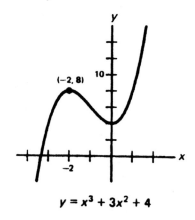

$y = x^3 + 3x^2 + 4$

7. $f'(x) = x^2 - 9 = (x - 3)(x + 3) = 0$

$x + 3$ $-$ $+$ $+$
$x - 3$ $-$ $-$ $+$

$$\begin{array}{c} + \quad | \quad - \quad | \quad + \\ -3 \qquad 3 \\ i \qquad d \qquad i \end{array}$$

max $(-3, 14)$; min $(3, -22)$
I: $y = -4$

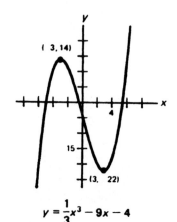

$y = \dfrac{1}{3}x^3 - 9x - 4$

Section 3.2

Solutions to Odd-Numbered Exercises

9. $f'(x) = 15x^4 - 15x^2 = 15x^2(x - 1)(x + 1)$

$15x^2$	+	+	+	+
$x - 1$	−	−	−	+
$x + 1$	−	+	+	+

$$\begin{array}{c|c|c|c|c}
+ & - & - & + \\
\hline
& -1 & 0 & 1 \\
i & d & d & i
\end{array}$$

max $(-1, 2)$; min $(1, -2)$;
I: $x = 0, y = 0$

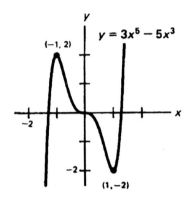

$y = 3x^5 - 5x^3$
$(-1, 2)$
$(1, -2)$

11. $f'(x) = 5(x - 2)^4$

$(x - 2)^4$	+	+

$$\begin{array}{c|c}
i & i \\
& 2
\end{array}$$

no max or min;
I: $x = 2, y = -32$

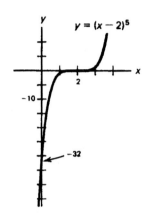

$y = (x - 2)^5$
-10
-32

13. $y' = 8x(x^2 - 1)^3$
$\quad = 8x'(x + 1)^3(x - 1)^3$

$8x$	−	−	+	+
$(x + 1)^3$	−	+	+	+
$(x - 1)^3$	−	−	−	+

$$\begin{array}{c|c|c|c}
- & + & - & + \\
\hline
-1 & 0 & 1 \\
d & i & d & i
\end{array}$$

min $(-1, 0)$, $(1, 0)$; max $(0, 1)$
I: $x = \pm 1, y = 1$; S: y-axis

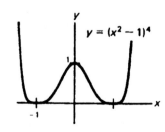

$y = (x^2 - 1)^4$
-1

15. $y' = \dfrac{x^2 - 8x}{(x - 4)^2} = 0$

x	−	+	+	+
$x - 8$	−	−	−	+
$(x - 4)^2$	+	+	+	+

$$\begin{array}{c|c|c|c}
+ & - & - & + \\
\hline
0 & 4 & 8 \\
i & d & d & i
\end{array}$$

max $(0, 0)$ min $(8, 16)$
I: $(0, 0)$ A: $x = 4, y = x + 4$, D: $x \neq 4$

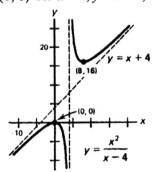

20
$y = x + 4$
$(8, 16)$
$(0, 0)$
-10
$y = \dfrac{x^2}{x - 4}$

Section 3.2

17. $y' = \dfrac{1}{(x+1)^2}$

$(x+1)^2$ +   +

i $\big|$ i

-1

no max or min; I: $(0, 0)$
A: $x = -1, y = 1$
D: $x \neq -1$

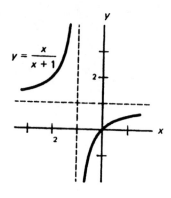

19. $y' = 1 - \dfrac{1}{x^2} = \dfrac{x^2 - 1}{x^2} = \dfrac{(x-1)(x+1)}{x^2}$

$x-1$	$-$	$-$	$-$	$+$
$x+1$	$-$	$+$	$+$	$+$
x^2	$+$	$+$	$+$	$+$

    + $-$ $-$ +

-1 0   1

i d d i

max $(-1, -2)$ min $(1, 2)$
A: $x = 0, y = x$ D: $x \neq 0$

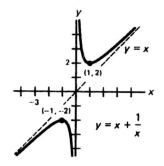

21. $y' = (1/2)x^{-1/2} = 1/(2\sqrt{x})$

y' $-$ +

no $\big|$ i
graph   0

min $(0, 0)$; D: $x \geq 0$

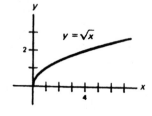

23. $y' = (2/3)x^{-1/3}$

y' $-$ +

d $\big|$ i
0

min $(0, 0)$ I: $(0, 0)$; S: y-axis
D: reals

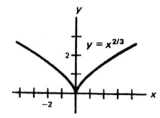

Section 3.3

KEY: up = concave upward; dn = concave downward; and same as in previous sections. Use steps as needed. Not all steps shown.

1. $y' = 2x - 4 = 0$; $x = 2$

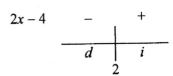

min $(2, -4)$
$y'' = 2$ always concave
upward and no points of inflection

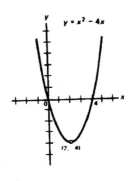

3. $y' = 4x^3 = 0$

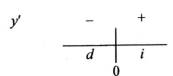

min $(0, 0)$
$y'' = 12x^2 = 0$

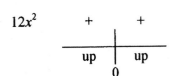

no pt of inflection

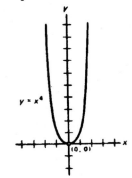

5. $y' = 3(1 - x)(1 + x)$

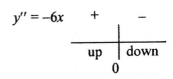

max $(1, 2)$ min $(-1, -2)$

$y'' = -6x$

pt. of inflection $(0, 0)$

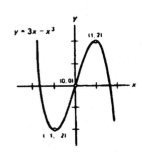

Section 3.3

7. $y' = -2x + 4$

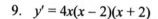

$$2x + 4 \qquad + \qquad -$$

i | d

2

max $(2, 3)$
$y'' = -2$ Thus graph concave
downward for all x; no points
of inflection.

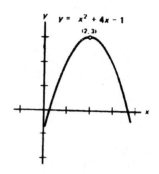

9. $y' = 4x(x - 2)(x + 2)$

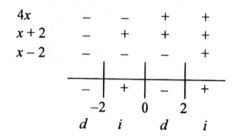

$4x$	$-$	$-$	$+$	$+$
$x + 2$	$-$	$+$	$+$	$+$
$x - 2$	$-$	$-$	$-$	$+$

$$- \qquad + \qquad - \qquad +$$

−2   0   2

d   i   d   i

min $(-2, -11), (2, 11)$; max $(0, 5)$
$y'' = 12x^2 - 16 = 4(\sqrt{3}x - 2)(\sqrt{3}x + 2)$

| $\sqrt{3}x - 2$ | $-$ | $-$ | $+$ |
| $\sqrt{3}x + 2$ | $-$ | $+$ | $+$ |

$$+ \qquad - \qquad +$$

$-2/\sqrt{3}$ $2/\sqrt{3}$

up   dn   up

inf. pts. $(-2/\sqrt{3}, -35/9), (2/\sqrt{3}, -35/9)$

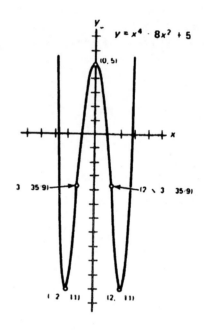

11. $y' = -2/x^3$

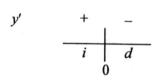

$$y' \qquad + \qquad -$$

i | d

0

no max or min
$y'' = 6/x^4$

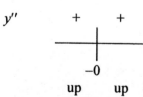

$$y'' \qquad + \qquad +$$

−0

up   up
no pts. of inflection

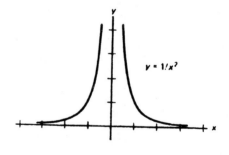

Section 3.3

13. $y' = 2/(x + 2)^2$

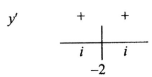

y' +   +

        i | i

       −2

no max or min

$y'' = -4/(x + 2)^3$

y'' +   −

      up | dn

       −2

no points of inflection

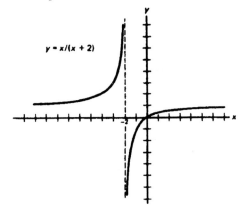

$y = x/(x + 2)$

15. $y' = 3/(x + 1)^4$

y' +   +

        i | i

       −1

no max or min

$y'' = -12/(x + 1)^5$

y'' +   −

      up | dn

       −1

no points of inflection

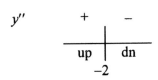

$$y = \frac{-1}{(x + 1)^3}$$

17. $y' = \dfrac{1 - x^2}{(x^2 + 1)^2} = \dfrac{(1 + x)(1 - x)}{(x^2 + 1)^2}$

$1 + x$	−	+	+
$1 - x$	+	+	−
$(x^2 + 1)^2$	+	+	+
	−	+	−
	-1		1
	d	i	d

max $(1, 1/2)$;   min $(-1, -1/2)$

$y'' = \dfrac{-2x(3 - x^2)}{(x^2 + 1)^3}$

$\quad = \dfrac{-2x(\sqrt{3} - x)(\sqrt{3} + x)}{(x^2 + 1)^3}$

$-2x$	+	+	−	−
$\sqrt{3} - x$	+	+	+	−
$\sqrt{3} + x$	−	+	+	+
$(x^2 + 1)^3$	+	+	+	+
	−	+	−	+
	$-\sqrt{3}$	0	$\sqrt{3}$	
	dn	up	dn	up

points of inflection: $(-\sqrt{3}, -\sqrt{3}/4)$, $(0, 0)$
and $(\sqrt{3}, \sqrt{3}/4)$

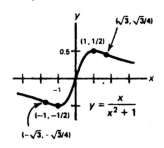

$y = \dfrac{x}{x^2 + 1}$

Section 3.3

19. $y' = 5/(x+3)^2$

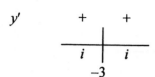

no max or min

$y'' = -10/(x+3)^3$

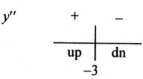

no points of inflection

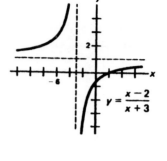

$$y = \frac{x-2}{x+3}$$

21. $y' = -8x/(x^2+4)^2$

$-8x$	+	−
$(x^2+4)^2$	+	+

	+	−
	i	0	d

max $(0, 1)$

$$y'' = \frac{-8(4-3x^2)}{(x^2+4)^3}$$

$$= \frac{-8(2+\sqrt{3}x)(2-\sqrt{3}x)}{(x^2+4)^3}$$

-8	−	−	−
$2+\sqrt{3}\,x$	−	+	+
$2-\sqrt{3}\,x$	+	+	−
$(x^2+4)^3$	+	+	+

	+	−	+
	$-2/\sqrt{3}$		$2/\sqrt{3}$
	up	dn	up

pts. of inflection $(-2/\sqrt{3}, 3/4)$; $(2/\sqrt{3}, 3/4)$

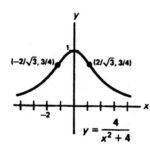

$$y = \frac{4}{x^2+4}$$

23. $y' = \dfrac{2-x}{x^3}$

$2-x$	+	+	−
x^3	−	+	+

	−	+	−
	0		2
d		i		d

max $(2, 1/4)$

$$y' = \frac{2(x-3)}{x^4}$$

	−	−	+
	+	+	+

	dn	dn	up
	0		3

pt of inflection $(3, 2/9)$

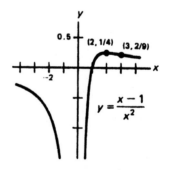

$$y = \frac{x-1}{x^2}$$

Section 3.3

Section 3.4

1.

```
                        12
Ans-(Ans²-150)/(
2Ans)
                     12.25
              12.24744898
              12.24744871
              12.24744871
■
```

3.

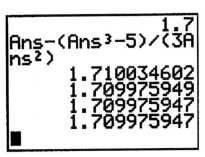

```
                       1.7
Ans-(Ans³-5)/(3A
ns²)
               1.710034602
               1.709975949
               1.709975947
               1.709975947
■
```

5.

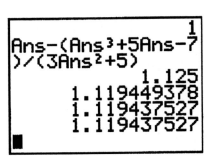

```
                        1
Ans-(Ans³+5Ans-7
)/(3Ans²+5)
                     1.125
              1.119449378
              1.119437527
              1.119437527
■
```

7.

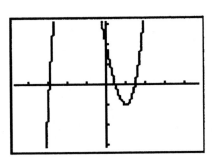

```
                       -3
Ans-(Ans³+Ans²-5
Ans+2)/(3Ans²+2A
ns-5)
                   -2.9375
              -2.935434556
              -2.935432332
■
```

```
                       .5
Ans-(Ans³+Ans²-5
Ans+2)/(3Ans²+2A
ns-5)
                .4615384615
                .4625976433
                .462598423
■
```

```
                      1.5
Ans-(Ans³+Ans²-5
Ans+2)/(3Ans²+2A
ns-5)
               1.473684211
               1.472834787
               1.472833909
■
```

Section 3.4

9.

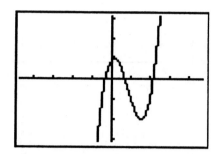

```
                          -.3
Ans-(2Ans³-5Ans²
+Ans+1)/(6Ans²-1
0Ans+1)
        -.3431718062
        -.3406739465
        -.3406653219
■
```

```
                  .7
Ans-(2Ans³-5Ans²
+Ans+1)/(6Ans²-1
0Ans+1)
        .6790849673
        .6789631883
        .6789631838
■
```

```
                 2.2
Ans-(2Ans³-5Ans²
+Ans+1)/(6Ans²-1
0Ans+1)
         2.16318408
         2.161704491
         2.161702138
■
```

11.

```
Plot1  Plot2  Plot3
\Y1■sin(2X)-cos(
X)-1
\Y2=
\Y3=
\Y4=
\Y5=
\Y6=
```

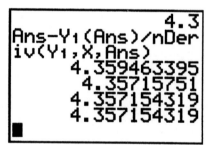

```
                   3
Ans-Y1(Ans)/nDer
iv(Y1,X,Ans)
         3.140397147
           3.1415923
         3.141592654
         3.141592654
■
```

```
                 4.3
Ans-Y1(Ans)/nDer
iv(Y1,X,Ans)
         4.359463395
          4.35715751
         4.357154319
         4.357154319
■
```

Section 3.4

13.

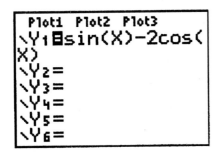 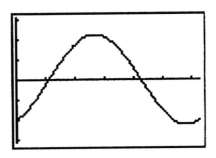

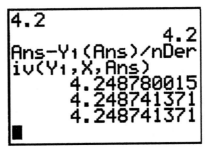

```
1.1
                1.1
Ans-Y₁(Ans)/nDer
iv(Y₁,X,Ans)
       1.107148841
       1.107148718
       1.107148718
■
```

```
4.2
                4.2
Ans-Y₁(Ans)/nDer
iv(Y₁,X,Ans)
       4.248780015
       4.248741371
       4.248741371
■
```

15.

```
                 2
Ans-(Ans^4-23)/(
4Ans^3)
          2.21875
       2.190495049
       2.189938915
       2.189938703
■
```

Section 3.5

1. $P = xy$ and $x + y = 56$
   $P = x(56 - x) = 56x - x^2$
   $dP/dx = 56 - 2x = 0$
   $x = 28, y = 28$

3. $V = lwh = (3 - 2x)^2 x$
   $dV/dx = x \cdot 2(3 - 2x)(-2) + (3 - 2x)^2$
   $= (3 - 2x)[-4x + (3 - 2x)]$
   $= (3 - 2x)(3 - 6x) = 0$
   max $x = 1/2$;
   dimensions $2 \times 2 \times 0.5$

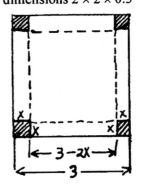

Sections 3.4 – 3.5

5. $P = 2x + y;\ A = xy$
$A = x(800 - 2x) = 800x - 2x^2$
$dA/dx = 800 - 4x = 0$
$x = 200,\ y = 400$

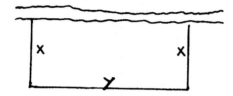

7. $A = \ell w;\ P = 2\ell + 2w = 36$
$\ell + w = 18$
$A = \ell(18 - \ell) = 18\ell - \ell^2$
$dA/D\ell = 18 - 2\ell = 0$
$\ell = 9,\ w = 9$
Since $A'' = -2 < 0,\ \ell = 9$ is a max.

9.

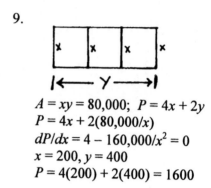

$A = xy = 80{,}000;\ P = 4x + 2y$
$P = 4x + 2(80{,}000/x)$
$dP/dx = 4 - 160{,}000/x^2 = 0$
$x = 200,\ y = 400$
$P = 4(200) + 2(400) = 1600$

11.

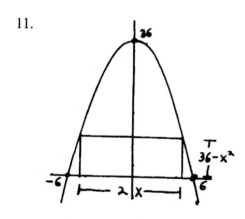

$A = \ell w = 2x(36 - x^2) = 72x - 2x^3$
$A' = 72 - 6x^2 = 0$
$x = \pm 2\sqrt{3},\ y = 24$
$A = (2)(2\sqrt{3})(24) = 96\sqrt{3}$
Since $A'' = -12x\Big|_{2\sqrt{3}} = -24\sqrt{3} < 0$

A is a max.

13. $A = 4xy;\ x^2 + 9y^2 = 16$
$A = 4y\sqrt{16 - 9y^2}$
$A' = 4y(1/2)(16 - 9y^2)^{-1/2}(-18y) + 4(16 - 9y^2)^{1/2}$
$= 4(16 - 9y^2)^{-1/2}[-9y^2 + 16 - 9y^2]$
$= \dfrac{4(16 - 18y^2)}{(16 - 9y^2)^{1/2}} = 0$
Thus, $y = 2\sqrt{2}/3,\ x = 2\sqrt{2}$;
$A = 4(2\sqrt{2}/3)(2\sqrt{2}) = 32/3$

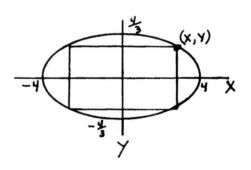

Section 3.5

15. $\dfrac{dP}{dR} = \dfrac{(R+4)^2(36) - 72R(R+4)}{(R+4)^4} = \dfrac{(R+4)[36(R+4) - 72R]}{(R+4)^4} = \dfrac{-36R + 144}{(R+4)^3} = 0$ thus $R = 4\ \Omega$

17. $A = 4xy;\ x^2 + y^2 = r^2$
$A = 4x(r^2 - x^2)^{1/2}$
$A' = 4(r^2 - x^2)^{1/2} + 4x(1/2)(r^2 - x^2)^{-1/2}(-2x)$
$\quad = 4(r^2 - x^2)^{-1/2}[(r^2 - x^2) - x^2]$
$\quad = \dfrac{4(r^2 - 2x^2)}{\sqrt{r^2 - x^2}} = 0$ thus $x = r\sqrt{2}/2,\ y = r\sqrt{2}/2$

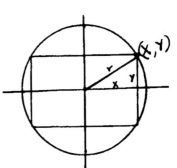

Since $A''(r\sqrt{2}/2) < 0$ and $x = y$, the max area is a square.

19. $m = y' = 6x - 6x^2$
$dm/dx = 6 - 12x = 0$
$x = 1/2$
Since $m'' = -12 < 0$, $x = 1/2$
gives a max slope and
$m = 6(1/2) - 6(1/2)^2 = 3/2$

21. $i = dq/dt = 4t^3 - 12t^2$
(minimize i)
$di/dt = 12t^2 - 24t$
$= 12t(t - 2) = 0$
$t = 0,\ t = 2$
At $t = 2$, $i = 4(2)^3 - 12(2)^2$
$= -16$, which is a min
because $d^2i/dt^2 = 24t - 24 = 24 > 0$ at $t = 2$

23. $V = x^2y = 4000;\ A = 4xy + x^2$
$A = 4x(4000/x^2) + x^2 = 16000/x + x^2$
$A' = -16000/x^2 + 2x = 0$
$2x = 16000/x^2$
$x^3 = 8000$
$x = 20;\ y = 10$
dimensions: $20 \times 20 \times 10$

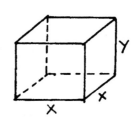

25. $P = 36.75x - (0.005x^3 + 0.45x^2 + 12.75x) = -0.005x^3 - 0.45x^2 + 24x$
$P' = -0.015x^2 - 0.9x + 24 = 0$
$x = \dfrac{0.9 \pm \sqrt{(-0.9)^2 - 4(-0.015)(24)}}{2(-0.015)} = \dfrac{0.9 \pm \sqrt{2.25}}{-0.03} = 20$

27. $V = \pi r^2 h;\ A = \pi r^2 + 2\pi rh = 24\pi$
$$h = \dfrac{24 - r^2}{2r}$$
$V = \pi r^2 (24 - r^2)/2r = \pi(12r - r^3/2)$
$V' = \pi(12 - 3r^2/2) = 0;\ 12 = 3r^2/2$
$r^2 = 8;\ r = 2\sqrt{2},$
$\quad h = (24 - 8)/(2\sqrt{2}) = 4\sqrt{2}$

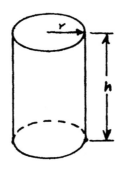

Section 3.5

29. $S = kwd^2$; $r^2 = d^2/4 + w^2/4$

$$d^2 = 4r^2 - w^2$$
$$S = kw(4r^2 - w^2) = 4kwr^2 - kw^3$$
$$dS/dw = 4kr^2 - 3kw^2 = 0$$
$$4r^2 - 3w^2 = 0$$
$$w^2 = 4r^2/3; \quad w = 2r/\sqrt{3}$$
$$d^2 = 4r^2 - 4r^2/3 = 8r^2/3$$
$$d = 2r\sqrt{2}/\sqrt{3} = 2r\sqrt{6}/3$$

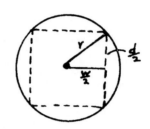

31. Use similar triangles:

$$\frac{8}{12} = \frac{x}{10-x}$$
$$80 - 8x = 12x$$

$$80 = 20x$$
$$x = 4$$

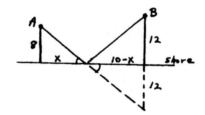

Section 3.6

1. $\dfrac{dy}{dt} = 6x\dfrac{dx}{dt}$, $\dfrac{dy}{dt} = 6(3)(2)$, $\dfrac{dy}{dt} = 36$.

3. $x^2 + 4^2 = 25$ and $x > 0$ implies $x = 3$,

$$2x\frac{dx}{dt} + 2y\frac{dy}{dt} = 0, \quad \frac{dx}{dt} = \frac{-y}{x}\frac{dy}{dt}, \quad \frac{dx}{dt} = \frac{-4}{3}\left(\frac{9}{2}\right) = -6$$

5. $dT/dt = 0.3°\text{C/s}$, find dR/dt when $T = 100°\text{C}$
$dR/dt = 0.002T \, dT/dt = 0.002(100)(0.3) = 0.06\,\Omega\text{/s}$

7. $A = s^2$; find dA/dt when $ds/dt = 0.08$ cm/min and $s = 12$ cm
$dA/dt = 2s \, ds/dt = 2(0.08)(12) = 1.92$ cm^2/min

9. $V = e^3$; find dV/dt when $ds/dt = 0.1$ cm/min and $e = 12$ cm
$dV/dt = 3e^2 \, de/dt = 3(12^2)(0.1) = 43.2$ cm^3/min

11. $A = \sqrt{3}\,s^2/4$; find dA/dt when $ds/dt = -0.04$ cm/min and $s = 8$ cm
$dA/dt = (\sqrt{3}/2)s \, ds/dt = (\sqrt{3}/2)(8)(-0.04) = -0.277$ cm^2/min

13. $h = 3r/4$; $V = (1/3)\pi r^2 h = (1/3)\pi r^2(3r/4) = \pi r^3/4$
Find dr/dt when $dV/dt = 5$ m^3/min, $h = 10$ m,
and $r = 40/3$ m; $dV/dt = (3/4)\pi r^2 dr/dt$;
$5 = (3/4)\pi(40/3)^2 dr/dt$; $dr/dt = 3/(80\pi)$ m/s

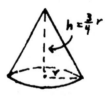

15. $P = I^2R$; find dP/dt when $I = 8A$, $dI/dt = -0.4$ A/s, $R = 75$ Ω, and $dR/dt = 5$ Ω/s.
$dP/dt = I^2 \, dR/dt + 2IR \, dI/dt = (8^2)(5) + 2(8)(75)(-0.4) = -160$ W/s

Sections 3.5 – 3.6

17. $r = 75$ mm (fixed) $= 7.5$ cm; $V = \pi r^2 h = \pi(7.5)^2 h$
Find dh/dt when $dV/dt = 90$ cm^3/s
$dV/dt = \pi(7.5)^2 dh/dt$; $dh/dt = 90/[(7.5)^2\pi] = 0.509$ cm/s

19. $z = 5$ m (fixed); $x^2 + y^2 = 25$
Find dy/dt when $dx/dt = -1$ m/s, $x = 4$ m, and $y = 3$ m.
$2x\, dx/dt + 2y\, dy/dt = 0$; $dy/dt = -x/y\, dx/dt = (-4/3)(-1) = 4/3$ m/s

21. $PV^{1.4} = k$; $P = kV^{-1.4}$; find dP/dt when $P = 60$ lb/in^2 · $V = 56$ in^3,
$dV/dt = -8$ in^3/min; $dP/dt = -1.4kV^{-2.4}\, dV/dt$
$= -1.4(PV^{1.4})V^{-2.4}\, dV/dt$
$= [-1.4P/V]\, dV/dt = [-1.4(60)/56](-8) = 12$ lb/in^2/min

23. $R = \dfrac{120R_2}{120 + R_2}$; find dR/dt when $dR_2/dt = -15$ Ω/s, $R_2 = 180$ Ω

$\dfrac{dR}{dt} = \dfrac{(120 + R_2)120dR_2/dt - 120R_2 dR_2/dt}{(120 + R_2)^2_{52}}$

$= \dfrac{(300)(120)(-15) - (120)(180)(-15)}{(120+180)^2} = -2.4$ Ω/s

25. $P = 30i^2$; $dP/di = 60i = 60(4) = 240$ AΩ

27. $V = \dfrac{0.02i}{(0.01)^2} = 200i$; find $di/dt = 0.04$ A/s; $dV/dt = 200\, di/dt = 200(0.04) = 8$ V/s

29. $\dfrac{h}{r} = \dfrac{8}{2} = 4$ $r = h/r$
$V = (1/3)\pi r^2 h = (1/3)\pi(h/4)^2 h = (1/48)\pi h^3$
$dV/dt = (\pi/16)h^2 dh/dt$

a) $2\pi = (\pi/16)(2^2)dh/dt$; $dh/dt = 8$ m/min
b) $2\pi = (\pi/16)(6^2)\, dh/dt$; $dh/dt = 8/9$ m/min

Section 3.7
1. $dy/dx = 10x - 24x^2$; $dy = (10x - 24x^2)dx$

3. $dy/dx = \dfrac{(2x - 1)(1) - (x + 3)(2)}{(2x - 1)^2} = \dfrac{-7}{(2x - 1)^2}$;

$dy = \dfrac{-7}{(2x - 1)^2}\, dx$

5. $dy/dt = 4(2t^2 + 1)^3(4t) = 16t(2t^2 + 1)^3$;
$dy = 16t(2t^2 + 1)^3 dt$

Sections 3.6 – 3.7

7. $ds/dt = -2(t^4 - t^{-2})^{-3}(4t^3 + 2t^{-3})$;

$$ds = -4t^{-3}(t^4 - t^{-2})^{-3}(2t^6 + 1)dt \text{ or } ds = \frac{-4t^3(2t^6 + 1)}{(t^6 - 1)^3}dt$$

9. $2x + 8y\, dy/dx = 0$; $dy/dx = -x/(4y)$; $dy = (-x/(4y))\, dx$

11. $3(x + y)^2(1 + y') = (1/2)x^{-1/2} + (1/2)y^{-1/2}y'$
    $3(x + y)^2 + [3(x + y)^2]y' = (1/2)x^{-1/2} + (1/2)y^{-1/2}y'$
    $6(x + y)^2 - x^{-1/2} = [y^{-1/2} - 6(x + y)^2]y'$

$$y' = \frac{6(x + y)^2 - x^{-1/2}}{y^{-1/2} - 6(x + y)^2}; dy = \frac{6(x + y)^2 - x^{-1/2}}{y^{-1/2} - 6(x + y)^2}dx$$

13. $dy = 32x^3\, dx = 32(3^3)(0.05) = 43.2$

15. $dV = (3r^2 - 6r)dr = [3 \cdot 4^2 - 6 \cdot 4](0.05) = 1.2$

17. $dV = 4\pi r^2\, dr = 4\pi(15)^2(0.1) = 282.7$

19. a) $dA = 2s\, ds = 2(12)(0.05) = 1.20$ cm^2
    b) $\Delta A = (12.05)^2 - 12^2 = 1.2025$ cm^2

    c) $\dfrac{dA}{A} \cdot 100\% = \dfrac{1.20}{144} \cdot 100\% = 0.833\%$

21. a) $V = (4/3)\pi r^3$; $dV = 4\pi r^2 dr = 4\pi(13)^2(0.04) = 84.9$ m^3

    b) steel $= (7800$ kg/m$^3)(84.9$ m$^3) = 662,000$ kg

23. $dP = 16d\, dd = 16(3.750)(0.005) = 0.3$ hp

25. $dv = (20/3)p^{-1/3}dP = (20/3)(125)^{-1/3}(3) = 4$ V

Section 3.7

Chapter 3 Review

1. $y = x(x - 4)(x + 4)$
   I: $x = 0, 4, -4;\ y = 0$

x	$-$	$-$	$+$	$+$
$x - 4$	$-$	$+$	$+$	$+$
$x + 4$	$-$	$-$	$-$	$+$

	$-$	$+$	$-$	$+$
	-4	0	4	
	b	a	b	a

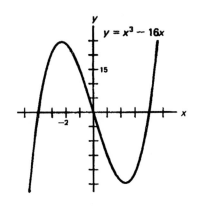

2. $y = \sqrt{-2 - 4x}$
   I: $x = -1/2$; D: $x \le -1/2$

	$-$	$+$
no	$-1/2$	a
graph		

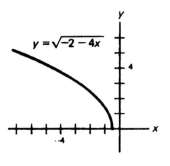

3. $y = (x + 2)(5 - x)(5 + x)$
   I: $x = -2, 5, -5;\ y = 50$

$x + 2$	$-$	$-$	$+$	$+$
$5 - x$	$+$	$+$	$+$	$-$
$5 + x$	$-$	$+$	$+$	$+$

	$+$	$-$	$+$	$-$
	-5	-2	5	
	a	b	a	b

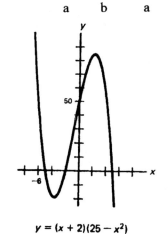

$y = (x + 2)(25 - x^2)$

4. $y = x(x + 4)(x - 1)^2$
   I: $x = 0, 1, -4;\ y = 0$

$(x - 1)^2$	$+$	$+$	$+$	$+$
$x + 4$	$-$	$+$	$+$	$+$
x	$-$	$-$	$+$	$+$

	$+$	$-$	$+$	$+$
	-4	0	1	
	a	b	a	a

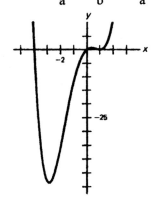

$y = (x^2 + 4x)(x - 1)^2$

Chapter 3 Review

5. $y = \pm\sqrt{x-4}$
   I: $x = 4$; D: $x \geq 4$
   S: x-axis

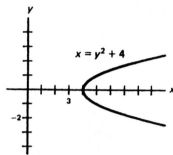

$x = y^2 + 4$

6. I: $x = 2$; $y = 1/2$
   D: $x \neq -4, 1$ A: $x = -4, x = 1, y = 0$

$x-2$	$-$	$-$	$+$	$+$
$x+4$	$-$	$+$	$+$	$+$
$x-1$	$-$	$-$	$-$	$+$

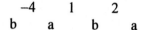

$-$ $+$ $-$ $+$
-4 1 2
b   a   b   a

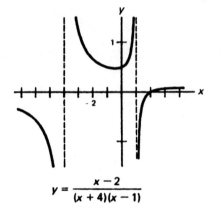

$$y = \frac{x-2}{(x+4)(x-1)}$$

7. I: $x = 0, y = 0$;
   D: $x \neq 1, -1$
   A: $x = \pm 1, y = 2$;
   S: y-axis

$2x^2$	$+$	$+$	$+$	$+$
$x+1$	$-$	$+$	$+$	$+$
$x-1$	$-$	$-$	$-$	$+$

$+$ $-$ $-$ $+$
-1 0 1
a   b   b   a

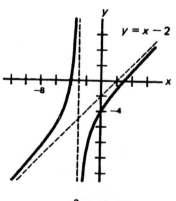

$$y = \frac{2x^2}{x^2-1}$$

8. $y = \dfrac{(x+4)(x-3)}{x+3}$; I: $x = -4, 3$; $y = -4$
   D: $x \neq -3$; A: $x = -3, y = x - 2$

$x+4$	$-$	$+$	$+$	$+$
$x-3$	$-$	$-$	$-$	$+$
$x+3$	$-$	$-$	$+$	$+$

$-$ $+$ $-$ $+$
-4 -3 3
b   a   b   a

$y = x - 2$

$$y = \frac{x^2+x-12}{x+3}$$

Chapter 3 Review

9. I: $x = 0, y = 0$; D: {reals}
    S: origin; A: $y = 0$

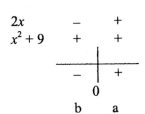

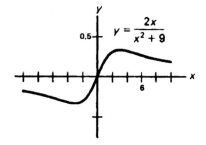

10. $y = \pm \sqrt{x/[(1-x)(x+4)]}$

    I: $x = 0, y = 0$
    D: $x < -4, 0 \le x < 1$
    S: x-axis A: $x = 1, x = -4$, $y = 0$
    $x = -4$ $y = 0$

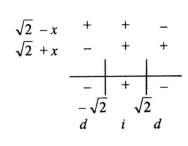

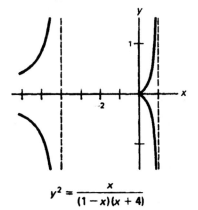

$$y^2 = \frac{x}{(1-x)(x+4)}$$

11. $y' = 6 - 3x^2$
    $= 3(\sqrt{2} - x)(\sqrt{2} + x)$

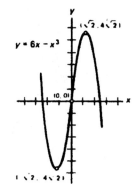

    max $(\sqrt{2}, 4\sqrt{2})$, min $(-\sqrt{2}, -4\sqrt{2})$
    $y'' = -6x$

    $-6x$ $\quad +\quad$ $\quad-$

         $\dfrac{\quad}{\text{up} \mid \text{dn}}$
              0

    point of inflection $(0, 0)$

Chapter 3 Review

12. $y' = 2x - 3$

y' $\dfrac{\quad - \quad \big| \quad + \quad}{d \quad \big| \quad i}$
        $3/2$

min $(3/2, -25/4)$
$y'' = 2$; concave up for all x;
no points of inflection

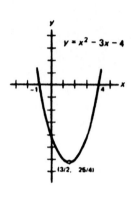

13. $y' = 3x^2$

y' $\dfrac{\quad + \quad \big| \quad + \quad}{i \quad \big| \quad i}$
        0

no max or min
$y'' = 6x$

$6x$ $\dfrac{\quad - \quad \big| \quad + \quad}{dn \quad \big| \quad up}$
        0

point of inflection $(0, -7)$

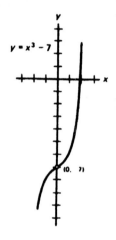

14. $y' = 6x^2 - 18x - 24 = 6(x^2 - 3x - 4)$
    $= 6(x - 4)(x + 1)$

$x + 1$ $\quad - \quad + \quad +$
$x - 4$ $\quad - \quad - \quad +$

$\dfrac{\quad + \quad \big| \quad - \quad \big| \quad + \quad}{\quad -1 \qquad 4 \quad}$
$\quad i \qquad d \qquad i$

max$(-1, 11)$; min $(4, -114)$
$y'' = 12x - 18$

y'' $\dfrac{\quad - \quad \big| \quad + \quad}{dn \quad \big| \quad up}$
        $3/2$

point of inf. $(3/2, -51.5)$

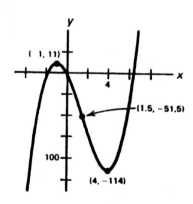

$y = 2x^3 - 9x^2 - 24x - 2$

Chapter 3 Review

Solutions to Odd-Numbered Exercises

15. $y' = -2/(x+1)^3$

$$
\begin{array}{c}
y' \quad \dfrac{+ \quad | \quad -}{i \quad | \quad d} \\[4pt]
\qquad -1
\end{array}
$$

no max or min

$y'' = 6/(x+1)^4$

$$
\begin{array}{c}
y'' \quad \dfrac{+ \quad | \quad +}{\text{up} \quad | \quad \text{dn}} \\[4pt]
\qquad -1
\end{array}
$$

no points of inflection

A: $y = 0$

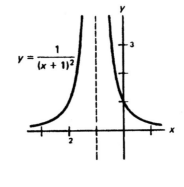

$$y = \frac{1}{(x+1)^2}$$

16. $y' = 10x/(x^2+4)^2$

$$
\begin{array}{c}
\begin{array}{c} 10x \\ (x^2+4)^2 \end{array}
\dfrac{\begin{array}{cc} - & + \\ + & + \end{array}}{\begin{array}{cc} - & + \end{array}} \\[4pt]
\qquad 0 \\
\qquad d \qquad i
\end{array}
$$

min $(0, -1/4)$

$$y'' = \frac{10(4-3x^2)}{(x^2+4)^3}$$

$$= \frac{10(2+\sqrt{3}x)(2-\sqrt{3}x)}{(x^2+4)^3}$$

$$
\begin{array}{c}
\begin{array}{c} 2+\sqrt{3}x \\ 2-\sqrt{3}x \\ (x^2+4)^3 \end{array}
\begin{array}{ccc} - & + & + \\ + & + & + \\ + & + & + \end{array} \\[6pt]
\dfrac{\quad}{\begin{array}{ccc} - & + & - \end{array}} \\[4pt]
-2/\sqrt{3} \quad 2/\sqrt{3} \\
\text{dn} \quad \text{up} \quad \text{dn}
\end{array}
$$

points of inflection $(2/\sqrt{3}, 1/16)$.

$(-2/\sqrt{3}, 1/16)$

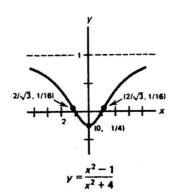

$$y = \frac{x^2-1}{x^2+4}$$

Chapter 3 Review

17. $y' = -10(x^2 + 1)^{-2}(2x)$

$$= \frac{-20x}{(x^2 + 1)^2}$$

	+	−
$-20x$	+	
$(x^2 + 1)^2$	+	+
	+	−

$$\begin{array}{cc} & 0 \\ i & d \end{array}$$

max $(0, 10)$

$$y'' = \frac{20(3x^2 - 1)}{(x^2 + 1)^3}$$

$$= \frac{20(\sqrt{3}x - 1)(\sqrt{3}x + 1)}{(x^2 + 1)^3}$$

$\sqrt{3}x - 1$	−	−	+
$\sqrt{3}x + 1$	−	+	+
$(x^2 + 1)^3$	+	+	+
	+	−	+

$$\begin{array}{ccc} -1/\sqrt{3} & & 1/\sqrt{3} \\ \text{up} & \text{dn} & \text{up} \end{array}$$

points of inflection $(-1/\sqrt{3}, 15/2)$
$(1/\sqrt{3}, 15/2)$
A: $y = 0$

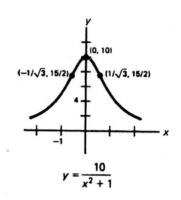

$(-1/\sqrt{3}, 15/2)$ (0, 10) $(1/\sqrt{3}, 15/2)$

$$y = \frac{10}{x^2 + 1}$$

18. $y' = \frac{-x - 2}{x^3}$

$-x - 2$	+	−	−
x^3	−	−	+
	−	+	−

$$\begin{array}{ccc} -2 & & 0 \\ d & i & d \end{array}$$

min$(-2, -1/4)$

$$y'' = \frac{2x + 6}{x^4}$$

$2x + 6$	−	+	+
x^4	+	+	+
	−	+	+

$$\begin{array}{ccc} -3 & & 0 \\ \text{dn} & \text{up} & \text{up} \end{array}$$

point of inflection $(-3, -2/9)$
A: $y = 0$

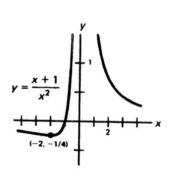

$$y = \frac{x + 1}{x^2}$$

$(-2, -1/4)$

Chapter 3 Review

19.

```
                     10
Ans-(Ans²-95)/(2
Ans)
                   9.75
         9.746794872
         9.746794345
         9.746794345
■
```

20.

```
                      5
Ans-(Ans³-120)/(
3Ans²)
         4.933333333
         4.932424316
         4.932424149
         4.932424149
■
```

21.

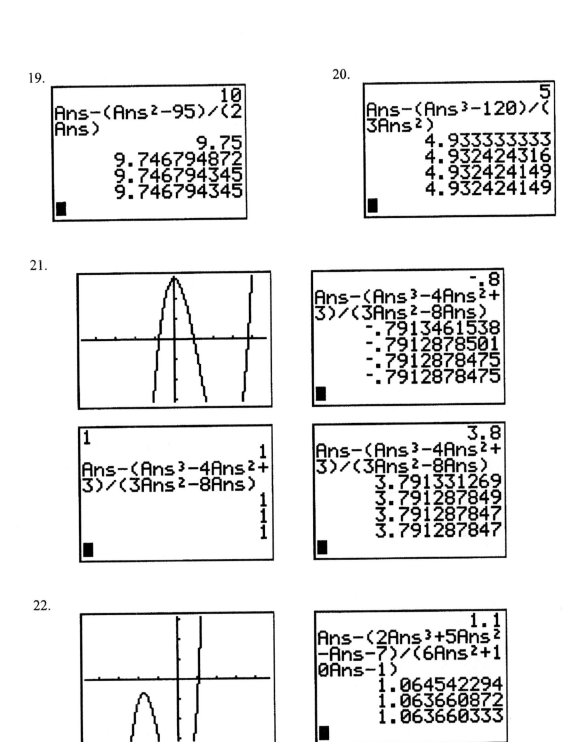

```
                    -.8
Ans-(Ans³-4Ans²+
3)/(3Ans²-8Ans)
         -.7913461538
         -.7912878501
         -.7912878475
         -.7912878475
■
```

```
1
                      1
Ans-(Ans³-4Ans²+
3)/(3Ans²-8Ans)
                      1
                      1
                      1
■
```

```
                    3.8
Ans-(Ans³-4Ans²+
3)/(3Ans²-8Ans)
         3.791331269
         3.791287849
         3.791287847
         3.791287847
■
```

22.

```
                    1.1
Ans-(2Ans³+5Ans²
-Ans-7)/(6Ans²+1
0Ans-1)
         1.064542294
         1.063660872
         1.063660333
■
```

23. $y' = 240 - 32t = 0$; $t = 7.5$ thus max ht y
$$= 240(7.5) - 16(7.5)^2 = 900 \text{ ft}$$

Chapter 3 Review

24. $a^2 + b^2 = 400$; $A = (1/2)ab$

$A = (1/2)a\sqrt{400 - a^2} = (1/2)a(400 - a^2)^{1/2}$

$A' = (1/2)a[(1/2)(400 - a^2)^{-1/2}(-2a)] + (1/2)(400 - a^2)^{1/2}$

$= (1/2)(400 - a^2)^{-1/2}[-a^2 + (400 - a^2)]$

$= \dfrac{400 - 2a^2}{2\sqrt{400 - a^2}} = 0$; thus $2a^2 = 400$ or $a = 10\sqrt{2}$

$b = \sqrt{400 - a^2} = 10\sqrt{2}$

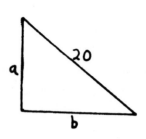

25. $dp/dr = 3 - 3r^2 = 0$

$r = 1\,\Omega$; $d^2p/dr^2 = -6r$

$= -6(1) = -6 < 0$

max power

26. $\dfrac{6-y}{y} = \dfrac{x}{8-x}$

$0 = 48 - 8y - 6x$ so $x = (24 - 4y)/3$

$A = xy = y(24 - 4y)/3 = 8y - 4y^2/3$

$dA/dy = 8 - 8y/3 = 0$

Thus, $y = 3$, $x = 4$; $A = (3)(4) = 12\ \text{m}^2$

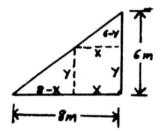

27. $A = (1/2)d\sqrt{36 - d^2}$

$dA/dd = (1/2)d(1/2)(36 - d^2)^{-1/2}(-2d) + (1/2)(36 - d^2)^{1/2}$

$= (1/2)(36 - d^2)^{-1/2}[-d^2 + (36 - d^2)]$

$= \dfrac{36 - 2d^2}{2\sqrt{36 - d^2}} = 0$ thus $d^2 = 18$; $d = 3\sqrt{2}$

gives max A and max V.

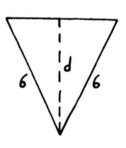

28. $d = \sqrt{(x-1)^2 + (\sqrt{x} - 0)^2} = \sqrt{x^2 - x + 1}$

$dd/dx = (1/2)(x^2 - x + 1)^{-1/2}(2x - 1)$

$= \dfrac{2x - 1}{2\sqrt{x^2 - x + 1}} = 0$

$2x - 1 = 0$; $x = 1/2$, $y = \sqrt{2}/2$

point $(1/2,\ \sqrt{2}/2)$

29. $V = (4/3)\pi r^3$; $dV/dt = 4\pi r^2\,dr$; $2 = 4\pi(12)^2\,dr$; $dr = 1/(288\pi)$ ft/s

30. $V = 0.06i/(0.01)^2$; $dV/dt = 600\,di/dt = 600(0.03) = 18$ V/s

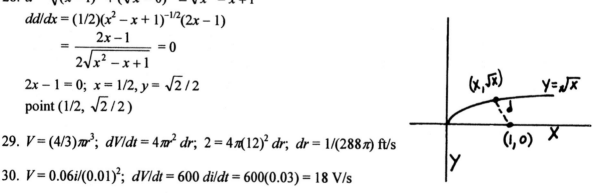

Chapter 3 Review

31. $i = \dfrac{1.5}{R + 0.3}$

$\dfrac{di}{dt} = \dfrac{-1.5}{(R+0.3)^2} \dfrac{dR}{dt}$

$= \dfrac{-1.5}{(8.3)^2}(0.4)$

$= -0.0087$ A/s

32. $A = \pi r^2$

$dA/dt = 2\pi r \, dr/dt$

$= 2\pi(8)(-0.05)$

$= -2.51$ cm^2/min

33. $A = \pi r^2$

$dA/dt = 2\pi r \, dr/dt$

$4 = 2\pi(2.5) \, dr/dt$

$dr/dt = 0.255$ km/d

34. $x = 12$ km (fixed)

$12^2 + y^2 = z^2$

$2y \, dy/dt = 2z \, dz/dt$

$\dfrac{dz}{dt} = \dfrac{y \, dy / dt}{z}$

$= \dfrac{10(3)}{2\sqrt{61}} = 1.92$ km/s

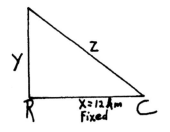

35. $dy/dx = 20x^4 - 18x^2 + 2$

$dy = (20x^4 - 18x^2 + 2)dx$

36. $y' = (-2/3)(3x - 5)^{-5/3}(3)$

$dy = -2(3x - 5)^{-5/3}dx$

37. $\dfrac{ds}{dt} = \dfrac{(5t+1)(6t) - (3t^2 - 4)(5)}{(5t+1)^2}$

$ds = \dfrac{15t^2 + 6t + 20}{(5t+1)^2} \, dt$

38. $2(x^2 + y^2)(2x + 2y \, y') = y' + 2$

$4x(x^2 + y^2) + 4y(x^2 + y^2)y' = y' + 2$

$4x(x^2 + y^2) - 2 = [1 - 4y(x^2 + y^2)]y'$

$dy = \dfrac{4x(x^2 + y^2) - 2}{1 - 4y(x^2 + y^2)} \, dx$

39. $ds = (6t - 5)dt$

$= [6(9.5) - 5](0.05) = 2.6$

40. $dy/dx = (-3/4)(8x + 3)^{-7/4}(8)$

$dy = -6(8x + 3)^{-7/4} \, dx$

$= -6[8(10) + 3]^{-7/4}(0.06)$

$= -0.000158$

41. $V = (4/3)\pi r^3$

$dV = 4\pi r^2 \, dr$

$= 4\pi(6)^2(0.1)$

$= 45.2$ in^3

42. $ds = (t^2 - 3)dt = (3^2 - 3)(0.05) = 0.3$ m

43. $V = e^3$; $dV = 3e^2 de = 3(192)^2(0.02) = 2211.84$ in$^3 = 9.58$ gal

44. $V = (4/3)\pi r^3$; $dV = 4\pi r^2 \, dr = 4\pi(4.00)^2(0.04) = 8.04$ cm^3

45. $F = k/x^2$; $dF = (-2k/x^3)dx = (-2k/(0.030)^3)(0.001) = -74.1 \, k$ N

Chapter 3 Review

CHAPTER 4

Section 4.1

37. $\sin\theta\cos 3\theta + \cos\theta\sin 3\theta =$
$\sin(\theta + 3\theta) = \sin 4\theta$

39. $\cos 4\theta\cos 3\theta + \sin 4\theta\sin 3\theta =$
$\cos(4\theta - 3\theta) = \cos\theta$

41. $\dfrac{\tan 3\theta + \tan 2\theta}{1 - \tan 3\theta \tan 2\theta} =$
$\tan(3\theta + 2\theta) = \tan 5\theta$

43. $\sin(\theta + \phi) + \sin(\theta - \phi) =$
$\sin\theta\cos\phi + \cos\theta\sin\phi + \sin\theta\cos\phi - \cos\theta\sin\phi = 2\sin\theta\cos\phi$

45. $2\sin\dfrac{x}{4}\cos\dfrac{x}{4} =$
$\sin 2\left(\dfrac{x}{4}\right) =$
$\sin\dfrac{x}{2}$

47. $1 - 2\sin^2 3x =$
$\cos 2(3x) =$
$\cos 6x$

49. $\sqrt{\dfrac{1 + \cos\dfrac{\theta}{4}}{2}} =$
$\cos\dfrac{1}{2}\left(\dfrac{\theta}{4}\right) = \cos\dfrac{\theta}{8}$

51. $\cos^2\dfrac{x}{6} - \sin^2\dfrac{x}{6} =$
$\cos 2\left(\dfrac{x}{6}\right) = \cos\dfrac{x}{3}$

53. $20\sin 4\theta\cos 4\theta =$
$10(2\sin 4\theta\cos 4\theta) =$
$10(\sin 2(4\theta)) =$
$10\sin 8\theta$

55. $4 - 8\sin^2\theta =$
$4(1 - 2\sin^2\theta) =$
$4\cos 2\theta$

57. $y = 2\cos x$
amp $= 2$
period $= 2\pi$

59. $y = 3\cos 6x$
amp $= 3$
period $= \dfrac{\pi}{3}$

61. $y = 2\sin 3\pi x$
amp $= 2$
period $= \dfrac{2\pi}{2\pi} = \dfrac{2}{3}$

63. $y = -\sin\left(4x - \dfrac{2\pi}{3}\right)$
amp $= 1$
period $= \dfrac{\pi}{2}$
phase $= \dfrac{-\pi}{6}$

65. $y = 3\sin\left(\dfrac{x}{2} - \dfrac{\pi}{4}\right)$
amp $= 3$
period $= \dfrac{2\pi}{\dfrac{1}{2}} = \dfrac{4}{\pi}$
phase $= \dfrac{-\pi}{2}$

67. $y = \tan 3x$
period $= \dfrac{\pi}{3}$

Section 4.1

Section 4.2

1. $y' = 7 \cos 7x$

3. $y' = -10 \sin 5x$

5. $y' = 6x^2 \cos x^3$

7. $y' = -24x \sin 4x^2$

9. $y' = -4 \cos(1 - x)$

11. $y' = 6x \cos(x^2 + 4)$

13. $y' = -4(10x + 1) \sin(5x^2 + x)$

15. $y' = -4(x^3 - x) \sin(x^4 - 2x^2 + 3)$

17. $y' = -6 \cos(3x - 1) \sin(3x - 1)$

19. $y' = 6 \sin^2(2x + 3) \cos(2x + 3)$

21. $y' = 4(2x - 5) \cos(2x - 5)^2$

23. $y' = -12x^2(x^3 - 4)^3 \sin(x^3 - 4)^4$

25. $y' = \cos 3x \cos x - 3 \sin x \sin 3x$

27. $y' = 5 \cos 5x \cos 6x - 6 \sin 5x \sin 6x$

29. $y' = -7 \sin 7x \cos 4x - 4 \cos 7x \sin 4x$

31. $y' = -3x^2 \sin(x^2 + 2x) \sin x^3 + 2(x + 1) \cos x^3 \cos(x^2 + 2x)$

33. $y' = 5(x^2 + 3x) \cos(5x - 2) + (2x + 3) \sin(5x - 2)$

35. $y' = \dfrac{5x \cos 5x - \sin 5x}{x^2}$

37. $y' = \dfrac{2x \cos 3x + 3(x^2 - 1) \sin 3x}{\cos^2 3x}$

39. $y' = 5 \cos 5x - 6 \sin 6x$

41. $y' = (2x - 3) \cos(x^2 - 3x) - 4 \sin 4x$

43. $y' = \dfrac{\cos^2 x + \sin^2 x}{\cos^2 x} = \dfrac{1}{\cos^2 x} = \sec^2 x$

45. $y' = -\sin x; \ y'' = -\cos x$

47. $y' = \cos x; \ y'' = -\sin x; \ y''' = -\cos x$

49. $y' = 5 \cos 5x - 6 \sin 6x; \ y'' = -25 \sin 5x - 36 \cos 6x$

51. $y' = 12 \cos 3x \big|_{x = \pi/18} = 12 \cos \pi/6 = 12(\sqrt{3}/2) = 6\sqrt{3}$

53. $y = -10 \sin 5x \big|_{x = \pi/10} = -10 \sin 5(\pi/10) = -10 \sin \pi/2 = -10$

Thus, $(\pi/10, 0); \ y - 0 = -10(x - \pi/10; \ 10x + y = \pi$

Section 4.3

1. $y' = 3 \sec^2 3x$

3. $y' = 7 \sec 7x \tan 7x$

5. $y' = -6x \csc^2(3x^2 - 7)$

7. $y' = -9 \csc(3x - 4) \cot(3x - 4)$

9. $y' = 10 \tan(5x - 2)\sec^2(5x - 2)$

11. $y' = 12 \cot^2 2x(-2 \csc^2 2x) = -24 \cot^2 2x \csc^2 2x$

Sections 4.2 – 4.3

13. $y' = (1/2)(x^2 + x)^{-1/2}(2x + 1) \sec \sqrt{x^2 + x} \, \tan \sqrt{x^2 + x}$

$\qquad = \dfrac{2x + 1}{2\sqrt{x^2 + x}} \sec \sqrt{x^2 + x} \tan \sqrt{x^2 + x}$

15. $y' = \dfrac{-3x \csc x \cot x - 3 \csc x}{9x^2} = \dfrac{-3 \csc x(\cot x + 1)}{9x^2}$

$\qquad\qquad\qquad\qquad\qquad = \dfrac{-\csc x(\cot x + 1)}{3x^2}$

17. $y' = 3 \sec^2 3x - 2x \sec(x^2 + 1) \tan (x^2 + 1)$

19. $y' = \sec^3 x + \sec x \tan^2 x = \sec x(\sec^2 x + \tan^2 x)$
$\qquad = \sec x[\sec^2 x + (\sec^2 x - 1)] = \sec x(2 \sec^2 x - 1)$

21. $y = \sin^2 x \dfrac{\cos x}{\sin x} = \sin x \cos x; \; y' = -\sin^2 x + \cos^2 x = \cos 2x$

23. $y' = x \sec x \tan x + \sec x = \sec x(x \tan x + 1)$

25. $y' = 2x + 2x^2 \tan x \sec^2 x + 2x \tan^2 x = 2x[1 + x \tan x \sec^2 x + \tan^2 x]$
$\qquad 2x[\sec^2 x + x \tan x \sec^2 x] = 2x \sec^2 x(x \tan x + 1)$

27. $y' = -3 \csc^2 3x - 3 \cot 3x \csc 3x \cot 3x = -3 \csc 3x(\csc^2 3x + \cot^2 3x)$
$\qquad = -3 \csc 3x[\csc^2 3x + (\csc^2 3x - 1)] = -3 \csc 3x (2 \csc^2 3x - 1)$

29. $y' = 3 \csc^2 3x \cos 3x + \sin 3x(-2 \csc 3x \cdot 3 \csc 3x \cot 3x)$
$\qquad = 3 \csc^2 3x \cos 3x - 6 \csc 3x \cot 3x$
$\qquad = 3 \csc 3x(\csc 3x \cos 3x - 2 \cot 3x)$
$\qquad = 3 \csc 3x\left\{\dfrac{\cos 3x}{\sin 3x} - 2 \cot 3x\right\} = 3 \csc 3x (\cot 3x - 2 \cot 3x)$
$\qquad = -3 \csc 3x \cot 3x$

31. $y' = 2(\sin x - \cos x)(\cos x + \sin x) = 2(\sin^2 x - \cos^2 x) = -2 \cos 2x$

33. $y' = 4(x + \sec^2 3x)^3(1 + 2 \sec 3x \cdot \sec 3x \tan 3x \cdot 3)$
$\qquad = 4(x + \sec^2 3x)^3(1 + 6 \sec^2 3x \tan 3x)$

35. $y' = 3(\sec x + \tan x)^2(\sec x \tan x + \sec^2 x)$
$\qquad = 3 \sec x(\sec x + \tan x)^2 (\sec x + \tan x) = 3 \sec x(\sec x + \tan x)^3$

37. $y' = \cos(\tan x) \cdot \sec^2 x$ $\qquad\qquad\qquad\qquad$ 39. $y' = -\sin x \sec^2(\cos x)$

41. $y' = 2 \sin(\cos x) \cos(\cos x) (-\sin x) = -2 \sin x \sin (\cos x) \cos(\cos x)$

Section 4.3

Section 4.4

1. $y = \sin 3x$

   $\arcsin y = 3x$

   $\dfrac{1}{3}\arcsin y = x$

3. $y = 4\cos x$

   $\dfrac{y}{4} = \cos x$

   $\arccos \dfrac{y}{4} = x$

5. $y = 5\tan \dfrac{x}{2}$

   $\dfrac{y}{5} = \tan \dfrac{x}{2}$

   $\arctan \dfrac{y}{5} = \dfrac{x}{2}$

   $2\arctan \dfrac{y}{5} = x$

7. $y = \dfrac{3}{2}\cot \dfrac{x}{4}$

   $\dfrac{2}{3}y = \cot \dfrac{x}{4}$

   $\operatorname{arccot}\dfrac{2}{3}y = \dfrac{x}{4}$

   $4\operatorname{arccot}\dfrac{2}{3}y = x$

9. $y = 3\sin(x-1)$

   $\dfrac{y}{3} = \sin(x-1)$

   $\arcsin \dfrac{y}{3} = x-1$

   $1 + \arcsin \dfrac{y}{3} = x$

11. $y = \dfrac{1}{2}\cos(3x+1)$

    $2y = \cos(3x+1)$

    $\arccos 2y = 3x+1$

    $-1 + \arccos 2y = 3x$

    $\dfrac{-1}{3} + \dfrac{1}{3}\arccos 2y = x$

13. $\arcsin\left(\dfrac{\sqrt{3}}{2}\right) =$

    $\dfrac{\pi}{3}$

15. $\tan\left(-\dfrac{1}{\sqrt{3}}\right) =$

    $\dfrac{-\pi}{6}$

17. $\arccos\left(-\dfrac{\sqrt{3}}{2}\right) =$

    $\dfrac{5\pi}{6}$

19. $\operatorname{arcsec}\sqrt{2} =$

    $\dfrac{\pi}{4}$

21. $\arctan \sqrt{3} =$

    $\dfrac{\pi}{3}$

23. $\arccos\left(\dfrac{1}{\sqrt{2}}\right) =$

    $\dfrac{\pi}{4}$

25. $\arcsin\left(-\dfrac{\sqrt{3}}{2}\right) =$

    $-\dfrac{\pi}{3}$

27. $\cos\left(\arctan \sqrt{3}\right) =$

    $\cos\left(\dfrac{\pi}{3}\right) = \dfrac{1}{2}$

29. $\sin\left[\arccos\left(-\dfrac{1}{\sqrt{2}}\right)\right] =$

    $\sin\left[\dfrac{3\pi}{4}\right] = \dfrac{1}{\sqrt{2}}$

31. $\tan[\arccos(-1)] =$

    $\tan \pi = 0$

33. $\sin\left[\arcsin \dfrac{\sqrt{3}}{2}\right] =$

    $\sin \dfrac{\pi}{3} = \dfrac{\sqrt{3}}{2}$

Section 4.4

35. $\cos\left[\arcsin\left(\dfrac{3}{5}\right)\right]$

Let $\theta = \arcsin\dfrac{3}{5}$

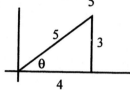

$\cos\theta = \dfrac{4}{5}$

37. $\tan\left[\arcsin(-0.1560)\right]$

Let $\theta = \arcsin(-0.1560)$

$\theta = -0.1566$

$\tan(-0.1566) = -0.1579$

39. $\cos(\arcsin x)$

Let $\theta = \arcsin x$

$\sin\theta = x$

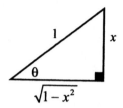

$\cos\theta = \sqrt{1-x^2}$

41. $\sin(\text{arcsec}\,x)$

Let $\theta = \text{arcsec}\,x$

$\sec\theta = x$

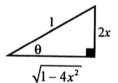

$\sin\theta = \dfrac{\sqrt{x^2-1}}{x}$

43. $\sec(\arccos x)$

Let $\theta = \arccos x$

$\cos\theta = x$

$\dfrac{1}{\sec\theta} = x$

$\dfrac{1}{x} = \sec\theta$

45. $\tan(\arctan x)$

Let $\theta = \arctan x$

$\tan\theta = x$

47. $\cos(\arcsin 2x)$

Let $\theta = \arcsin 2x$

$\sin\theta = 2x$

$\cos\theta = \sqrt{1-4x^2}$

Section 4.4

49. $\sin\left(2\arcsin x\right)$
Let $\theta = 2\arcsin x$

$\dfrac{\theta}{2} = \arcsin x$

$\sin\dfrac{\theta}{2} = x$

$\pm\sqrt{\dfrac{1-\cos\theta}{2}} = x$

$\dfrac{1-\cos\theta}{2} = x^2$

$1-\cos\theta = 2x^2$

$1-2x^2 = \cos\theta$

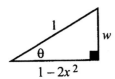

$w^2 = 1-\left(1-2x^2\right)^2$

$w^2 = 1-\left(1-4x^2+4x^4\right)$

$w^2 = 4x^2-4x^4$

$w^2 = 4x^2\left(1-x^2\right)$

$w = 2x\sqrt{1-x^2}$

$\sin\left(2\arcsin x\right) = \sin\left(2\left[\dfrac{\theta}{2}\right]\right)$

$= \sin\theta$

Thus from the triangle

$\sin\theta = \dfrac{w}{1}$

$\sin\theta = 2x\sqrt{1-x^2}$

Section 4.5

1. $y' = \dfrac{5}{\sqrt{1-25x^2}}$

3. $y' = \dfrac{3}{1+9x^2}$

5. $y' = -\dfrac{-1}{\left|1-x\right|\sqrt{(1-x)^2-1}} = \dfrac{1}{\left|1-x\right|\sqrt{x^2-2x}}$

7. $y' = 3\cdot\dfrac{-1}{\sqrt{1-(x-1)^2}} = \dfrac{-3}{\sqrt{2x-x^2}}$

9. $y' = 2\cdot\dfrac{-1(6x)}{1+9x^4} = \dfrac{-12x}{1+9x^4}$

11. $y' = 5\cdot\dfrac{3x^2}{\left|x^3\right|\sqrt{x^6-1}} = \dfrac{15}{\left|x\right|\sqrt{x^6-1}}$

Sections 4.4 – 4.5

13. $y' = 3 \arcsin^2 x \cdot \dfrac{1}{\sqrt{1-x^2}} = \dfrac{3 \arcsin^2 x}{\sqrt{1-x^2}}$

15. $y' = 2(2 \arccos 3x) \dfrac{-1(3x)}{\sqrt{1-9x^2}} = \dfrac{-12 \arccos 3x}{\sqrt{1-9x^2}}$

17. $y' = 12 \arctan^3 \sqrt{x} \dfrac{1}{1+x}(1/2)x^{-1/2} = \dfrac{6 \arctan^3 \sqrt{x}}{\sqrt{x}(1+x)}$

19. $y' = \dfrac{1}{\sqrt{1-x^2}} - \dfrac{1}{\sqrt{1-x^2}} = 0$

21. $y' = (1/2)(1-x^2)^{-1/2}(-2x) + \dfrac{1}{\sqrt{1-x^2}} = \dfrac{-x}{\sqrt{1-x^2}} + \dfrac{1}{\sqrt{1-x^2}} = \dfrac{1-x}{\sqrt{1-x^2}}$

23. $y' = \dfrac{3x}{\sqrt{1-9x^2}} + \arcsin 3x$

25. $y' = \dfrac{x}{1+x^2} + \arctan x$

27. $y' = x \cdot \dfrac{1}{\sqrt{1-x^2}} + \arcsin x + (1/2)(1-x^2)^{-1/2}(-2x)$

$\quad = \dfrac{x}{\sqrt{1-x^2}} + \arcsin x - \dfrac{x}{\sqrt{1-x^2}} = \arcsin x$

29. $y' = \dfrac{\arcsin x - x \cdot \dfrac{1}{\sqrt{1-x^2}}}{\arcsin^2 x} \cdot \dfrac{\sqrt{1-x^2}}{\sqrt{1-x^2}} = \dfrac{\sqrt{1-x^2}\arcsin x - x}{\sqrt{1-x^2}\arcsin^2 x}$

31. $y' = \dfrac{1}{\sqrt{1-x^2}}\Bigg|_{x=1/2} = 2/\sqrt{3}$

33. $y' = \dfrac{x}{x^2+1} + \arctan \Bigg|_{x=-1} = -\dfrac{1}{2} - \dfrac{\pi}{4} = \dfrac{-\pi-2}{4}$

35. If $y = \arccos u$, then $u = \cos y$ where $0 \le y \le \pi$.
So $du/dx = -\sin y \, dy/dx$

Then $\dfrac{dy}{dx} = \dfrac{-1}{\sin y}\dfrac{du}{dx}$ but $\sin y = \sqrt{1 - \cos y}$ since $0 \le y \le \pi$.

$\dfrac{dy}{dx} = -\dfrac{1}{\sqrt{1-u^2}}\dfrac{du}{dx}$

Section 4.5

Section 4.6
The graphs are in the text answer section.

1. $y = 4^x$

x	-1	0	1	2
y	0.25	1	4	16

3. $y = \left(\dfrac{1}{3}\right)^x$

x	-2	-1	0	1
y	9	3	1	$\dfrac{1}{3}$

5. $y = 4^{-x}$

x	-2	-1	0	1
y	16	4	1	0.25

7. $y = \left(\dfrac{4}{3}\right)^{-x}$

x	-2	-1	0	1
y	1.8	1.3	1	0.75

9. $3^2 = 9$

$\log_3 9 = 2$

11. $5^3 = 125$

$\log_5 125 = 3$

13. $9^{1/2} = 3$

$\log_9 3 = \dfrac{1}{2}$

15. $10^{-5} = 0.00001$

$\log_{10} 0.00001 = -5$

17. $\log_5 25 = 2$

$5^2 = 25$

19. $\log_{25} 5 = \dfrac{1}{2}$

$25^{1/2} = 5$

21. $\log_2 \left(\dfrac{1}{4}\right) = -2$

$2^{-2} = \dfrac{1}{4}$

23. $\log_{10} 0.01 = -2$

$10^{-2} = 0.01$

The graphs for 25, 27, and 29 appear in the text answer section.

25. $y = \log_4 x$

x	0.25	1	4	16
y	-1	0	1	2

27. $y = \log_{10} x$

x	0.1	1	10	100
y	-1	0	1	2

29. $y = \log_{1/4} x$

x	4	1	0.25	0.06
y	-1	0	1	2

31. $\log_4 x = 3$

$4^3 = x$

$x = 64$

33. $\log_9 3 = x$

$9^x = 3$

$\left(3^2\right)^x = 3^1$

$2x = 1$

$x = \dfrac{1}{2}$

35. $\log_2 8 = x$

$2^x = 8$

$2^x = 2^3$

$x = 3$

Section 4.6

37. $\log_{25} 5 = x$
$25^x = 5$
$(5^2)^x = 5^1$
$2x = 1$
$x = \dfrac{1}{2}$

39. $\log_x 25 = 2$
$x^2 = 25$
$x = 5$
(Base cannot be negative.)

41. $\log_{1/2}\left(\dfrac{1}{8}\right) = x$
$\left(\dfrac{1}{2}\right)^x = \dfrac{1}{8}$
$\left(\dfrac{1}{2}\right)^x = \left(\dfrac{1}{2}\right)^3$
$x = 3$

43. $\log_{12} x = 2$
$12^2 = x$
$x = 144$

45. $\log_x 9 = \dfrac{2}{3}$
$x^{2/3} = 9$
$x = 9^{3/2}$
$x = 27$

47. $\log_x\left(\dfrac{1}{8}\right) = -\dfrac{3}{2}$
$x^{-3/2} = \dfrac{1}{8}$
$x = \left(\dfrac{1}{8}\right)^{-2/3}$
$x = \left[\left(\dfrac{1}{8}\right)^{1/3}\right]^{-2}$
$x = \left(\dfrac{1}{2}\right)^{-2}$
$x = \dfrac{1^{-2}}{2^{-2}}$
$x = 4$

49. $\log_2 5x^3 y = \log_2 5 + \log_2 x^3 + \log_2 y = \log_2 5 + 3\log_2 x + \log_2 y$

51. $\log_b \dfrac{y^3 \sqrt{x}}{z^2} = \log_b y^3 \sqrt{x} - \log_b z^2 = \log_b y^3 + \log_b \sqrt{x} - $
$\left(\log_b z^2\right) = 3\log_b y + \dfrac{1}{2}\log_b x - 2\log_b z$

53. $\log_b \sqrt[3]{\dfrac{x^2}{y}} = \dfrac{1}{3}\log_b \dfrac{x^2}{y} = \dfrac{1}{3}\left(\log_b x^2 - \log_b y\right) = $
$\dfrac{1}{3}\left(2\log_b x - \log_b y\right) = \dfrac{2}{3}\log_b x - \dfrac{1}{3}\log_b y$

55. $\log_2 \dfrac{1}{x}\sqrt{\dfrac{y}{z}} = \log_2 \dfrac{1}{x} + \log_2 \sqrt{\dfrac{y}{z}} = -\log_2 x + \dfrac{1}{2}\log_2 \dfrac{y}{z} = $
$-\log_2 x + \dfrac{1}{2}\left(\log_2 y - \log_2 z\right) = -\log_2 x + \dfrac{1}{2}\log_2 y - \dfrac{1}{2}\log_2 z$

57. $\log_b \dfrac{z^3 \sqrt{x}}{\sqrt[3]{y}} = \log_b z^3 \sqrt{x} - \log_b \sqrt[3]{y} = \log_b z^3 + \log_b \sqrt{x} - $
$\left(\dfrac{1}{3}\log_b y\right) = 3\log_b z + \dfrac{1}{2}\log_b x - \dfrac{1}{3}\log_b y$

Section 4.6

59. $\log_b \dfrac{x^2(x+1)}{\sqrt{x+2}} = \log_b x^2(x+1) - \log_b \sqrt{x+2} =$

$\log_b x^2 + \log_b (x+1) - \dfrac{1}{2}\log_b (x+2) = 2\log_b x + \log_b (x+1) - \dfrac{1}{2}\log_b (x+2)$

61. $\log_b x + 2\log_b y = \log_b x + \log_b y^2 = \log_b xy^2$

63. $\log_b x + 2\log_b y - 3\log_b z = \log_b x + \log_b y^2 - \log_b z^3 = \log_b \dfrac{xy^2}{z^3}$

65. $\log_3 x + \dfrac{1}{3}\log_3 y - \dfrac{1}{2}\log_3 z = \log_3 x + \log_3 \sqrt[3]{y} - \log_3 \sqrt{z} = \log_3 \dfrac{x\sqrt[3]{y}}{\sqrt{z}}$

67. $2\log_{10} x - \dfrac{1}{2}\log_{10}(x-3) - \log_{10}(x+1) =$

$\log_{10} x^2 - \log_{10} \sqrt{x-3} - \log_{10}(x+1) = \log_{10} \dfrac{x^2}{(x+1)\sqrt{x-3}}$

69. $5\log_b x + \dfrac{1}{3}\log_b (x-1) - \log_b (x+2) =$

$\log_b x^5 + \log_b \sqrt[3]{x-1} - \log_b (x+2) = \log_b \dfrac{x^5\sqrt[3]{x-1}}{x+2}$

71. $\log_{10} x + 2\log_{10}(x-1) - \dfrac{1}{3}[\log_{10}(x+2) + \log_{10}(x-5)] =$

$\log_{10} x + \log_{10}(x-1)^2 - \dfrac{1}{3}[\log_{10}(x+2)(x-5)] = \log_{10} \dfrac{x(x-1)^2}{\sqrt[3]{(x+2)(x-5)}}$

73. $\log_b b^3 = 3$ 75. $\log_3 9 = \log_3 3^2 = 2$ 77. $\log_5 125 = \log_5 5^3 = 3$

79. $\log_2 \dfrac{1}{4} = \log_2 (2^{-2}) = -2$ 81. $\log_{10} 0.001 = \log_{10} 10^{-3} = -3$

83. $\log_3 1 = \log_3 3^0 = 0$ 85. $6^{\log_6 5} = 5$

87. $25^{\log_5 6} = \left(5^2\right)^{\log_5 6} = 5^{2\log_5 6} = 5^{\log_5 6^2} = 36$ 89. $4^{\log_2\left(\frac{1}{5}\right)} = 2^{2\log_2\left(\frac{1}{5}\right)}$

$$= 2^{\log_2\left(\frac{1}{5}\right)^2}$$

$$= \left(\dfrac{1}{5}\right)^2$$

$$= \dfrac{1}{25}$$

Section 4.6

1. $y' = \dfrac{1}{4x-3}\log e(4) = \dfrac{4\log e}{4x-3}$

3. $y' = \dfrac{1}{3x}\log_2 e(3) = \dfrac{\log_2 e}{x}$

5. $y' = \dfrac{1}{2x^3-3}(6x^2) = \dfrac{6x^2}{2x^3-3}$

7. $y' = \dfrac{1}{\tan 3x}3\sec^2 3x = \dfrac{3\sec^2 3x}{\tan 3x}$
   or $3\sec 3x\csc 3x$

9. $y' = \dfrac{x\cos x+\sin x}{x\sin x}$

11. $y' = \dfrac{1}{\sqrt{3x-2}}(1/2)(3x-2)^{-1/2}(3) = \dfrac{3}{2(3x-2)}$

13. $y' = \dfrac{x^2+1}{x^3}\cdot\dfrac{(x^2+1)(3x^2)-x^3(2x)}{(x^2+1)^2} = \dfrac{x^2+1}{x^3}\cdot\dfrac{x^2(x^2+3)}{(x^2+1)^2} = \dfrac{x^2+3}{x(x^2-1)}$

15. $y' = \sec^2(\ln x)\cdot\dfrac{1}{x} = \dfrac{\sec^2(\ln x)}{x}$

17. $y' = \dfrac{1}{\ln x}\cdot\dfrac{1}{x} = \dfrac{1}{x\ln x}$

19. $y' = \dfrac{1}{1+\ln^2 x^2}\cdot\dfrac{1}{x^2}(2x) = \dfrac{2}{x(1+\ln^2 x^2)}$

21. $y' = \dfrac{1}{\arccos^2 x}(2\arccos x)\dfrac{-1}{\sqrt{1-x^2}} = \dfrac{-2}{(\arccos x)\sqrt{1-x^2}}$

23. $\ln y = \ln(3x+2)+2\ln(6x-1)+\ln(x-4)$

$\dfrac{1}{y}\cdot y' = \dfrac{3}{3x+2}+\dfrac{12}{6x-1}+\dfrac{1}{x-4}$

$y' = (3x+2)(6x-1)^2(x-4)\left\{\dfrac{3}{3x+2}+\dfrac{12}{6x-1}+\dfrac{1}{x+4}\right\}$

25. $\ln y = \ln(x+1)+\ln(2x+1)-\ln(3x-4)-\ln(1-8x)$

$\dfrac{1}{y}\cdot y' = \dfrac{1}{x+1}+\dfrac{2}{2x+1}-\dfrac{3}{3x-4}+\dfrac{1}{1-8x}$

$y' = \dfrac{(x+1)(2x+1)}{(3x-4)(1-8x)}\left\{\dfrac{1}{x+1}-\dfrac{3}{3x-4}+\dfrac{2}{2x+1}+\dfrac{1}{1-8x}\right\}$

27. $\ln y = \ln x^x = x\ln x$; $y'/y = x(1/x)+\ln x$; $y' = x^x(1+\ln x)$

29. $\ln y = \ln x^{2/x} = (2/x)\ln x$; $y'/y = \dfrac{2}{x}\cdot\dfrac{1}{x}-\dfrac{2\ln x}{x^2}$; $y' = 2x^{2/x}\dfrac{(1-\ln x)}{x^2}$

31. $\ln y = x \ln(\sin x);\ y'/y = x \dfrac{1}{\sin x}(\cos x) + \ln(\sin x);\ y' = (\sin x)^x [x \cot x + \ln(\sin x)]$

33. $\ln y = x^2 \ln (1 + x);\ y'/y = x^2 \dfrac{1}{1+x} + 2x \ln(1 + x)$

$$y' = (1+x)^{x^2}\left\{ \dfrac{x^2}{1+x} + 2x \ln(1+x) \right\}$$

35. $m = y'\Big|_{x=1} = \dfrac{1}{x}\Big|_{x=1} = 1$ Thus, $y = x - 1$

37. $m = y'\Big|_{x=\pi/6} = \cot x\Big|_{x=\pi/6} = \sqrt{3}$;   at $x = \pi/6,\ y = -\ln 2$

$y + \ln 2 = \sqrt{3}(x - \pi/6)$ or $y = \sqrt{3}(x - \pi/6) - \ln 2$

Section 4.8

1. $y' = 5e^{5x}$

3. $y' = 12x^2 e^{x^3}$

5. $y' = \dfrac{(3)10^{3x}}{\log e}$

7. $y' = \dfrac{-6}{e^{6x}}$

9. $y' = \dfrac{e\sqrt{x}}{2\sqrt{x}}$

11. $y' = (\cos x)e^{\sin x}$

13. $y' = 6x(e^{x^2-1})(2x) + 6e^{x^2-1} = 6e^{x^2-1}(2x^2 + 1)$

15. $y' = (\cos x)(e^{3x^2})(6x) + (e^{3x^2})(-\sin x) = e^{3x^2}(6x \cos x - \sin x)$

17. $y' = \dfrac{1}{\cos e^{5x}}(-\sin e^{5x})(e^{5x})(5) = \dfrac{-5e^{5x} \sin e^{5x}}{\cos e^{5x}} = -5e^{5x} \tan e^{5x}$

19. $y' = e^x + e^{-x}$

21. $y' = \dfrac{(3e^x - x)(4x) - (2x^2)(3e^x - 1)}{(3e^x - x)^2} = \dfrac{-6e^x x^2 + 2x^2 - 4x^2 + 12xe^x}{(3e^x - x)^2} = \dfrac{12xe^x - 6e^x x^2 - 2x^2}{(3e^x - x)^2}$

$y' = \dfrac{6xe^x(2 - x) - 2x^2}{(3e^x - x)^2}$

23. $y' = \dfrac{6e^{3x}}{1 + e^{6x}}$

25. $y' = 3(\arccos^2 e^{-2x})\dfrac{-1}{\sqrt{1 - e^{-4x}}}(-2) = \dfrac{6e^{-2x}\arccos^2 e^{-2x}}{\sqrt{1 - e^{-4x}}}$

27. $y' = xe^x + e^x - e^x = xe^x$

29. $y = x^2;\ y' = 2x$

Sections 4.7 – 4.8

1. $\displaystyle\lim_{x\to 0}\frac{\sin x}{5x}$

$\displaystyle =\lim_{x\to 0}\frac{\cos x}{5}=\frac{1}{5}$

3. $\displaystyle\lim_{x\to 1}\frac{\ln x}{x^2-5x+4}$

$\displaystyle =\lim_{x\to 1}\frac{\dfrac{1}{x}}{2x-5}$

$\displaystyle =\lim_{x\to 1}\frac{1}{2x^2-5x}=\frac{-1}{3}$

5. $\displaystyle\lim_{x\to 0}\frac{\ln(\cos x)}{x^2}$

$\displaystyle =\lim_{x\to 0}\frac{\dfrac{1}{\cos x}\cdot -\sin x}{2x}$

$\displaystyle =\lim_{x\to 0}\frac{-\tan x}{2x}$

$\displaystyle =\lim_{x\to 0}\frac{-\sec^2 x}{2}=\frac{-1}{2}$

7. $\displaystyle\lim_{x\to 0}\frac{x-\sin x}{x^3}$

$\displaystyle =\lim_{x\to 0}\frac{1-\cos x}{3x^2}$

$\displaystyle =\lim_{x\to 0}\frac{\sin x}{6x}$

$\displaystyle =\lim_{x\to 0}\frac{\cos x}{6}=\frac{1}{6}$

9. $\displaystyle\lim_{x\to 0} x\cot x$

$\displaystyle =\lim_{x\to 0}\frac{x}{\tan x}$

$\displaystyle =\lim_{x\to 0}\frac{x}{\sec^2 x}=1$

11. $\displaystyle\lim_{x\to\infty}\frac{5x^2+7x-1}{2x^2-x+9}$

$\displaystyle =\lim_{x\to\infty}\frac{10x+7}{4x-1}$

$\displaystyle =\lim_{x\to\infty}\frac{10}{4}=\frac{5}{2}$

13. $\displaystyle\lim_{x\to\infty}\frac{\ln x}{\sqrt{x}}$

$\displaystyle =\lim_{x\to\infty}\frac{\dfrac{1}{x}}{\dfrac{1}{2}x^{\frac{-1}{2}}}$

$\displaystyle =\lim_{x\to\infty}\frac{2}{\sqrt{x}}=0$

15. $\displaystyle\lim_{t\to\infty} t^2 e^{-5t}$

$\displaystyle =\lim_{t\to\infty}\frac{t^2}{e^{5t}}$

$\displaystyle =\lim_{t\to\infty}\frac{2t}{5e^{5t}}$

$\displaystyle =\lim_{t\to\infty}\frac{2}{25e^{5t}}=0$

17. $\displaystyle\lim_{x\to\infty} 5x^{-2}e^{x^2}$

$\displaystyle =\lim_{x\to\infty}\frac{5e^{x^2}}{x^2}$

$\displaystyle =\lim_{x\to\infty}\frac{5(2x)e^{x^2}}{2x}$

$\displaystyle =\lim_{x\to\infty} 5e^{x^2}$

does not exist

19. $\displaystyle\lim_{x\to 0}(\csc x-\cot x)$

$\displaystyle =\lim_{x\to 0}\left(\frac{1}{\sin x}-\frac{\cos x}{\sin x}\right)$

$\displaystyle =\lim_{x\to 0}\frac{1-\cos x}{\sin x}$

$\displaystyle =\lim_{x\to 0}\frac{\sin x}{\cos x}=0$

21. $\displaystyle\lim_{x\to\frac{\pi}{2}}\frac{\sec x}{\tan x}$

$=\displaystyle\lim_{x\to\frac{\pi}{2}}\frac{\sec x\tan x}{\sec^2 x}$

$=\displaystyle\lim_{x\to\frac{\pi}{2}}\frac{\tan x}{\sec x}$

$=\displaystyle\lim_{x\to\frac{\pi}{2}}\frac{\sec^2 x}{\sec x\tan x}$

$=\displaystyle\lim_{x\to\frac{\pi}{2}}\frac{\sec x}{\tan x}$

This is the original problem.

First use algebra to simplify:

$\displaystyle\lim_{x\to\frac{\pi}{2}}\frac{\sec x}{\tan x}$

$=\displaystyle\lim_{x\to\frac{\pi}{2}}\frac{1}{\cos x}\cdot\frac{\cos x}{\sin x}$

$=\displaystyle\lim_{x\to\frac{\pi}{2}}\frac{1}{\sin x}=1$

Section 4.10

1. I: $x = 3\pi/4 + 2n\pi,\ x = 7\pi/4 + 2n\pi$
    $y = 0$
    $y' = \cos x - \sin x = 0;\ \tan x = 1$
    $x = \pi/4 = n\pi,$ when n is even,
    $y = 1/\sqrt{2} + 1/\sqrt{2} = \sqrt{2}$
    $y'' = -\sin x - \cos x = -1/\sqrt{2} - 1/\sqrt{2} = -\sqrt{2} < 0$
    Thus $(\pi/4 + 2n\pi,\ \sqrt{2}\,)$ are relative maxima.
    When n is odd, $y = -1/\sqrt{2} - 1/\sqrt{2} = -\sqrt{2};$
    $y'' = 1/\sqrt{2} + 1/\sqrt{2} = \sqrt{2} > 0$
    Thus $(\pi/4 + (2n + 1)\pi,\ -\sqrt{2}\,)$ are relative minima.
    Points of inflection: $y'' = 0$ or $\tan x = -1;\ x = 3\pi/4 + n\pi,$
    for n even, $y = 1/\sqrt{2} - 1/\sqrt{2} = 0.$ For n odd, $y = -1/\sqrt{2} + 1/\sqrt{2} = 0.$
    Thus $(3\pi/4 + n\pi,\ 0)$ are points of inflection.

3. I: $x = 1;$ no y-intercept
    $y' = \dfrac{x(1/x) - \ln x}{x^2} = \dfrac{1 - \ln x}{x^2} = 0;$
    $\ln x = 1;$
    $x = e,\ y = 1/e$

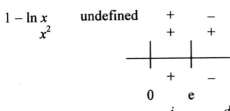

$\begin{array}{ccc} \dfrac{1 - \ln x}{x^2} & + & - \\[4pt] & + & + \end{array}$

undefined

$\dfrac{1-\ln x}{x^2}$ undefined $+$ $-$
                                     $+$ $+$

         0 e
         i d Thus, $(e,\ 1/e)$ max

Sections 4.9 – 4.10

$$y'' = \frac{x^2(-1/x) - (1 - \ln x)(2x)}{x^4} = \frac{x(-3 + 2\ln x)}{x^4} = \frac{-3 + 2\ln x}{x^3} = 0$$

$-3 + 2\ln x = 0; \quad \ln x = 3/2; \quad x = e^{3/2}$

$f''(x)$ $-$ $+$

0   *dn* $e^{3/2}$ up

Thus $(e^{3/2}, 3e^{-3/2}/2)$ is point of inflection.

5. I: $y = 2$; no x-intercept
$y' = e^x - e^{-x} = 0$; $x = -x$;
$x = 0, y = 2$;
$y'' = e^x + e^{-x} > 0$ for all x;
Thus $(0, 2)$ min, graph is concave
upward and no points of inflection.

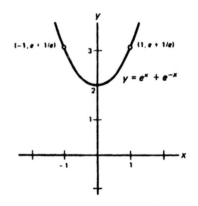

7. I: $x = 0$; $y = 0$
$y' = -x^2 e^{-x} + 2xe^{-x}$
$= xe^{-x}(2 - x) = 0$
Note: $e^{-x} > 0$

x	$-$	$+$	$+$
$2 - x$	$+$	$+$	$-$

$f'(x)$ $-$ 0 $+$ 2 $-$
         d   *i*   *d*

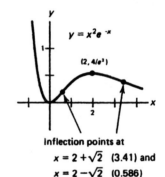

Inflection points at
$x = 2 + \sqrt{2}$ (3.41) and
$x = 2 - \sqrt{2}$ (0.586)

$(0, 0)$ min; $(2, 4/e^2)$ max

$y'' = e^{-x}(2 - 2x) - e^{-x}(2x - x^2) = e^{-x}(2 - 4x + x^2) = 0$;
$x = 2 \pm \sqrt{2}$

$x - (2 + \sqrt{2})$	$-$	$-$	$+$
$x - (2 - \sqrt{2})$	$-$	$+$	$+$

         $(2 - \sqrt{2})$ $(2 + \sqrt{2})$

$(3.41, 17.8)$ and $(0.586, -1.8)$ are points of inflection.

Section 4.10

9. no intercepts

$x = 0$ is a vertical asymptote

$y' = \dfrac{e^x}{(1-e^x)^2}$ $y' > 0$ for all x except 0.

no max or min.

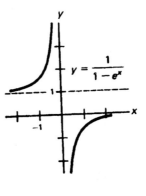

$y = \dfrac{1}{1-e^x}$

$$y'' = \frac{(1-e^x)^2 e^x - e^x[2(1-e^x)(-e^x)]}{(1-e^x)^4}$$

$$= \frac{e^x(1-e^x)[(1-e^x)+2e^x]}{(1-e^x)^4} = \frac{e^x(1+e^x)}{(1-e^x)^3};$$

$y'' > 0$ for all x except 0.

no points of inflection.

11. $y' = x(-2e^{-2x}) + e^{-2x} = e^{-2x}(1-2x) = 0;$

Note: $e^{nx} > 0$ for all n and x and will not be in sign charts.

$1 - 2x$ +   −

$$\begin{array}{c c}
 & \\
\hline
i & 1/2 \; d
\end{array}$$

Thus, $(1/2, \; 1/(2e))$ max

13. $y' = \dfrac{2x^{2x}x - 2x^2 e^{2x}}{e^{4x}} = \dfrac{2xe^{2x}(1-x)}{e^{4x}} = \dfrac{2x(1-x)}{e^{2x}} = 0$

$2x$ −   +   +

$1-x$ +   +   −

$$\begin{array}{ccc}
d & i & d \\
\hline
0 & & 1
\end{array}$$

Thus, $(0, 0)$ is a min; $(1, 1/e^2)$ is a max.

15. $y' = \dfrac{x^2(1/x) - (\ln x)(2x)}{x^4} = \dfrac{x(1-2\ln x)}{x^4} = \dfrac{1-2\ln x}{x^3} = 0$

$\begin{array}{l} 1 - 2\ln x \\ x^3 \end{array}$ +   − $1 - 2\ln x = 0$

                    +   + $\ln x = 1/2$

                                         $x = e^{1/2}$

$$\begin{array}{ccc}
 & 0 \; i & e^{1/2} \; d
\end{array}$$

not
defined

Thus $(e^{1/2}, \; 1/(2e))$ is a max.

Section 4.10

1. $\sec x \cot x = \csc x$

$$\sec x \cot x = \frac{1}{\cot x} \cdot \frac{\cos x}{\sin x}$$

$$= \frac{1}{\sin x}$$

$$= \csc x$$

2. $\sec^2 \theta + \tan^2 \theta + 1 = \dfrac{2}{\cos^2 \theta}$

$$\sec^2 \theta + \tan^2 \theta + 1 = \sec^2 \theta + \sec^2 \theta$$

$$= 2 \sec^2 \theta$$

$$= \frac{2}{\cos^2 \theta}$$

3. $\dfrac{\cos \theta}{\cos \theta + \sin \theta} = \dfrac{\cot \theta}{1 + \cot \theta}$

$$\frac{\cot \theta}{1 + \cot \theta} = \frac{\dfrac{\cos \theta}{\sin \theta}}{1 + \dfrac{\cos \theta}{\sin \theta}}$$

$$= \frac{\dfrac{\cos \theta}{\sin \theta}}{\dfrac{\sin \theta + \cos \theta}{\sin \theta}}$$

$$= \frac{\cos \theta}{\sin \theta + \cos \theta}$$

4. $\cos\left(\theta - \dfrac{3\pi}{2}\right) = -\sin \theta$

$$\cos\left(\theta - \frac{3\pi}{2}\right) = \cos \theta \cos \frac{3\pi}{2} + \sin \theta \sin \frac{3\pi}{2}$$

$$= \cos \theta(0) + \sin \theta(-1)$$

$$= -\sin \theta$$

5. $\left(\sin \dfrac{1}{2} x + \cos \dfrac{1}{2} x\right)^2 = 1 + \sin x$

$$\left(\sin \frac{1}{2} x + \cos \frac{1}{2} x\right)^2 =$$

$$\sin^2 \frac{1}{2} x + 2 \sin \frac{1}{2} x \cos \frac{1}{2} x + \cos^2 \frac{1}{2} x$$

$$= 1 + \sin 2\left(\frac{1}{2} x\right)$$

$$= 1 + \sin x$$

6. $2 \cos^2 \dfrac{\theta}{2} = \dfrac{1 + \sec \theta}{\sec \theta}$

$$2 \cos^2 \frac{\theta}{2} = \cos \theta + 1$$

$$= \frac{1}{\sec \theta} + \frac{\sec \theta}{\sec \theta}$$

$$= \frac{1 + \sec \theta}{\sec \theta}$$

7. $\dfrac{2 \cot \theta}{1 + \cot^2 \theta} = \sin 2\theta$

$$\frac{2 \cot \theta}{1 + \cot^2 \theta} = \frac{2 \cot \theta}{\csc^2 \theta}$$

$$= 2 \frac{\cos \theta}{\sin \theta} \div \frac{1}{\sin^2 \theta}$$

$$= 2 \sin \theta \cos \theta$$

$$= \sin 2\theta$$

8. $\csc x - \cot x = \tan \dfrac{1}{2} x$

$$\csc x - \cot x = \frac{1}{\sin x} = \frac{\cos x}{\sin x}$$

$$= \frac{1 - \cos x}{\sin x}$$

$$= \tan \frac{x}{2}$$

9. $\tan 2x = \dfrac{2\cos x}{\csc x - 2\sin x}$

$\dfrac{2\cos x}{\csc x - 2\sin x} = \dfrac{2\cos x}{\dfrac{1}{\sin x} - 2\sin x}$

$\qquad = \dfrac{2\cos x}{\dfrac{1 - 2\sin^2 x}{\sin x}}$

$\qquad = \dfrac{2\cos x}{\dfrac{\cos 2x}{\sin x}}$

$\qquad = \dfrac{2\cos x}{1} \cdot \dfrac{\sin x}{\cos 2x}$

$\qquad = \dfrac{\sin 2x}{\cos 2x}$

$\qquad = \tan 2x$

10. $\tan^2 \dfrac{x}{2} + 1 = 2\tan \dfrac{x}{2}\csc x$

$\tan^2 \dfrac{x}{2} + 1 = \sec^2 \dfrac{x}{2}$

$\qquad = \dfrac{1}{\cos^2 \dfrac{x}{2}}$

$\qquad = \dfrac{1}{\dfrac{1 + \cos x}{2}}$

$\qquad = \dfrac{2}{1 + \cos x}$

$\qquad = \dfrac{2(1 - \cos x)}{(1 + \cos x)(1 - \cos x)}$

$\qquad = \dfrac{2(1 - \cos x)}{1 - \cos^2 x}$

$\qquad = \dfrac{2(1 - \cos x)}{\sin^2 x}$

$\qquad = 2\left(\dfrac{1 - \cos x}{\sin x}\right)\dfrac{1}{\sin x}$

$\qquad = 2\tan \dfrac{x}{2}\csc x$

11. $\sin \theta \cos \theta =$

$\dfrac{1}{2} \cdot 2\sin \theta \cos \theta =$

$\dfrac{1}{2}\sin 2\theta$

12. $\cos^2 3\theta - \sin^2 3\theta =$

$\cos 2(3\theta) =$

$\cos 6\theta$

13. $\dfrac{1 + \cos 4\theta}{2} =$

$\cos^2 \dfrac{4\theta}{2} =$

$\cos^2 2\theta$

14. $1 - 2\sin^2 \dfrac{\theta}{3} =$

$\cos 2\left(\dfrac{\theta}{3}\right) =$

$\cos \dfrac{2\theta}{3}$

15. $\cos 2x \cos 3x - \sin 2x \sin 3x =$
$\cos (2x + 3x) = \cos 5x$

16. $\sin 2x \cos x - \cos 2x \sin x =$
$\sin(2x - x) = \sin x$

The graphs of 17 – 24 appear in the text answer section.

Chapter 4 Review

25. $y' = 2x \cos (x^2 + 3)$

26. $y' = -8 \sin 8x$

27. $y' = 3 \cos^2(5x - 1)[-\sin(5x - 1)](5) = -15 \cos^2(5x - 1) \sin(5x - 1)$

28. $y' = \sin 3x(-2 \sin 2x) + \cos 2x (3 \cos 3x) = -2 \sin 2x \sin 3x + 3 \cos 2x \cos 3x$

29. $y' = 3 \sec^2(3x - 2)$

30. $y' = 4 \sec(4x + 3) \tan (4x + 3)$

31. $y' = -12x \csc^2 6x^2$

32. $y' = -2(16x + 1) \csc(8x^2 + x) \csc(8x^2 + x) \cot(8x^2 + x) = -2(16x + 1)\csc^2(8x^2 + x) \cot (8x^2 + x)$

33. $y' = \sec^2 x \cos x + \sin x(2 \sec x \cdot \sec x \tan x)$

$$= \frac{1}{\cos^2 x} \cos x + 2 \sin x \frac{1}{\cos^2 x} \frac{\sin x}{\cos x} = \frac{1}{\cos x} + \frac{2 \sin^2 x}{\cos^3 x}$$

$$= \frac{\cos^2 x + 2\sin^2 x}{\cos^3 x} = \frac{\cos^2 x + 2(1 - \cos^2 x)}{\cos^3 x}$$

$$= \frac{2 - \cos^2 x}{\cos^3 x} = \frac{2}{\cos^3 x} - \frac{1}{\cos x} = 2 \sec^3 x - \sec x$$

34. $y' = 2x - 2 \csc x(-\csc x \cot x) = 2x + 2 \csc^2 x \cot x$

35. $y' = \sec^2(\sec x) \sec x \tan x$

36. $y' = \frac{(1 + \sin x)(-\sin x) - \cos x(\cos x)}{(1 + \sin x)^2} = \frac{-\sin x - \sin^2 x - \cos^2 x}{(1 + \sin x)^2}$

$$= \frac{-(\sin x + 1)}{(1 + \sin x)^2} = \frac{-1}{1 + \sin x}$$

37. $y' = 3(1 - \sin x)^2(-\cos x) = -3 \cos x(1 - \sin x)^2$

38. $y' = 2(1 + \sec 4x) \sec 4x \tan 4x(4) = 8(1 + \sec 4x) \sec 4x \tan 4x$

39. $y = \frac{1}{2} \sin \frac{3x}{4}$

$2y = \sin \frac{3x}{4}$

$\arcsin 2y = \frac{3x}{4}$

$\frac{4}{3} \arcsin 2y = x$

40. $y = 5 \tan(1 - 2x)$

$\frac{y}{5} = \tan(1 - 2x)$

$\arctan(y/5) = 1 - 2x$

$2x = 1 - \arctan(y/5)$

$x = \frac{1}{2}(1 - \arctan(y/5))$

Chapter 4 Review

Solutions to Odd-Numbered Exercises

41. $\arcsin\left(\dfrac{1}{\sqrt{2}}\right)$

Quadrant 1

$\dfrac{\pi}{4}$

42. $\arctan\left(-\dfrac{1}{\sqrt{3}}\right)$

Quadrant IV

$-\dfrac{\pi}{6}$

43. $\text{arcsec}(-1)$

π

44. $\arccos\left(-\dfrac{1}{2}\right)$

Quadrant II

$\dfrac{2\pi}{3}$

45. $\sin\left[\arccos\left(-\dfrac{1}{2}\right)\right] =$

$\sin\left[\dfrac{2\pi}{3}\right] = \dfrac{\sqrt{3}}{2}$

46. $\tan(\arctan\sqrt{3}) =$

$\tan\left(\dfrac{\pi}{3}\right) = \sqrt{3}$

47. $\sin(\text{arccot } x) =$

Let $\theta = \text{arccot } x$

$\sin\theta = \dfrac{1}{\sqrt{x^2+1}}$

$= \dfrac{\sqrt{x^2+1}}{\sqrt{x^2+1}\sqrt{x^2+1}}$

$= \dfrac{\sqrt{x^2+1}}{x^2+1}$

48. $y' = \dfrac{3x^2}{\sqrt{1-x^6}}$

49. $y' = \dfrac{3}{1+9x^2}$

50. $y' = -3\dfrac{1}{\sqrt{1-(1/(2x))^2}}\dfrac{-1}{2x^2} = \dfrac{(-3/2)(-1/x^2)}{\dfrac{\sqrt{4x^2-1}}{2|x|}} = \dfrac{3}{2x^2}\dfrac{2|x|}{\sqrt{4x^2-1}} = \dfrac{3}{|x|\sqrt{4x^2-1}}$

Chapter 4 Review

51. $y' = \dfrac{2}{\left|4x\right|\sqrt{16x^2-1}}(4) = \dfrac{2}{\left|x\right|\sqrt{16x^2-1}}$

52. $y' = 2\arcsin 3\sqrt{x}\,\dfrac{(3/2)x^{-1/2}}{\sqrt{1-9x}} = \dfrac{3\arcsin 3\sqrt{x}}{\sqrt{x}\sqrt{1-9x}} = \dfrac{3\arcsin 3\sqrt{x}}{\sqrt{x-9x^2}}$

53. $y' = \dfrac{x}{\sqrt{1-x^2}} + \arcsin x$

The graphs of 54 and 55 appear in the text answer section.

56. $\log_9 x = 2$

$9^2 = x$

$x = 81$

57. $\log_x 8 = 3$

$x^3 = 8$

$x = 2$

58. $\log_2 32 = x$

$2^x = 32$

$2^x = 2^5$

$x = 5$

59. $\log_4 6x^2 y =$

$\log_4 6 + \log_4 x^2 + \log_4 y =$

$\log_4 6 + 2\log_4 x + \log_4 y$

60. $\log_3 \dfrac{5x\sqrt{y}}{z^3} =$

$\log_3 5 + \log_3 x + \log_3 \sqrt{y} - \log_3 z^3 =$

$\log_3 5 + \log_3 x + \dfrac{1}{2}\log_3 y - 3\log_3 z$

61. $\log \dfrac{x^2(x+1)^3}{\sqrt{x-4}} =$

$\log x^2 + \log(x+1)^3 - \log\sqrt{x-4} =$

$2\log x + 3\log(x+1) - \dfrac{1}{2}\log(x-4)$

62. $\ln \dfrac{[x(x-1)]^3}{\sqrt{x+1}} =$

$\ln[x(x-1)]^3 - \ln\sqrt{x+1} =$

$3\ln x(x-1) - \dfrac{1}{2}\ln(x+1) =$

$3\ln x + 3\ln(x-1) - \dfrac{1}{2}\ln(x+1)$

63. $\log_2 x + 3\log_2 y - 2\log_2 z =$

$\log_2 x + \log_2 y^3 - \log_2 z^2 =$

$\log_2 \dfrac{xy^3}{z^2}$

64. $\dfrac{1}{2}\log(x+1) - 3\log(x-2) =$

$\log\sqrt{x+1} - \log(x-2)^3 =$

$\log \dfrac{\sqrt{x+1}}{(x-2)^3}$

65. $4\ln x - 5\ln(x+1) - \ln(x+2) =$

$\ln x^4 - \ln(x+1)^5 - \ln(x+2) =$

$\ln \dfrac{x^4}{(x+1)^5(x+2)}$

66. $\dfrac{1}{2}[\ln x + \ln(x+2)] - 2\ln(x-5) =$

$\dfrac{1}{2}\ln x + \dfrac{1}{2}\ln(x+2) - \ln(x-5)^2 =$

$\ln\sqrt{x} + \ln\sqrt{x+2} - \ln(x-5)^2 =$

$\ln\sqrt{x}\sqrt{x+2} - \ln(x-5)^2 =$

$\ln \dfrac{\sqrt{x(x+2)}}{(x-5)^2}$

Chapter 4 Review

67. $\log 1000 = 3$

68. $\log 10^{x^2} = x^2$

69. $\ln e^2 = 2$

70. $\ln e^x = x$

71. $y' = \dfrac{6x^2}{2x^3 - 4} = \dfrac{3x^2}{x^3 - 2}$

72. $y' = \dfrac{4 \log_3 e}{4x + 1}$

73. $y = 2 \ln x - \ln(x^2 + 3); \ \ y' = \dfrac{2}{x} - \dfrac{2x}{x^2 + 3} = \dfrac{6}{x(x^2 + 3)}$

74. $y' = -\sin(\ln x) \cdot (1/x) = (-1/x) \sin (\ln x)$

75. $\ln y = (1/2) \ln(x + 1) + \ln(3x - 4) - 2 \ln x - \ln(x + 2)$

$$y'/y = \frac{1}{2} \cdot \frac{1}{x+1} + \frac{3}{3x-4} - \frac{2}{x} - \frac{1}{x+2}$$

$$y' = \frac{\sqrt{x+1}(3x-4)}{x^2(x+2)} \left\{ \frac{1}{2(x+1)} + \frac{3}{3x-4} - \frac{2}{x} - \frac{1}{x+2} \right\}$$

76. $\ln y = (1 - x) \ln x; \ \ y'/y = \dfrac{1-x}{x} - \ln x;$

$$y' = x^{1-x}[(1-x)/x - \ln x]$$

77. $y' = 2xe^{x^2} + 5$

78. $y' = \dfrac{3(8^{3x})}{\log_8 e}$

79. $y' = -\csc^2 e^{2x}(e^{2x})(2) = -2e^{2x} \csc^2 e^{2x}$

80. $y' = e^{\sin x^2} (\cos x^2)(2x) = 2x \cos x^2 e^{\sin x^2}$

81. $y' = \dfrac{1}{\sqrt{1 - e^{-8x}}} (e^{-4x})(-4) = \dfrac{-4e^{-4x}}{\sqrt{1 - e^{-8x}}}$

82. $y' = x^3 e^{-4x}(-4) + e^{-4x}(3x^2) = -4e^{-4x} x^3 + 3x^2 e^{-4x} = e^{-4x} x^2 (3 - 4x)$

83. I: $x = 0, y = 0$

$y' = -xe^{-x} + e^{-x} = e^{-x}(1 - x) = 0; \ x = 1$

1 − x   +   −

           i   1   d

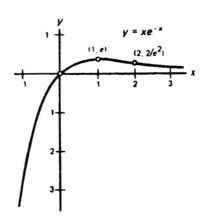

Thus, (1, 1/e) max

$y'' = xe^{-x} - 2e^{-x} = e^{-x}(x - 2) = 0; \ x = 2$

x − 2   −   +

      dn   2   up

Thus $(2, 2/e^2)$ is point of inflection.

Chapter 4 Review

84. I: $y = 1$; $y' = -2xe^{-x^2}$

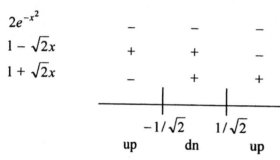

$-2x$ +   −

i   0   d

Thus, (0, 1) is max

$$y'' = -4x^2e^{-x^2} - 2e^{-x^2}$$
$$= -2e^{-x^2}(1 - 2x^2) = 0$$
$$x^2 = 1/2; \quad x = \pm 1/\sqrt{2}$$

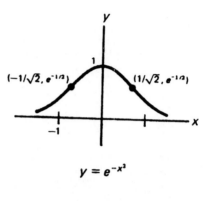

$(-1/\sqrt{2}, e^{-1/2})$ $(1/\sqrt{2}, e^{-1/2})$

$y = e^{-x^2}$

$2e^{-x^2}$ −   −   −

$1 - \sqrt{2}x$ +   +   −

$1 + \sqrt{2}x$ −   +   +

$-1/\sqrt{2}$ $1/\sqrt{2}$

up   dn   up

Thus $(-1/\sqrt{2}, e^{-1/2})$ and $(1/\sqrt{2}, e^{-1/2})$ are points of inflection.

85.
$$\lim_{x \to 2} \frac{\ln(x-1)}{\sqrt{x-2}}$$
$$= \lim_{x \to 2} \frac{\frac{1}{(x-1)}}{0.5(x-2)}$$
$$= \lim_{x \to 2} \frac{2\sqrt{x-2}}{x-1}$$
$$= 0$$

86.
$$\lim_{x \to \infty} xe^{-3x}$$
$$= \lim_{x \to \infty} \frac{x}{e^{3x}}$$
$$= \lim_{x \to \infty} \frac{1}{3e^{3x}}$$
$$= 0$$

87.
$$\lim_{x \to 0} \frac{x\sin x}{1 - \cos x}$$
$$= \lim_{x \to 0} \frac{x\cos x + \sin x}{\sin x}$$
$$= \lim_{x \to 0} \frac{-x\sin x + \cos x + \cos x}{\cos x}$$
$$= \lim_{x \to 0} \frac{-x\sin x + 2\cos x}{\cos x}$$
$$= 2$$

88.
$$\lim_{x \to \pi/2} \frac{\cos^2 x}{\sin x - 1}$$
$$= \lim_{x \to \pi/2} \frac{-\cos^2 x(1 + \sin x)}{(1 - \sin x)(1 + \sin x)}$$
$$= \lim_{x \to \pi/2} \frac{-\cos^2 x(1 + \sin x)}{1 - \sin x^2}$$
$$= \lim_{x \to \pi/2} \frac{-\cos^2 x(1 + \sin x)}{\cos x^2}$$
$$= \lim_{x \to \pi/2} -(1 + \sin x)$$
$$= -2$$

89. $V_L = L \, di/dt = 2(-100 \sin 5t)(5) = -1000 \sin 5t$

Chapter 4 Review

90. $p = dW/dt = 120 \cos 3t \, (-\sin 3t)(3) = -360 \sin 3t \cos 3t = -180 \sin 6t$

91. $y = 2 \ln x;$ $\quad m = y' = 2/x \Big|_{x = 1} = 2;$ $\quad y - 0 = 2(x - 1);$
$$y = 2x - 2$$

92. $v = ds/dt = e^{\sin t} \, (\cos t) = \cos t \, e^{\sin t}$

93. $v = 4e^{-t};$ $\quad a = -4e^{-t} \Big|_{t = 1/2} = -4e^{-1/2} = -2.43$

Chapter 4 Review

CHAPTER 5

Section 5.1

1. $\dfrac{x^8}{8} + C$

3. $\dfrac{x^9}{3} + C$

5. $4x + C$

7. $\displaystyle\int 9x^{5/6}\,dx = (54/11)x^{11/6} + C$

9. $\displaystyle\int 6x^{-3}\,dx = -3x^{-2} + C = -3/x^2 + C$

11. $(5/3)x^3 - 6x^2 + 8x + C$

13. $\displaystyle\int (3x^2 - x + 5x^{-3})\,dx = x^3 - (1/2)x^2 - 5/(2x^2) + C$

15. $\displaystyle\int (4x^4 - 12x^2 + 9)\,dx = (4/5)x^5 - 4x^3 + 9x + C$

17. $\displaystyle\int (6x + 2)^{1/2}\,dx = (1/6)\int (6x + 2)^{1/2} \cdot 6\,dx = (1/6)(2/3)(6x + 2)^{3/2} + C = (1/9)(6x + 2)^{3/2} + C$

19. $\displaystyle\int 8x(x^2 + 3)^3\,dx = 4\int (x^2 + 3)^3 \cdot 2x\,dx = (x^2 + 3)^4 + C$

21. $\displaystyle\int (5x^2 - 1)^{1/3} \cdot x\,dx = (1/10)\int (5x^2 - 1)^{1/3} \cdot 10x\,dx$
$$= (3/40)(5x^2 - 1)^{4/3} + C$$

23. $\displaystyle\int (x^2 - 1)^4 x\,dx = (1/2)\int (x^2 - 1)^4 \cdot 2x\,dx$
$$= (1/10)(x^2 - 1)^5 + C$$

25. $\displaystyle\int (x^2 + 1)^{-1/2} \cdot 2x\,dx = 2(x^2 + 1)^{1/2} + C$

27. $\displaystyle\int (3x^2 + 2)(x^3 + 2x)^3\,dx = \int u^3\,du = \dfrac{u^4}{4} + C$ let $u = x^3 + 2x$
$$= (1/4)(x^3 + 2x)^4 + C \quad du = (3x^2 + 2)dx$$

29. $\displaystyle\int (x^3 - 4)^{-2} x^2\,dx = (1/3)\int (x^3 - 4)^{-2} 3x^2\,dx = \dfrac{-1}{3(x^3 - 4)} + C$

31. $\displaystyle\int (5x^2 - x)^{1/2} (10x - 1)\,dx = (2/3)(5x^2 - x)^{3/2} + C$

33. $\displaystyle\int (x^2 + x)^{-1/2} (2x + 1)\,dx = 2(x^2 + x)^{1/2} + C$

35. $\displaystyle\int (2x + 3)^2\,dx = (1/2)\int (2x + 3)^2 \cdot 2\,dx = (1/6)(2x + 3)^3 + C$

37. $(1/2)\displaystyle\int (2x - 1)^4 \cdot 2\,dx = (1/10)(2x - 1)^5 + C$

Section 5.1

39. $\int (x^6 + 3x^4 + 3x^2 + 1)\, dx = (1/7)x^7 + (3/5)x^5 + x^3 + x + C$

41. $2\int (x^2 + 1)^3 \cdot 2x\, dx = (1/2)(x^2 + 1)^4 + C$

43. $2\int (5x^3 + 1)^4 15x^2\, dx = (2/5)(5x^3 + 1)^5 + C$

45. $2\int (x^3 + 1)^{-1/2} 3x^2\, dx = 4(x^3 + 1)^{1/2} + C$

47. $2\int (x^3 + 3x)^{-1/3} (3x^2 + 3)\, dx = 3(x^3 + 3x)^{2/3} + C$

49. $\int x^{-3}(x - 1)\, dx = \int (x^{-2} - x^{-3})dx = -1/x + 1/(2x^2) + C$

Section 5.2

1. $y = \int 3x\, dx = (3/2)x^2 + C$

   $1 = (3/2)(0) + C$

   $C = 1$ and $y = (3/2)x^2 + 1$

3. $y = \int (3x^2 + 3)\, dx = x^3 + 3x + C$

   $2 = (-1)^3 + 3(-1) + C$

   $C = 6$ and $y = x^3 + 3x + 6$

5. $y = (1/2)\int (x^2 - 3)^2\, 2x\, dx$

   $= (1/6)(x^2 - 3)^3 + C$

   $7/6 = (1/6)(2^2 - 3)^3 + C$

   $C = 1$

   $y = (1/6)(x^2 - 3)^3 + 1$

7. $v = \int a\, dt = \int 3t\, dt = (3/2)t^2 + C_1$

   $40 = (3/2)(4)^2 + C_1;\ \ C_1 = 16$

   $v = (3/2)t^2 + 16$

   $s = \int v\, dt = \int [(3/2)t^2 + 16]\, dt$

   $s = (1/2)t^3 + 16t + C_2$

   $86 = (1/2)(2)^3 + 16(2) + C_2$

   $C_2 = 50$

   $s = (1/2)t^3 + 16t + 50$

9. $v = \int -32\, dt = -32t + C_1$

   Since $v = 0$ when $t = 0$, $C_1 = 0$.

   $v = -32t$

   $s = \int -32t\, dt = -16t^2 + C_2$

   Since $s = 100$ when $t = 0$,

   $C_2 = 100;\ s = -16t^2 + 100$

   At $t = 2$, $s = -16(2)^2 + 100 =$

   36 ft from the ground, thus ball

   has fallen 64 ft. Object hits

   ground at $s = 0$; $0 = -16t^2 + 100$

   $t = 5/2\ s$ so $v = -32(5/2) = -80$ ft/s (down)

11. $\Delta s = (1/2)\, a(\Delta t)^2$

    $3000 = (1/2)\, a(30)^2$

    $a = 8$ ft/s

    $v = \int a\, dt$

    $= \int 8\, dt = 8t + C$

    At $t = 0$, $v = 0$

    Thus, $C = 0$

    $v = 8t \Big|_{t = 30} = 240$ ft/s

13. $v = \int -9.80 \, dt = -9.80t + C_1$

   At $t = 0$, $v = 25$, so $C_1 = 25$

   $v = -9.80t + 25$

   $s = \int v \, dt = \int (-9.80t + 25) \, dt$; $s = -4.90t^2 + 25t + C_2$

   At $t = 0$, $s = 0$ Thus $C_2 = 0$; Thus $s = -4.90t^2 + 25t$

   a) Max height occurs, $v = 0$; Thus $t = 2.55$ s or $s = -4.90(2.55)^2 + 25(2.55) = 31.9$ m

   b) Ball hits ground when $s = 0$, so $0 = -4.90t^2 + 25t = t(-4.90t + 25)$

   $t = 0$ s or $t = 5.10$ s

   c) Ball hits the ground with speed v

   $= -9.80t + 25 \Big|_{t = 5.1} = -25$ m/s

15. $v = \int a \, dt = \int -32 \, dt = -32t + C_1$, At $t = 0$, $v = 30$, so $C_1 = 30$

   and $v = -32t + 30$, $s = \int v \, dt = \int (-32t + 30) dt$

   $s = -16t^2 + 30t + C_2$

   a) At $t = 0$, $s = 200$ ft, so $C_2 = 200$ and $s = -16t^2 + 30t + 200$

   b) Hits ground $s = 0$, so $0 = -16t^2 + 30t + 200$ or $t = 4.59$ s

17. $\theta = \int \omega \, dt = \int (80 - 12t + 3t^2) dt = t^3 - 6t^2 + 80t + C_1$

   At $t = 0$, $\theta = 0$, so $C_1 = 0$. $\theta = t^3 - 6t^2 + 80t \Big|_{t = 3s} = 213$ revolutions

19. $V_C = (1/10^{-4}) \int \left(\frac{1}{2} t^{1/2} + 0.2 \right) dt$

   $= (1/10^{-4}) \left[\frac{1}{3} t^{3/2} + (1/5)t \right] + C$

   At $t = 0$, $V = 100$, so $C = 100$

   $V_C = \frac{1}{10^{-4}} \left[(1/3)t^{3/2} + t/5 \right] + 100 \Big|_{t = 0.16}$

   $= 633$ V

21. $q = \int i \, dt = \int t(t^2 + 1)^{1/2} \, dt = (1/3)(t^2 + 1)^{3/2} + C$

   At $t = 0$, $q = 0$, so $C = -1/3$ and

   $q = (1/3)(t^2 + 1)^{3/2} - 1/3 \Big|_{t = 1 s} = (1/3)(2^{3/2} - 1) = \frac{2\sqrt{2} - 1}{3}$

Section 5.2

23. $v = \int a\, dt = \int -32dt = -32t + C_1$; At $t = 0$, $C_1 = v_o$

$\quad s = \int v\, dt = \int(-32t + v_o)dt$

$\quad s = -16t^2 + v_o t + C_2$; At $t = 0$, $C_2 = s_o$.

Thus, $s = -16t^2 + v_o t + s_o$.

Similarly for the other form of this equation.

Section 5.3

1. $\int x\, dx = x^2/2 + C$; $A = F(2) - F(0) = 2 - 0 = 2$

3. $\int 2x^2\, dx = (2/3)x^3 + C$; $A = F(3) - F(1) = 18 - 2/3 = 17\dfrac{1}{3}$

5. $\int(3x^2 - 2x)\, dx = x^3 - x^2 + C$; $A = F(2) - F(1) = 4 - 0 = 4$

7. $\int 3x^{-2}\, dx = -3/x + C$; $A = F(2) - F(1) = -3/2 - (-3) = 3/2$

9. $\int(3x - 2)^{1/2}\, dx = (1/3)(2/3)(3x - 2)^{3/2} + C$; $A = F(2) - F(1) = 14/9$

11. $\int(4x - x^3)\, dx = 2x^2 - (1/4)x^4 + C$; $A = F(2) - F(0) = 4 - 0 = 4$

13. $\int(1 - x^4)\, dx = x - x^5/5 + C$; $A = F(1) - F(-1) = 4/5 + 4/5 = 8/5$

15. $\int(2x + 1)^{1/2}\, dx = (1/2)(2/3)(2x + 1)^{3/2} + C$; $A = F(12) - F(4) = 125/3 - 9 = 98/3$

17. $\int(x - 1)^{1/3}\, dx = (3/4)(x - 1)^{4/3} + C$; $A = F(1) - F(0) = 0 - (-3/4) = 3/4$

19. $\int(1/x^2)\, dx = -1/x + C$; $A = F(5) - F(1) = -1/5 - (-1) = 4/5$

21. $\int(2x - 1)^{-2}\, dx = -1/[2(2x - 1)] + C$; $A = F(0) - F(-3) = 1/2 - 1/14 = 3/7$

23. $\int(9 - x^2)\, dx = 9x - x^3/3 + C$; $A = F(3) - F(-3) = 18 - (-18) = 36$

25. $\int(2x - x^2)\, dx = x^2 - x^3/3 + C$; $A = F(2) - F(0) = 4/3 - 0 = 4/3$

27. $\int(x^2 - x^3)\, dx = x^3/3 - x^4/4 + C$; $A = F(1) - F(0) = 1/12 - 0 = 1/12$

29. $\int(x^2 - x^4)\, dx = x^3/3 - x^5/5 + C$; $2A = 2[F(1) - F(0)] = 2[2/15 - 0] = 4/15$

Sections 5.2 – 5.3

Section 5.4

1. $(5/2)x^2\Big|_0^1 = 5/2 - 0 = 5/2$

3. $(x^3/3 + 3x)\Big|_1^2 = (8/3 + 6) - (1/3 + 3) = 16/3$

5. $x^4/4 + x\Big|_2^0 = 0 - (4 + 2) = -6$

7. $x^3/3 + x^2/2 + 2x\Big|_{-1}^1 = (1/3 + 1/2 + 2) - (-1/3 + 1/2 - 2) = 14/3$

9. $2x^{3/2} + 2x^{1/2}\Big|_4^9 = (2 \cdot 9^{3/2} + 2 \cdot 9^{1/2}) - (2 \cdot 4^{3/2} + 2 \cdot 4^{1/2}) = 40$

11. $\int_1^9 (x^{1/2} + 3x^{-1/2})dx = (2/3)x^{3/2} + 6x^{1/2}\Big|_1^9 = [(2/3)9^{3/2} + 6 \cdot 9^{1/2}] - [(2/3) \cdot 1^{3/2} + 6 \cdot 1^{1/2}] = 88/3$

13. $(1/15)(3x + 4)^5\Big|_1^2 = (1/15)[10^5 - 7^5] = 5546.2$

15. $(1/3)(2x + 4)^{3/2}\Big|_0^{16} = (1/3)[36^{3/2} - 4^{3/2}] = 208/3$

17. $(1/2)(x^2 - 3)^4\Big|_1^2 = (1/2)[(2^2 - 3)^4 - (1^2 - 3)^4] = -15/2$

19. $(-3/10)(1 - x^2)^{5/2}\Big|_{-1}^0 = (-3/10)[(1 - 0^2)^{5/3} - (1 - (-1)^2)^{5/3}] = -3/10$

21. $(1/3)(x^2 + 1)^{3/2}\Big|_0^1 = (1/3)[2^{3/2} - 1^{3/2}] = (2\sqrt{2} - 1)/3$

23. $6(x^2 + 9)^{1/2}\Big|_0^4 = 6[(4^2 + 9)^{1/2} - (0^2 + 9)^{1/2}] = 12$

25. $2(x^3 + x)^{1/2}\Big|_1^2 = 2\sqrt{10} - 2\sqrt{2}$

Chapter 5 Review

1. $(5/3)x^3 - (1/2)x^2 + C$

2. $(3/8)x^8 + x^2 + 4x + C$

3. $(4/3)x^{9/2} + C$

4. $(12/5)x^{5/3} + C$

5. $\int 3x^{-5}dx = -3/(4x^4) + C$

6. $\int x^{-3/2}dx = -2/\sqrt{x} + C$

Section 5.4 – Chapter 5 Review

7. $\dfrac{1}{2}\int(3x^4+2x-1)^3 \cdot 2(6x^3+1)\,dx = (1/8)(3x^4+2x-1)^4 + C$

8. $\dfrac{1}{2}\int(7x^2+8x+2)^{3/5} \cdot 2(7x+4)\,dx = (5/16)(7x^2+8x+2)^{8/5} + C$

9. $\int(x^2+5x)^{-1/2}(2x+5)\,dx = 2(x^2+5x)^{1/2} + C$

10. $\int(5x^3+4x)^{-2/3}(15x^2+4)\,dx = 3(5x^3+4x)^{1/3} + C$

11. $y = \int 3x^2\,dx = x^3 + C$

$-3 = 1 + C$ so $C = -4$ and $y = x^3 - 4$

12. $v = \int -32\,dt = -32t + C_1$ At $t = 0$, $v = 25$, so $C_1 = 25$

Thus $v = -32t + 25$

$s = \int v\,dt = \int(-32t+25)dt$ Thus $s = -16t^2 + 25t + C_2$

At $t = 0$, $s = 100$, so $C_2 = 100$ and thus $s = -16t^2 + 25t + 100$

13. $R = \int(0.009T^2 + 0.02T - 0.7)dt = 0.003T^3 + 0.01T^2 - 0.7T + C_1$

At $T = 0$, $R = 0.2$, then $C_1 = 0.2$

$R = 0.003T^3 + 0.01T^2 - 0.7T + 0.2$;   find R when $T = 30$

$R = 0.003(30)^3 + 0.01(30)^2 - 0.7(30) + 0.2 = 69.2\ \Omega$

14. $q = \int i\,dt = \int \dfrac{3t^2+1}{\sqrt{t^3+t+2}}\,dt = \int(t^3+t+2)^{-1/2}(3t^2+1)dt$

$$= 2(t^3+t+2)^{1/2} + C .$$

At $t = 0$, $q = 2\sqrt{2}$ so $C = 0$ Thus $q = 2(t^3+t+2)^{1/2}\Big|_{t=0.2}$

$$= 2\sqrt{2.208} = 2.97\ C$$

15. $\int(x^2+1)\,dx = x^3/3 + x + C$; $A = F(2) - F(0) = (8/3 + 2) - 0 = 14/3$

16. $\int(8+6x^2)\,dx = 8x - 2x^3 + C$; $A = F(1) - F(0) = (8-2) - 0 = 6$

17. $\int x^{-5}\,dx = -1/(4x^4) + C$; $A = F(2) - F(1) = (-1/64) - (-1/4) = 15/64$

18. $2\int(x^2+1)^{-2} \cdot 2x\,dx = -2/(x^2+1) + C$; $A = F(1) - F(0) = -1 - (-2) = 1$

Chapter 5 Review

19. $(1/5)\int(5x+6)^{1/2}\cdot 5\,dx = (2/15)(5x+6)^{3/2} + C; \quad A = F(6) - F(0)$

$\qquad\qquad = (2/15)[(5\cdot 6 + 6)^{3/2} - (50+6)^{3/2}] = (144 - 4\sqrt{6})/5$

20. $(1/2)\int(x^2+4)^{1/2}\cdot 2x\,dx = (1/3)(x^2+4)^{3/2} + C;$

$\quad A = F(2) - F(0) = (1/3)[(2^2+4)^{3/2} - (0^2+4)^{3/2}] = (16\sqrt{2} - 8)/3$

21. $x^4/4 + (2/3)x^3 + (1/2)x^2 \Big|_0^1 = (1/4 + 2/3 + 1/2) - 0 = 17/12$

22. $(3/5)x^5 - (1/2)x^4 + (7/2)x^2 \Big|_1^2 = [(3/5)2^5 - (1/2)2^4 + (7/2)2^2] - (3/5 - 1/2 + 7/2) = 108/5$

23. $(1/2)\int_0^2 2x(x^2+1)^2\,dx = (x^2+1)^3/6 \Big|_0^2 = 5^3/6 - 1/6 = 62/3$

24. $(5/3)\int_1^2 (x^3+2)^{-2}\cdot 3x^2\,dx = -5/[3(x^3+2)]\Big|_1^2 = -5/30 - (-5/9) = 7/18$

25. $(2/3)(x^2+x)^{3/2}\Big|_1^2 = \dfrac{12\sqrt{6}}{3} - \dfrac{4\sqrt{2}}{3} = \dfrac{12\sqrt{6} - 4\sqrt{2}}{3}$

26. $(1/3)\int_0^3 (x^3+1)^{-1/2}\cdot 3x^2\,dx = (2/3)(x^3+1)^{1/2}\Big|_0^3 = \dfrac{4\sqrt{7} - 2}{3}$

27. $\int_2^1 (3x^2 - x + x^{-2})\,dx = x^3 - (1/2)x^2 - 1/x\Big|_2^1 = (1 - 1/2 - 1) - (8 - 2 - 1/2) = -6$

28. $(3/4)\int_0^{1/2} (2x^2+1/2)^{-1/2}\cdot 4x\,dx = (3/2)(2x^2+1/2)^{1/2}\Big|_0^{1/2} = \dfrac{6 - 3\sqrt{2}}{4}$

Chapter 5 Review

CHAPTER 6

Section 6.1

1.
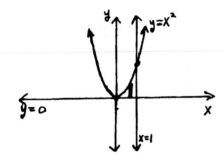

$$A = \int_0^1 x^2 \, dx = x^3/3 \Big|_0^1 = 1/3$$

3.
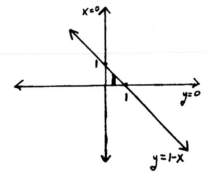

$$A = \int_0^1 (1-x) \, dx = x - x^2/2 \Big|_0^1 = 1/2$$

5.
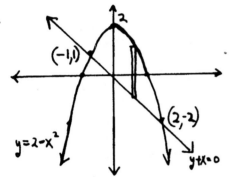

$$A = \int_{-1}^2 [(2-x^2) - x(-x)] \, dx$$

$$= 2x - x^3/3 = x^2/2 \Big|_{-1}^2 = \frac{9}{2}$$

7.
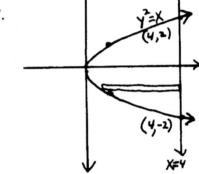

$$A = \int_{-2}^2 (4 - y^2) \, dy = 4y - \frac{y^3}{3} \ \Big|_{-2}^2 = \frac{32}{3}$$

9.
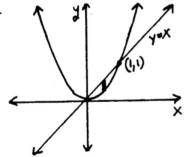

$$A = \int_0^1 (x - x^2) \, dx = x^2/2 - x^3/3 \Big|_0^1 = 1/6$$

11.
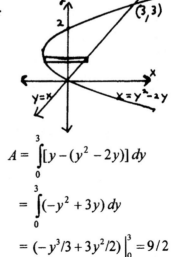

$$A = \int_0^3 [y - (y^2 - 2y)] \, dy$$

$$= \int_0^3 (-y^2 + 3y) \, dy$$

$$= (-y^3/3 + 3y^2/2) \Big|_0^3 = 9/2$$

Section 6.1

13. $A = \int\limits_{-1}^{0}(x^3 - x)\,dx + \int\limits_{0}^{1}[0 - (x^3 - x)]\,dx$

$\quad = x^4/4 - x^2/2 \Big|_{-1}^{0} + (-x^4/4 + x^2/2)\Big|_{0}^{1} = 1/2$

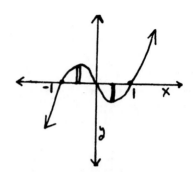

15.

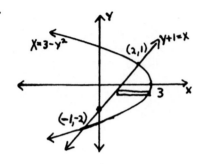

$A = \int\limits_{-2}^{1}[(3 - y^2) - (y + 1)]\,dy$

$\quad = \int\limits_{-2}^{1}(-y^2 - y + 2)\,dy$

$\quad = (-y^3/3 - y^2/2 + 2y)\Big|_{-2}^{1} = 9/2$

17.

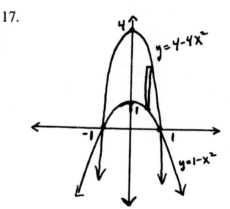

$A = \int\limits_{-1}^{1}[(4 - 4x^2) - (1 - x^2)]\,dx$

$\quad = \int\limits_{-1}^{1}(3 - 3x^2)\,dx$

$\quad = (3x - x^3)\Big|_{-1}^{1} = 4$

19.

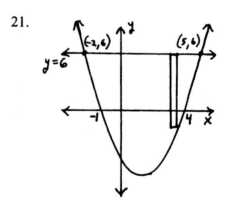

$A = \int\limits_{-1}^{1}[(2 - y^2) - y^4]\,dy$

$\quad = (2y - y^3/3 - y^5/5)\Big|_{-1}^{1} = 44/15$

21.

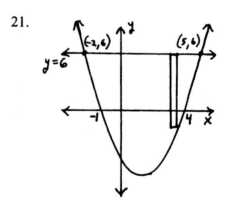

$A = \int\limits_{-2}^{5}[6 - (x^2 - 3x - 4)]\,dx$

$\quad = (-x^3/3 + 3x^2/2 + 10x)\Big|_{-2}^{5} = 343/6$

Section 6.1

23.

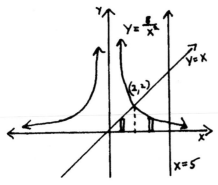

$A = \int\limits_0^2 x\, dx + \int\limits_2^5 (8/x^2)\, dx$

$= x^2/2 \Big|_0^2 + (-8/x) \Big|_2^5 = 22/5$

25.

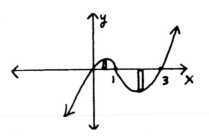

$A = \int\limits_0^1 (x^3 - 4x^2 + 3x)]\, dx +$

$\qquad \int\limits_1^3 [0 - (x^3 - 4x^2 + 3x)]\, dx$

$= (x^4/4 - 4x^3/3 + 3x^2/2) \Big|_0^1 +$

$\qquad (-x^4/4 + 4x^3/3 + 3x^2/2) \Big|_1^3 = 37/12$

Section 6.2

1.

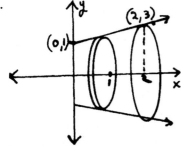

$V = \pi \int\limits_0^2 y^2\, dx = \pi \int\limits_0^2 (x+1)^2\, dx$

$= \pi \int\limits_0^2 (x^2 + 2x + 1)\, dx$

$= \pi(x^3/3 + x^2 + 1) \Big|_0^2 = 26\pi/3$

3.

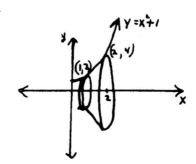

$V = \pi \int\limits_1^2 y^2\, dx = \pi \int\limits_1^2 (x^2 + 1)^2\, dx$

$= \pi \int\limits_1^2 (x^4 + 2x^2 + 1)\, dx$

$= \pi(x^5/5 + 2x^3/3 + x) \Big|_1^2 = 178\pi/15$

Sections 6.1 – 6.2

5.

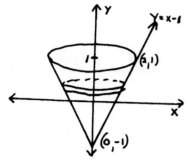

$$V = \pi \int_{-1}^{1} x^2 \, dy = \pi \int_{-1}^{1} (y+1)^2 \, dy$$

$$= \pi \int_{-1}^{1} (y^2 + 2y + 1) \, dy$$

$$= \pi (y^3/3 + y^2 + y) \Big|_{-1}^{1} = 8\pi/3$$

7.

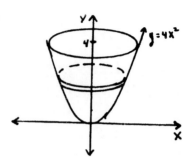

$$V = \pi \int_{0}^{4} x^2 \, dy = \pi \int_{0}^{4} (y/4) \, dy$$

$$= \pi (y^2/8) \Big|_{0}^{4} = 2\pi$$

9.

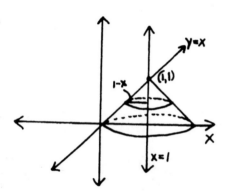

$$V = \pi \int_{0}^{1} (1-x)^2 \, dy = \pi \int_{0}^{1} (1-y)^2 \, dy$$

$$= \pi \int_{0}^{1} (1 - 2y + y^2) \, dy$$

$$= \pi (y - y^2 + y^3/3) \Big|_{0}^{1} = \frac{\pi}{3}$$

11.

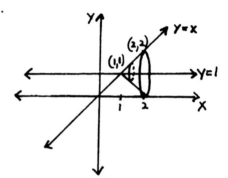

$$V = \pi \int_{1}^{2} (1-y^2) \, dx = \pi \int_{1}^{2} (1-x)^2 \, dx$$

$$= \pi \int_{1}^{2} (1 - 2x + x^2) \, dx$$

$$= \pi (x - x^2 + x^3/3) \Big|_{1}^{2} = \pi/3$$

13.

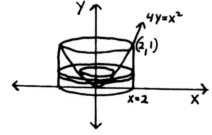

$$V = \pi \int_{0}^{1} (2^2 - x^2) \, dy = \pi \int_{0}^{1} (4 - 4y) \, dy$$

$$= 4\pi (y - y^2/2) \Big|_{0}^{1} = 2\pi$$

Section 6.2

15.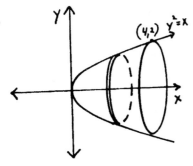

$$V = \pi \int_0^4 y^2\,dx = \pi \int_0^4 x\,dx$$

$$= \pi (x^2/2) \Big|_0^4 = 8\pi$$

17.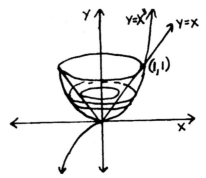

$$V = \pi \int_0^1 [(y^{1/3})^2 - y^2]\,dy$$

$$= \pi \int_0^1 (y^{2/3} - y^2)\,dy$$

$$= \pi (3y^{5/3}/5 - y^3/3) \Big|_0^1 = 4\pi/15$$

19.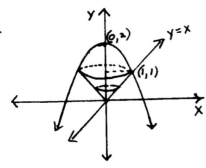

$$V = \pi \int_0^1 y^2\,dy + \pi \int_1^2 (\sqrt{2-y})^2\,dy$$

$$= \pi (y^3/3) \Big|_0^1 + \pi (2y - y^2/2) \Big|_1^2$$

$$= \pi/3 + \pi/2 = 5\pi/6$$

21.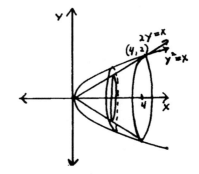

$$V = \pi \int_0^4 [(\sqrt{x})^2 - (x/2)^2]\,dx$$

$$= \pi \int_0^4 (x - x^2/4)\,dx$$

$$= \pi (x^2/2 - x^3/12) \Big|_0^4 = 8\pi/3$$

23. $V = 2\pi \int_0^1 [(3-x^2)^2 - (x^2+1)^2]\,dx$

$$= 2\pi \int_0^1 [(9 - 6x^2 + x^4) - (x^4 + 2x^2 + 1)]\,dx$$

$$= 2\pi \int_0^1 (8 - 8x^2)\,dx = 2\pi (8x - 8x^3/3) \Big|_0^1 = 32\pi/3$$

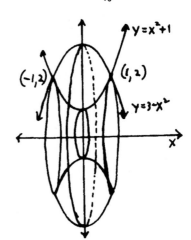

Section 6.2

25. $V = 2\pi \int_0^5 y^2 \, dx = 2\pi \int_0^5 \frac{225 - 9x^2}{25} \, dx$

$= 2\pi \int_0^5 (9 - 9x^2/25) \, dx$

$= 2\pi(9x - 3x^3/25) \Big|_0^5 = 60\pi$

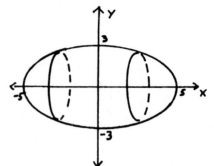

27. intersection of $y = 2$ and
$$9x^2 + 25y^2 = 225$$
$$9x^2 + 25(2)^2 = 225$$
$$9x^2 = 125$$
$$x = \pm 5\sqrt{5}/3$$

$V = 2\pi \int_0^{5\sqrt{5}/3} (y^2 - 2^2) \, dx$

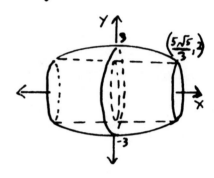

$= 2\pi \int_0^{5\sqrt{5}/3} \left\{ \frac{225 - 9x^2}{25} - 2^2 \right\} \cdot dx = 2\pi \int_0^{5\sqrt{5}/3} (9 - 9x^2/25 - 4) \, dx$

$= 2\pi \int_0^{5\sqrt{5}/3} (5 - 9x^2/25) \, dx = 2\pi(5x - 3x^3/25) \Big|_0^{5\sqrt{5}/3} = \frac{100\pi\sqrt{5}}{9}$ in³

29. $V = 2\pi \int_0^r x^2 \, dy = 2\pi \int_0^r (r^2 - y^2) \, dy$

$= 2\pi(r^2 y - y^3/3) \Big|_0^r = 2\pi(r^3 - r^3/3) = \frac{4}{3}\pi r^3$

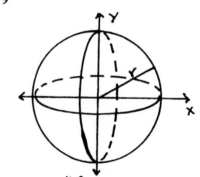

Section 6.3

1.

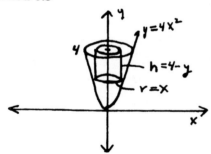

$V = 2\pi \int_0^1 x(4 - 4x^2) \, dx$

$= 2\pi \int_0^4 (4x - 4x^3) \, dx$

$= 2\pi(2x^2 - x^4) \Big|_0^1 = 2\pi$

3.

$V = 2\pi \int_0^2 xy \, dx = 2\pi \int_0^2 x(x^2/4) \, dx$

$= 2\pi \int_0^2 (x^3/4) \, dx = 2\pi(x^4/16) \Big|_0^2 = 2\pi$

Sections 6.2 – 6.3

5.

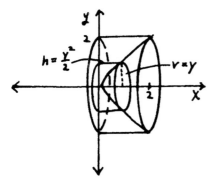

$$V = 2\pi \int_0^2 xy\,dy$$

$$= 2\pi \int_0^2 y(y^2/2)\,dy$$

$$= 2\pi \int_0^2 (y^3/2)\,dy$$

$$= 2\pi(y^4/8)\Big|_0^2 = 4\pi$$

7.

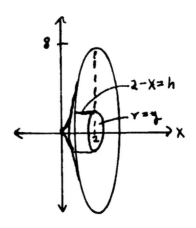

$$V = 2\pi \int_0^8 y(2-x)\,dy$$

$$= 2\pi \int_0^8 y(2-y^{1/3})\,dy$$

$$= 2\pi \int_0^8 (2y - y^{4/3})\,dy$$

$$= 2\pi(y^2 - 3/7)y^{7/3}\Big|_0^8 = 128\pi/7$$

9.

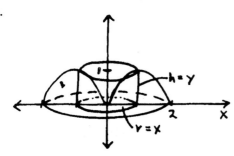

$$V = 2\pi \int_0^2 xy\,dy$$

$$= 2\pi \int_0^2 x(2x - x^2)\,dx$$

$$= 2\pi \int_0^2 (2x^2 - x^3)\,dx$$

$$= 2\pi(2x^3/3 - x^4/4)\Big|_0^2 = 8\pi/3$$

11.

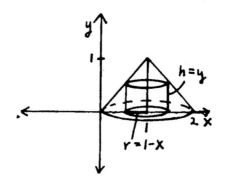

$$V = 2\pi \int_0^1 (1-x)y\,dx$$

$$= 2\pi \int_0^1 (1-x)x\,dx$$

$$= 2\pi \int_0^1 (x - x^2)\,dx$$

$$= 2\pi(x^2/2 - x^3/3)\Big|_0^1 = \pi/3$$

Section 6.3

13.

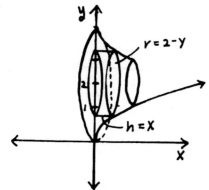

$$V = 2\pi \int_0^1 x(2-y)\,dy$$

$$= 2\pi \int_0^1 y^2(2-y)\,dy$$

$$= 2\pi \int_0^1 (2y^2 - y^3)\,dy$$

$$= 2\pi(2y^3/3 - y^4/4)\Big|_0^1 = 5\pi/6$$

15.

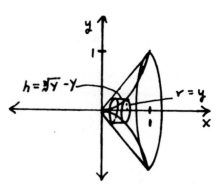

$$V = 2\pi \int_0^1 y(y^{1/3} - y)\,dy$$

$$= 2\pi \int_0^1 (y^{4/3} - y^2)\,dy$$

$$= 2\pi(3y^{7/3}/7 - y^3/3)\Big|_0^1 = 4\pi/21$$

17.

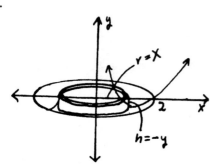

$$V = 2\pi \int_1^2 x(-y)\,dx$$

$$= -2\pi \int_1^2 x(x^2 - 3x + 2)\,dx$$

$$= -2\pi \int_1^2 (x^3 - 3x^2 + 2x)\,dx$$

$$= -2\pi(x^4/4 - x^3 + x^2)\Big|_1^2 = \pi/2$$

19.

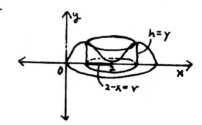

$$V = 2\pi \int_0^2 y(2-x)\,dx$$

$$= 2\pi \int_0^2 x(x-2)^2(2-x)\,dx$$

$$= 2\pi \int_0^2 (-x^4 + 6x^3 - 12x^2 + 8x)\,dx$$

$$= 2\pi(-x^5/5 + 3x^4/2 - 4x^3 + 4x^2)\Big|_0^2$$

$$= 16\pi/5$$

Section 6.3

Section 6.4

1. $M_o = (3)(-5) + (7)(3) + (4)(6) = 30$; $m = 3 + 7 + 4 = 14$
   $\bar{x} = 30/14 = 15/7$

3. $M_o = (24)(-15) + (15)(-9) + (12)(3) + (9)(6) = -405$
   $m = 24 + 15 + 12 + 9 = 60$; $\bar{x} = -405/60 = -6.75$

5. Given $\bar{x} = 0$, find $(x, 0)$. $M_y = (6)(9) + (18)(-2) + (3)(x)$; $m = 3$
   $\bar{x} = \dfrac{M_y}{m}$ so $0 = \dfrac{3x + 18}{3}$ Thus $x = -6$.

7. Given $\bar{x} = 3$, find $(x, 0)$. $M_y = (24)(-8) + (36)(12) + (9)(x) = 240 + 9x$
   $m = 24 + 36 + 9 = 69$; $\bar{x} = M_y/m$; $3 = \dfrac{240 + 9x}{69}$; $x = -11/3$

9. Given $\bar{x} = 0$, find m. $M_y = (6)(-3) + (9)(12) + m(-6) = 90 - 6m$
   $\bar{x} = M_y/m$; $0 = \dfrac{90 - 6m}{m}$; $m = 15$

11. Given $\bar{x} = 3$, find m. $M_y = (25)(-6) + (45)(8) + (40)(10) + m(-4) = 610 - 4m$
    $m = 25 + 45 + 40 + m = 110 + m$; $\bar{x} = M_y/m$; $3 = \dfrac{610 - 4m}{110 + m}$; $m = 40$

13. $M_y = 75000(0) + 50000(18) + 25000(30) = 1,650,000$
    $m = 75000 + 50000 + 25000 = 150,000$; $\bar{x} = \dfrac{1,650,000}{150,000} = 11$

    Thus locate airport 11 miles north of Flatville.

15. $m = 6 + 3 + 12 = 21$; $M_x = 6(4) + 3(2) + 12(3) = 66$;
    $M_y = 6(1) + 3(6) + 12(3) = 60$; $\bar{x} = M_y/m = 60/21 = 20/7$
    $\bar{y} = M_x/m = 66/21 = 22/7$ Thus center of mass is $(20/7, 22/7)$.

17. $m = 8 + 16 + 20 + 36 = 80$; $M_x = 8(12) + 16(8) + 20(-4) + 36(-20) = -576$
    $M_y = 8(8) + 16(-12) + 20(-16) + 36(4) = -304$
    $\bar{x} = M_y/m -304/80 = -3.8$; $\bar{y} = M_x/m = -576/80 = -7.2$
    Thus center of mass is $(-3.8, -7.2)$.

19. Given $\bar{x} = 0$ and $\bar{y} = 0$, find (x, y); $m = 6 + 9 + 10 = 25$
    $M_x = 6(2) + 9(8) + 10y = 84 + 10y$; $M_y = 6(4) + 9(-5) + 10x = -21 + 10x$
    $\bar{x} = M_y/m$; $0 = \dfrac{-21 + 10x}{25}$; $x = 2.1$; $\bar{y} = M_x/m = \dfrac{84 + 10y}{25}$; $y = -8.4$
    Thus the point is $(2.1, -8.4)$.

Section 6.4

21. Given $\bar{x} = -1$ and $\bar{y} = -2$, find m'. $m = 15 + 25 + 40 + m' = 80 + m'$

$M_y = 15(10) + 25(-6) + 40(8) + m'(-5) = 320 - 5m'$

$M_x = 15(3) + 25(-1) + 40(-2) + m'(-3) = -60 - 3m'$

$\bar{x} = M_y/m; \quad -1 = \dfrac{320 - 5m'}{80 + m'}; \quad m' = 100$

23. Place $A(1250)$ at $(0, 0)$, $B(820)$ at $(6, -3)$, and $C(520)$ at $(-2, -8)$; find $(\bar{x}, \bar{y})$, $m = 820 + 520 + 1250 = 2590$;

$M_x = 820(-3) + 520(-8) + 1250(0) = -6620$

$M_y = 820(6) + 520(-2) + 1250(0) = 3880$

$\bar{x} = M_y/m = 3880/2590 = 1.50; \quad \bar{y} = M_x/m = -6620/2590 = -2.56$

Thus best location: 1.50 mi east and 2.56 mi. south of A.

Section 6.5

1. $\dfrac{\displaystyle\int_0^{20} x \, dx}{\displaystyle\int_0^{20} dx} = \dfrac{\left.\dfrac{x^2}{2}\right|_0^{20}}{\left. x \right|_0^{20}} = \dfrac{200}{20} = 10$ Thus center is 10 cm from either end.

3. $\bar{x} = \dfrac{M_o}{m} = \dfrac{\displaystyle\int_0^{10} (0.1x)x \, dx}{\displaystyle\int_0^{10} 0.1x \, dx} = \dfrac{\displaystyle\int_0^{10} x^2 \, dx}{\displaystyle\int_0^{10} x \, dx} = \dfrac{\left.\dfrac{x^3}{3}\right|_0^{10}}{\left.\dfrac{x^2}{2}\right|_0^{10}} = \dfrac{\dfrac{1000}{3}}{50} = \dfrac{20}{3}$ cm from lighter end

5. $\bar{x} = \dfrac{M_o}{m} = \dfrac{\displaystyle\int_0^{12} (4 + x^2) \, dx}{\displaystyle\int_0^{12} (4 + x^2) \, dx} = \dfrac{\left.\dfrac{1}{4}(x^2 + 4)^2\right|_0^{12}}{\left.(4x + x^3/3)\right|_0^{12}} = \dfrac{5472}{624} = 8.77$ cm from lighter end

7. $\bar{x} = \dfrac{M_o}{m} = \dfrac{\displaystyle\int_0^{6} (kx)x \, dx}{\displaystyle\int_0^{6} kx \, dx} = \dfrac{\displaystyle\int_0^{6} x^2 \, dx}{\displaystyle\int_0^{6} x \, dx} = \dfrac{\left.\dfrac{x^3}{3}\right|_0^{6}}{\left.\dfrac{x^2}{2}\right|_0^{6}} = \dfrac{72}{18} = 4$ cm from given end

9. center of rectangle is $(4, 6)$; $A = 32$
center of square is $(6, 2)$; $A = 16$
$m = 32 + 16 = 48$
$M_y = 32(4) + 16(6) = 224$
$M_x = 32(6) + 16(2) = 224$
$\bar{x} = \dfrac{M_y}{m} = \dfrac{224}{48} = 4\dfrac{2}{3}; \quad \bar{y} = \dfrac{M_x}{m} = \dfrac{224}{48} = 4\dfrac{2}{3}$

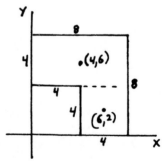

Sections 6.4 – 6.5

11. center of rectangle 1 is $(2, 8)$; $A = 32$
center of rectangle 2 is $(8, 2)$; $A = 56$
center of rectangle 3 is $(14, -4)$; $A = 32$
$m = 32 + 56 + 32 = 120$
$M_y = 32(2) + 56(8) + 32(14) = 960$
$M_x = 32(8) + 56(2) + 32(-4) = 240$
$\bar{x} = \dfrac{M_y}{m} = \dfrac{960}{120} = 8$; $\bar{y} = \dfrac{M_x}{120} = 2$

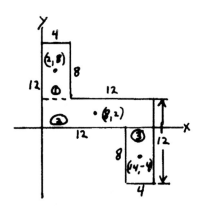

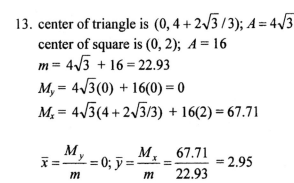

13. center of triangle is $(0, 4 + 2\sqrt{3}/3)$; $A = 4\sqrt{3}$
center of square is $(0, 2)$; $A = 16$
$m = 4\sqrt{3} + 16 = 22.93$
$M_y = 4\sqrt{3}(0) + 16(0) = 0$
$M_x = 4\sqrt{3}(4 + 2\sqrt{3}/3) + 16(2) = 67.71$

$\bar{x} = \dfrac{M_y}{m} = 0$; $\bar{y} = \dfrac{M_x}{m} = \dfrac{67.71}{22.93} = 2.95$

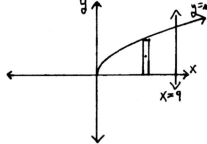

15. $A = \displaystyle\int_0^9 \sqrt{x}\, dx = \dfrac{2}{3}x^{3/2}\Big|_0^9 = 18$

$\bar{x} = \dfrac{\displaystyle\int_0^9 x\sqrt{x}\, dx}{18} = \dfrac{\dfrac{2}{5}\cdot x^{5/2}\Big|_0^9}{18} = \dfrac{97.2}{18} = 5.4$

$\bar{y} = \dfrac{\dfrac{1}{2}\displaystyle\int_0^9 (\sqrt{x})^2\, dx}{18} = \dfrac{\dfrac{1}{2}\cdot\dfrac{x^2}{2}\Big|_0^9}{18} = \dfrac{20.25}{18} = \dfrac{9}{8}$

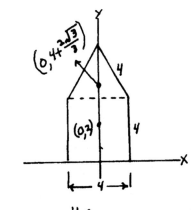

17. $A = \displaystyle\int_0^2 -(x^2 - 2x)\, dx = -(x^3/3 - x^2)\Big|_0^2 = \dfrac{4}{3}$

$\bar{x} = \dfrac{\displaystyle\int_0^2 x[0 - (x^2 - 2x)]\, dx}{4/3} = \dfrac{\displaystyle\int_0^2 (-x^3 + 2x^2)\, dx}{4/3}$

$= (3/4)(-x^4 + 2x^3/3)\Big|_0^2 = 1$

$\bar{y} = \dfrac{\dfrac{1}{2}\displaystyle\int_0^2 [0^2 - (x^2 - 2x)^2]\, dx}{4/3} = 3/8\displaystyle\int_0^2 -(x^4 - 4x^3 + 4x^2)\, dx$

$= (-3/8)(x^5/5 - x^4 + 4x^3/3\Big|_0^2 = -2/5$

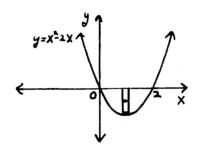

Section 6.5

19. $A = \int_{-2}^{2} (4-x^2)\,dx = (4x - x^3/3)\Big|_{-2}^{2} = 32/3$

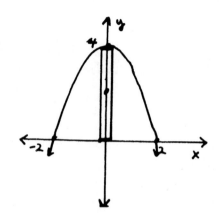

$\bar{x} = 0$ from sketch

$$\bar{y} = \frac{\dfrac{1}{2}\displaystyle\int_{-2}^{2}(4-x^2)^2\,dx}{32/3} = \frac{3}{64}\int_{-2}^{2}(16-8x^2+x^4)\,dx$$

$\quad = (3/64)(16x - 8x^3/3 + x^5/5)\Big|_{-2}^{2} = 8/5$

Thus $(0, 8/5)$.

21. $A = \int_{0}^{1}(x-x^3)\,dx = (x^2/2 - x^4/4)\Big|_{0}^{1} = 1/4$

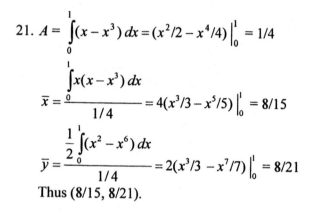

$$\bar{x} = \frac{\displaystyle\int_{0}^{1}x(x-x^3)\,dx}{1/4} = 4(x^3/3 - x^5/5)\Big|_{0}^{1} = 8/15$$

$$\bar{y} = \frac{\dfrac{1}{2}\displaystyle\int_{0}^{1}(x^2-x^6)\,dx}{1/4} = 2(x^3/3 - x^7/7)\Big|_{0}^{1} = 8/21$$

Thus $(8/15, 8/21)$.

23. $\bar{x} = 0$; $A = (1/2)\pi r^2 = \pi/2$

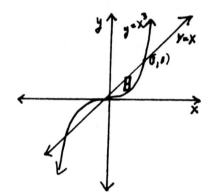

$$\bar{y} = \frac{\dfrac{1}{2}\displaystyle\int_{-1}^{1}(\sqrt{1-x^2})^2\,dx}{\pi/2} = \frac{1}{\pi}\int_{-1}^{1}(1-x^2)\,dx$$

$\quad = (1/\pi)(x - x^3/3)\Big|_{-1}^{1} = \dfrac{4}{3\pi}$

25. $\bar{x} = \dfrac{\displaystyle\int_{0}^{1}x(x^3)^2\,dx}{\displaystyle\int_{0}^{1}(x^3)^2\,dx} = \dfrac{x^8/8\Big|_{0}^{1}}{x^7/7\Big|_{0}^{1}} = \dfrac{1/8}{1/7} = \dfrac{7}{8}$

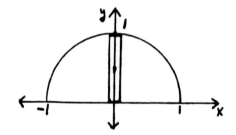

27. $\bar{x} = \dfrac{\displaystyle\int_{0}^{3}x(3-x)^2\,dx}{\displaystyle\int_{0}^{3}(3-x)^2\,dx} = \dfrac{(9x^2/2 - 2x^3 + x^4/4)\Big|_{0}^{3}}{(9x - 3x^2 + x^3/3)\Big|_{0}^{3}} = \dfrac{27/4}{9} = \dfrac{3}{4}$; $\bar{y} = 0$

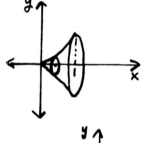

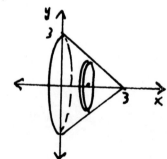

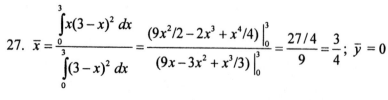

Section 6.5

$$29. \quad \bar{y} = \frac{\int_0^1 y(y)\,dy}{\int_0^1 y\,dy} = \frac{y^3/3\,\big|_0^1}{y^2/2\,\big|_0^1} = \frac{2}{3};\ \bar{x} = 0$$

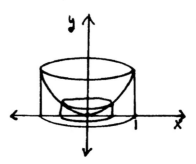

Section 6.6

1. $I_y = 9(3)^2 + 12(5)^2 + 15(3)^2 = 516;\ m = 36;\ R = \sqrt{516/36} = 3.79$

3. $I_y = 15(3)^2 + 10(6)^2 + 18(9)^2 + 12(1)^2 = 1965;\ m = 55;\ R = \sqrt{1965/55} = 5.98$

5. $I_x = 9(-2)^2 + 12(4)^2 + 15(7)^2 = 963;\ m = 36;\ R = \sqrt{963/36} = 5.17$

7. $I_x = 9(-9)^2 + 5(-2)^2 + 8(0)^2 + 10(1)^2 = 759;\ m = 32;\ R = \sqrt{759/32} = 4.87$

9. $I_y = 5\int_0^2 x^2(4 - x^2)\,dx = 5\int_0^2 (4x^2 - x^4)\,dx = 5(4x^3/3 - x^5/5)\big|_0^2 = 64/3$

$\qquad m = 5\int_0^2 (4 - x^2)\,dx = 5(4x - x^3/3)\big|_0^2 = 80/3;\ R = \sqrt{(64/3)/(80/3)} = 0.894$

11. $I_x = 4\int_0^1 y^2(\sqrt{y} - y)\,dy = 4\int_0^1 (y^{5/2} - y^3)\,dy = 4(2y^{7/2}/7 - y^4/4)\big|_0^1 = 1/7$

$\qquad m = 4\int_0^1 (\sqrt{y} - y)\,dy = 4(2y^{3/2}/3 - y^2/2)\big|_0^1 = \frac{2}{3};\ R = \sqrt{(1/7)/(2/3)} = 0.463$

13. $I_y = 3\int_0^2 x^2[(5 - x^2) - 1]\,dx = 3\int_0^2 (4x^2 - x^4)\,dx = 3(4x^3/3 - x^5/5)\big|_0^2 = 64/5$

$\qquad m = 3\int_0^2 [(5 - x^2) - 1]\,dx = 3(4x - x^3/3)\big|_0^2 = 16\ \ R = \sqrt{(64/5)/16} = 0.894$

15. $I_y = 2\int_1^2 x^2(1/x^2)\,dx = 2\int_1^2 dx = 2x\,\big|_1^2 = 2;$

$\qquad m = 2\int_1^2 x^{-2}\,dx = 2(-1/x)\,\big|_1^2 = 1;\ R = \sqrt{2/1} = 1.41$

17. $m = 2\pi(15)\int_0^2 x(3x)\,dx = 30\pi\int_0^2 3x^2\,dx = 30\pi x^3\big|_0^2 = 240\pi$

$\qquad I_y = 2\pi(15)\int_0^2 x^3(3x)\,dx = 30\pi\int_0^2 3x^4\,dx = 18\pi x^5\,\big|_0^2 = 576\pi\ ;\ R = \sqrt{576\pi/240\pi} = 1.55$

Sections 6.5 – 6.6

19. $I_x = 2\pi \int_0^{16} y^3 (2 - y^{1/2}/2)\, dy = 2\pi \int_0^{16} (2y^3 - y^{7/2})\, dy = 2\pi(y^4/2 - y^{9/2}/9)\Big|_0^{16} = 7282\pi = 2.29 \times 10^4$

$m = 2\pi \int_0^{16} y(2 - y^{1/2}/2)\, dy = 2\pi \int_0^{16} (2y - y^{3/2}/2)\, dy = 2\pi(y^2 - y^{5/2}/5)\Big|_0^{16} = 102.4\pi = 322$;

$R = \sqrt{7282\pi / 102.4\pi} = 8.43$

21. $I_y = 2\pi(12) \int_0^3 x^3(9 - x^2)\, dx = 24\pi \int_0^3 (9x^3 - x^5)\, dx = 24\pi(9x^4/4 - x^6/6)\Big|_0^3 = 1458\pi$;

$m = 2\pi(12) \int_0^3 x(9 - x^2)\, dx = 24\pi \int_0^3 (9x - x^3)\, dx = 24\pi(9x^2/2 - x^4/4)\Big|_0^3 = 486\pi$;

$R = \sqrt{1458\pi / 486\pi} = 1.73$

23. $I_y = 2\pi(15) \int_0^4 x^3(4x - x^2)\, dx = 30\pi \int_0^4 (4x^4 - x^5)\, dx = 30\pi(4x^5/5 - x^6/6)\Big|_0^4 = 4096\pi$;

$m = 2\pi(15) \int_0^4 x(4x - x^2)\, dx = 30\pi \int_0^4 (4x^2 - x^3)\, dx = 30\pi(4x^3/3 - x^4/4)\Big|_0^4 = 640\pi$;

$R = \sqrt{4096\pi / 640\pi} = 2.53$

Section 6.7

1. $W = \int_0^3 (x^3 - x)\, dx = (x^4/4 - x^2/2)\Big|_0^3 = 63/4$

3. $20 = k(10)$; $k = 2$; $W = \int_0^5 2x\, dx = x^2\Big|_0^5 = 25$ in.-lb

5. $150 = 4k$; $k = 75/2$; $W = \int_0^6 (75x/2)\, dx = (75x^2/4)\Big|_0^6 = 657$ N · cm or 6.75 J

7. $W = \int_{0.01}^{0.05} 3.62 \times 10^{-16} x^{-2}\, dx = 3.62 \times 10^{-16}(-1/x)\Big|_{0.01}^{0.05} = 2.896 \times 10^{-14}$ J

9. a) $W = \int_{40}^{50} 2x\, dx = x^2\Big|_{40}^{50} = 900$ ft-lb

    b) $W = \int_{25}^{50} 2x\, dx = x^2\Big|_{40}^{50} = 1875$ ft-lb

    c) $W = \int_0^{50} 2x\, dx = x^2\Big|_0^{50} = 2500$ ft-lb

Sections 6.6 – 6.7

11. $W = \int\limits_{0}^{12} 62.4\pi(4)^2 x\,dx = 998.4\pi(x^2/2)\;\Big|_{0}^{12} = 225{,}800$ ft-lb

13. $W = \int\limits_{10}^{22} 62.4\pi(4)^2 x\,dx = 998.4\pi(x^2/2)\;\Big|_{10}^{22} = 602{,}200$ ft-lb

15. $\dfrac{5}{12} = \dfrac{r}{12 - x}$ thus $r = \dfrac{5(12 - x)}{12}$

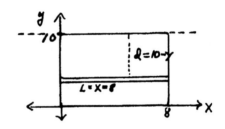

$F = 62.4\pi\dfrac{25(12 - x)^2}{144}\Delta x$

$W = \int\limits_{0}^{12} 10.83\pi x(12 - x)^2\,dx$

$= 10.83\pi \int\limits_{0}^{12}(x^3 - 24x^2 + 144x)\,dx = 10.83\pi(x^4/4 - 8x^3 + 72x^2)\;\Big|_{0}^{12}$

$= 58{,}810$ ft-lb

17. $F = 62.4\int\limits_{0}^{10}(10 - y)(8)\,dy$

$= (62.4)(8)(10y - y^2/2)\;\Big|_{0}^{10}$

$= 25{,}000$ lb

19. $F = 9800\int\limits_{0}^{3}(3 - y)(8)\,dy$

$= 9800(8)(3y - y^2/2)\;\Big|_{0}^{3}$

$= 352{,}800$ N

Note: $\rho g = 9800$ kg/(m^2 s^2)
3/4 full; $A = 24$ m^2;
depth $= 3$ m

21. $F = 870(9.80)\int\limits_{-5}^{0}(-y)(2)\sqrt{25 - y^2}\,dy$

$= 8526\int\limits_{-5}^{0}(25 - y^2)^{1/2}(-2y)\,dy$

$= 8526(2/3)(25 - y^2)^{3/2}\;\Big|_{-5}^{0} = 710{,}500$ N

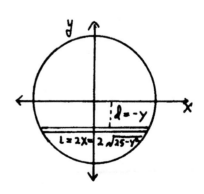

Section 6.7

23. $F = 62.4 \int_0^2 (2-y)(2y)\,dy$

$\qquad = 62.4 \int_0^2 (4y - 2y^2)\,dy$

$\qquad = 62.4(2y^2 - 2y^3/3) \Big|_0^2 = 166.4$ lb

$\dfrac{4}{2} = \dfrac{L}{Y}$ Thus $L = 2y$.

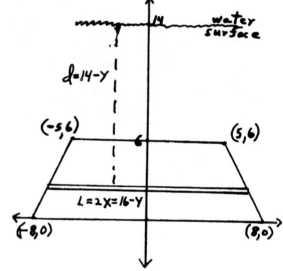

25. $F = 62.4 \int_0^6 (14-y)(16-y)\,dy$

$\qquad = 62.4 \int_0^6 (224 - 30y + y^2)\,dy$

$\qquad = 62.4(224y - 15y^2 - y^3/3) \Big|_0^6 = 54{,}660$ lb

Note: equation of line through (5, 6)
and (8, 0) is $y = -2(x - 8)$
Thus $L = 2x = 16 - y$.

27. $F = 62.4 \int_6^9 (9-y)(18)\,dy + 62.4 \int_0^6 (9-y)(3y)\,dy$

$\qquad = 1123.2 \int_6^9 (9-y)\,dy + 187.2 \int_0^6 (9y - y^2)\,dy$

$\qquad = 1123.2(9y - y^2/2) \Big|_6^9 + 187.2(9y^2/2 - y^3/3) \Big|_0^6 = 5054 + 16{,}848 = 21{,}902$ lb

Note: equation of line through (0, 0) and (18, 6) is $y = x/3$.

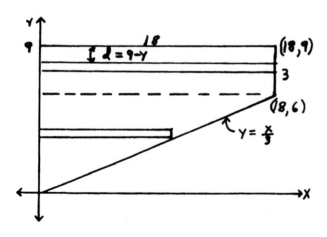

Section 6.7

Solutions to Odd-Numbered Exercises

29. $y_{av} = \dfrac{1}{3-1} \displaystyle\int_1^3 x^2 \, dx = (1/2)(x^3/3)\Big|_1^3 = 13/3$

31. $y_{av} = \dfrac{1}{10-5} \displaystyle\int_5^{10} \dfrac{dx}{\sqrt{x-1}} = (1/5)\int_5^{10}(x-1)^{-1/2} \, dx = (2/5)(x-1)^{1/2}\Big|_5^{10} = 2/5$

33. $I_{av} = \dfrac{1}{0.5-0.1} \displaystyle\int_{0.1}^{0.5}(6t - t^2) \, dt = (2.5)(3t^2 - t^3/3)\Big|_{0.1}^{0.5} = 1.70 \, A$

Chapter 6 Review

1.

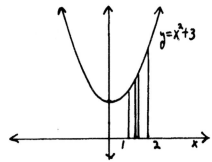

$A = \displaystyle\int_1^2 (x^2 + 3) \, dx = (x^3/3 + 3x)\Big|_1^2 = 16/3$

2.

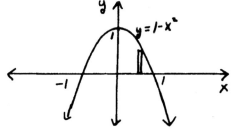

$A = \displaystyle\int_0^1 (1 - x^2) \, dx = (x - x^3/3)\Big|_0^1 = \dfrac{2}{3}$

3.

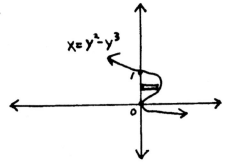

$A = \displaystyle\int_0^1 (y^2 - y^3) \, dy$

$= (y^3/3 - y^4/4)\Big|_0^1 = 1/12$

4.

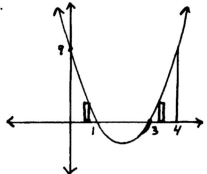

$A = \displaystyle\int_0^1 (3x^2 - 12x + 9) \, dx$

$+ \displaystyle\int_3^4 (3x^2 - 12x + 9) \, dx$

$= 4 + 4 = 8$

Section 6.7 – Chapter 6 Review

5.

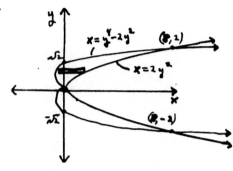

$$A = 2\int_0^2 [2y^2 - (y^4 - 2y^2)]\, dy$$

Note: S: x-axis

$$= 2\int_0^2 (4y^2 - y^4)\, dy$$

$$= 2(4y^3/3 - y^5/5)\Big|_0^2 = 128/15$$

7.
$$V = 2\pi\int_0^2 y(4 - y^2)\, dy = 2\pi\int_0^2 (4y - y^3)\, dy$$

$$= 2\pi(2y^2 - y^4/4)\Big|_0^2 = 8\pi$$

8. (See # 7 diagram)

$$V = \pi\int_0^4 (\sqrt{x})^2\, dx = \pi(x^2/2)\Big|_0^4 = 8\pi$$

9.

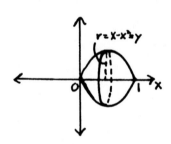

$$V = \pi\int_0^1 (x - x^2)^2\, dx = \pi\int_0^1 (x^2 - 2x^3 + x^4)\, dx$$

$$= \pi(x^3/3 - x^4/2 + x^5/5)\Big|_0^1 = \pi/30$$

6.

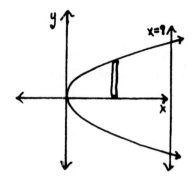

$$A = 2\int_0^9 \sqrt{x}\, dx = 2(2x^{3/2}/3)\Big|_0^9 = 36$$

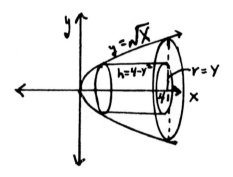

Chapter 6 Review

10. $V = 2\pi \int_0^2 x[(3x - x^2) - x]\,dx = 2\pi \int_0^2 (2x^2 - x^3)\,dx$

$= 2\pi(2x^3/3 - x^4/4)\Big|_0^2 = \dfrac{8\pi}{3}$

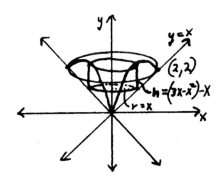

11.

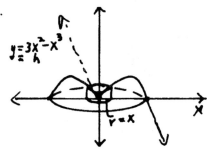

$V = 2\pi \int_0^3 x(3x^2 - x^3)\,dx$

$= 2\pi \int_0^3 (3x^3 - x^4)\,dx$

$= 2\pi(3x^4/4 - x^5/5)\Big|_0^3 = 243\pi/10$

12.

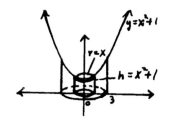

$V = 2\pi \int_0^3 x(x^2 + 1)\,dx$

$= 2\pi \int_0^3 (x^3 + x)\,dx$

$= 2\pi(x^4/4 - x^2/2)\Big|_0^3 = \dfrac{99\pi}{2}$

13.

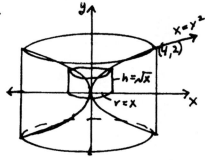

$V = (2)2\pi \int_0^4 x\sqrt{x}\,dx$

$= 4\pi \int_0^4 x^{3/2}\,dx$

$= 4\pi(2x^{5/2}/5)\Big|_0^4 = 256\pi/5$

14.

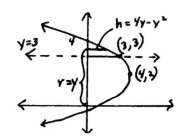

$V = 2\pi \int_0^3 y(4y - y^2)\,dy$

$= 2\pi \int_0^3 (4y^2 - y^3)\,dy$

$= 2\pi(4y^3/3 - y^4/4)\Big|_0^3 = \dfrac{63\pi}{2}$

Chapter 6 Review

15. $M_o = 12(-4) + 20(9) + 24(12) = 420$; $m = 12 + 20 + 24 = 56$
    $\bar{x} = M_o/m = 420/56 = 7.5$

16. $m = 24 + 36 + 30 = 90$; $M_x = 24(-3) + 36(-15) + 30(0) = -612$
    $M_y = 24(11) + 36(-4) + 30(-7) = -90$; $\bar{x} = -90/90 = -1$; $\bar{y} = -612/90 = -6.8$

17. center of top rectangle is $(8, 10.5)$; $A = 40$
    center of bottom rectangle is $(10, 4)$; $A = 160$
    $M_y = 40(8) + 160(10) = 1920$
    $M_x = 40(10.5) + 160(4) = 1060$
    $m = 40 + 160 = 200$; $\bar{x} = 1920/200 = 9.6$; $\bar{y} = 1060/200 = 5.3$

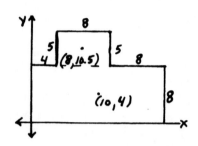

18. $\bar{x} = \dfrac{\displaystyle\int_0^4 x(5x)\,dx}{40} = \dfrac{(5x^3/3)\big|_0^4}{40} = 2\dfrac{2}{3}$

$\bar{y} = \dfrac{\dfrac{1}{2}\displaystyle\int_0^4 (5x)^2\,dx}{40} = \dfrac{(25x^3/3)\big|_0^4}{80} = 6\dfrac{2}{3}$

$A = \dfrac{1}{2}(4)(20) = 40$

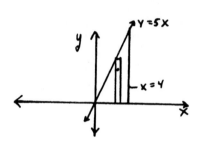

19. $A = \displaystyle\int_0^3 [(6x - x^2) - 3x]\,dx = \int_0^3 (3x - x^2)\,dx$

$= (3x^2/2 - x^3/3)\big|_0^3 = 9/2$

$\bar{x} = \dfrac{\displaystyle\int_0^3 x[(6x - x^2) - 3x]\,dx}{9/2}$

$= \dfrac{2}{9}\displaystyle\int_0^3 (3x^2 - x^3)\,dx$

$= (2/9)(x^3 - x^4/4)\big|_0^3 = 3/2$

$\bar{y} = \dfrac{\dfrac{1}{2}\displaystyle\int_0^3 [(6x - x^2)^2 - (3x)^2]\,dx}{9/2} = \dfrac{1}{9}\displaystyle\int_0^3 (27x^2 - 12x^3 + x^4)\,dx$

$= (1/9)(9x^3 - 3x^4 + x^5/3)\big|_0^3 = 27/5$

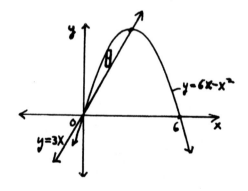

Chapter 6 Review

Solutions to Odd-Numbered Exercises

20. $A = \displaystyle\int_0^6 -(x^2 - 6x)\,dx$

$\quad = -(x^3/3 - 3x^2)\Big|_0^6 = 36$

Note: $\bar{x} = 3$

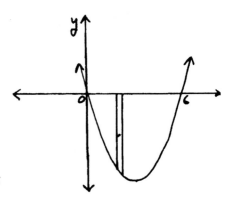

$\bar{y} = \dfrac{\dfrac{1}{2}\displaystyle\int_0^6 [0^2 - (x^2 - 6x)^2]\,dx}{36} = \dfrac{1}{72}\displaystyle\int_0^6 (-x^4 + 12x^3 - 36x^2)\,dx$

$\quad = (1/72)(-x^5/5 + 3x^4 - 12x^3)\Big|_0^6 = -3.6$ Thus $(3, -3.6)$.

21. $\bar{y} = \dfrac{\displaystyle\int_0^2 y(y/2)^2\,dy}{\displaystyle\int_0^2 (y/2)^2\,dy} = \dfrac{(y^4/16)\Big|_0^2}{(y^3/12)\Big|_0^2}$

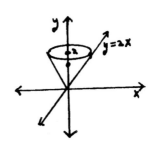

$\quad = \dfrac{\dfrac{1}{8}}{\dfrac{1}{12}} = \dfrac{3}{2}$ Thus $(0, 3/2)$.

22. $\bar{x} = \dfrac{\displaystyle\int_0^1 x(x^2)^2\,dx}{\displaystyle\int_0^1 (x^2)^2\,dx} = \dfrac{\displaystyle\int_0^1 x^5\,dx}{\displaystyle\int_0^1 x^4\,dx} = \dfrac{(x^6/6)\Big|_0^1}{(x^5/5)\Big|_0^1}$

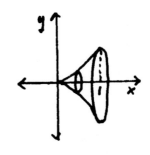

$\quad = \dfrac{5}{6}$ Thus $(5/6, 0)$.

23. $\bar{y} = \dfrac{\displaystyle\int_0^4 y(y^2 - 4y)^2\,dy}{\displaystyle\int_0^4 (y^2 - 4y)^2\,dy} = \dfrac{\displaystyle\int_0^4 (y^5 - 8y^4 + 16y^3)\,dy}{\displaystyle\int_0^4 (y^4 - 8y^3 + 16y^2)\,dy}$

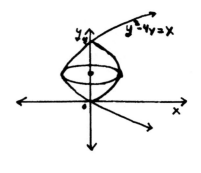

$\quad = \dfrac{(y^6/6 - 8y^5/5 + 4y^4)\Big|_0^4}{(y^5/5 - 2y^4 + 16y^3/3)\Big|_0^4} = \dfrac{68.2\overline{6}}{34.1\overline{3}} = 2; \bar{x} = 0$

24. $I_x = 10(2)^2 + 6(7)^2 + 8(-4)^2 = 462; \ m = 24; \ R = \sqrt{462/24} = 4.39$

Chapter 6 Review

25. $I_y = 1\int_0^4 x^2(3x)\,dx = (3/4)x^4\Big|_0^4 = 192; \; m = 1\int_0^4 3x\,dx = (3x^2/2)\Big|_0^4 = 24$

   $R = \sqrt{192/24} = 2.83$

26. $I_x = 1\int_0^{12} y^2(4 - y/3)\,dy = (4y^3/3 - y^4/12)\Big|_0^{12} = 576; \; m = 24 \text{ from } \#25$

   $R = \sqrt{576/24} = 4.90$

27. $I_x = 4\int_0^1 y^2(1 - y^2)\,dy = 4(y^3/3 - y^5/5)\Big|_0^1 = 8/15$

   $m = 4\int_0^1 (1 - y^2)\,dy = 4(y - y^3/3)\Big|_0^1 = 8/3;$

   $\sqrt{(8/15)/(8/3)}\,; \; R = 0.447$

28. $I_x = 2\pi(4)\int_0^1 y^3(1 - y^{1/3})\,dy = 8\pi\int_0^1 (y^3 - y^{10/3})\,dy = 8\pi(y^4/4 - 3y^{13/3}/13)\Big|_0^1 = 2\pi/13$

   $m = 2\pi(4)\int_0^1 y(1 - y^{1/3})\,dy = 8\pi\int_0^1 (y - y^{4/3})\,dy = 8\pi(y^2/2 - 3y^{7/3}/7)\Big|_0^1 = 4\pi/7$

   $R = \sqrt{2\pi/13/4\pi/7} = 0.519$

29. $I_y = 2\pi(4)\int_0^1 x^3(x^3)\,dx = 8\pi\int_0^1 x^6\,dx = 8\pi(x^7/7)\Big|_0^1 = 8\pi/7$

   $m = 8\pi\int_0^1 x^3 \cdot x\,dx = 8\pi\int_0^1 x^4\,dx = 8\pi(x^5/5)\Big|_0^1 = 8\pi/5$

   $R = \sqrt{(8\pi/7)/(8\pi/5)} = 0.845$

30. $I_y = 2\pi(3)\int_1^4 x^3(1/x)\,dx = 6\pi(x^3/3)\Big|_1^4 = 126\pi;$

   $m = 6\pi\int_1^4 x(1/x)\,dx = 6\pi x\Big|_1^4 = 18\pi; \; R = \sqrt{126\pi/18\pi} = 2.65$

31. $16 = 4k; \; k = 4; \; W = \int_0^{10} 4x\,dx = 2x^2\Big|_0^{10} = 200 \text{ in.-lb}$

32. $W = \int_{0.01}^{0.02} 5.24 \times 10^{-18} x^{-2}\,dx = 5.24 \times 10^{-18}(x^{-1}/-1)\Big|_{0.01}^{0.02} = 2.62 \times 10^{-16} \text{ J}$

Chapter 6 Review

138

33. $W = \int_0^{200} 4x \, dx + 250(200) = 2x^2 \Big|_0^{200} + 50{,}000 \ = 130{,}000$ ft-lb

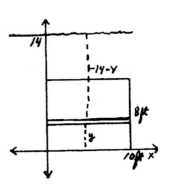

34. $F = 62.4 \int_0^8 (14 - y)(10) \, dy$

$= 624(14y - y^2/2) \Big|_0^8 = 49{,}920$ lb

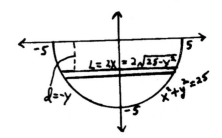

35. $F = 9800 \int_{-5}^0 (-y)(2\sqrt{25 - y^2}) \, dy$

$= 9800(2/3)(25 - y^2)^{3/2} \Big|_{-5}^0$

816,700 N

36. $V_{av} = \dfrac{1}{3 - 0} \int_0^3 (t^2 + 3t + 2) \, dt = (1/3)(t^3/3 + 3t^2/2 + 2t) \ \Big|_0^3 = 9.5$ V

37. $I_{av} = \dfrac{1}{9 - 4} \int_4^9 4t^{3/2} \, dt = (1/5)(8t^{5/2}/5) \Big|_4^9 = 67.52$ A

38. $P_{av} = \dfrac{1}{3 - 1} \int_1^3 2t^3 \, dt = (1/2)(t^4/2) \Big|_1^3 = 20$ W

Chapter 6 Review

CHAPTER 7

Section 7.1

1. $u = 3x + 2$; $du = 3\ dx$; $(1/3) \int (3x+2)^{1/2} \cdot 3\ dx = (2/9)(3x+2)^{3/2} + C$

3. $u = x + 4$; $du = dx$; $\int (4+x)^{-1/2}\ dx = 2(4+x)^{1/2} + C$

5. $u = x^2 + 4x$; $du = (2x+4)\ dx$; $(1/2) \int (x^2+4x)^{3/4}(x+2)(2)\ dx = (2/7)(x^2+4x)^{7/4} + C$

7. $u = \cos x$; $du = -\sin x\ dx$; $- \int \cos^3 x\ \sin x(-dx) = (-1/4)\cos^4 x + C$

9. $u = \tan 4x$; $du = 4\sec^2 4x\ dx$; $(1/4) \int \tan^3 4x\ \sec^2 4x \cdot 4\ dx = (1/16)\tan^4 4x + c$

11. $u\cos 4x + 1$; $du = -4\sin 4x\ dx$; $(-1/4) \int \sin 4x\ (\cos 4x + 1)(-4)dx = (-1/8)(\cos 4x + 1)^2 + C$

13. $u = 9 + \sec x$; $du = \sec x \tan x\ dx$; $\int (9 + \sec x)^{1/2}\sec x \tan x\ dx = (2/3)(9 + \sec x)^{3/2} + C$

15. $u = 1 + e^{2x}$; $du = 2e^{2x}dx$; $(1/2)\int (1+e^{2x})^{1/2}2e^{2x}dx = (1/3)(1 + e^{2x})^{3/2} + C$

17. $u = 1 + e^{x^2}$; $du = 2xe^{x^2}\ dx$; $(1/2)\int (1+e^{x^2})^{-1/2}2xe^{x^2}\ dx = (1 + e^{x^2})^{1/2} + C$

19. $u = \ln(3x-5)$; $du = \dfrac{3\ dx}{3x-5}$; $\dfrac{1}{3}\int \dfrac{\ln(3x-5)}{3x-5}3\ dx = (1/6)\ln^2|3x-5| + C$

21. $u = \ln x$; $du = (1/x)\ dx$; $\int \dfrac{dx}{x\ln^2 x} = \dfrac{\ln^{-2}x\ dx}{x} = -\dfrac{1}{\ln x} + C$

23. $u = \text{Arcsin } 3x$; $du = \dfrac{3\ dx}{\sqrt{1-9x^2}}$; $\dfrac{1}{3}\int \dfrac{3\ \text{Arcsin } 3x}{\sqrt{1-9x^2}}dx = \dfrac{1}{6}\text{Arcsin}^2 3x + C$

25. $u = \sin x$; $du = \cos x\ dx$; $\int \csc^{-4}x \cot x\ dx = \int \sin^4 x \cdot \dfrac{\cos x}{\sin x}dx = \int \sin^3 x \cos x\ dx = (1/4)\sin^4 x + C$

27. $u = \text{Arctan } x$; $du = \dfrac{dx}{1+x^2}$; $\int \dfrac{\text{Arctan}^2 x\ dx}{1+x^2} = (1/3)\text{Arctan}^3 x + C$

29. $u = x^2 - 9$; $du = 2x\ dx$; $\dfrac{1}{2}\int_3^5 2x\sqrt{x^2-9}\ dx = \dfrac{1}{3}(x^2-9)^{3/2}\Big|_3^5 = \dfrac{64}{3}$

31. $u = 1 + e^{3x}; \; du = 3e^{3x} \, dx; \; \dfrac{1}{3}\int_0^1 (1 + e^{3x})^{-1/2} 3e^{3x} \, dx = \dfrac{2}{3}(1 + e^{3x})^{1/2}\Big|_0^1 = (2/3)(\sqrt{1 + e^3}) - \sqrt{2})$

33. $u = \ln(2x - 1); \; du = \dfrac{1}{2x-1} \cdot 2 \, dx; \; \int_1^2 \dfrac{\ln(2x-1)}{2x-1} \, dx = \dfrac{1}{2}\int_1^2 \dfrac{2\ln(2x-1)\,dx}{2x-1} = (1/4)\ln^2(2x-1)\Big|_1^2 = \dfrac{\ln^2 3}{4}$

35. $A = \displaystyle\int_0^{\pi/2} \sin^2 x \cos x \, dx = (1/3)\sin^3 x \; \Big|_0^{\pi/2} = 1/3$

Section 7.2

1. $u = 3x + 2; \; du = 3 \, dx; \; \dfrac{1}{3}\int \dfrac{3\,dx}{3x + 2} = (1/3)\ln|3x + 2| + C$

3. $u = 1 - 4x; \; du = -4 \, dx; \; -\dfrac{1}{4}\int \dfrac{-4\,dx}{1 - 4x} = (-1/4)\ln|1 - 4x| + C$

5. $u = 1 - x^2; \; du = -2x \, dx; \; -2\int \dfrac{2x\,dx}{1 - x^2} = -2\ln|1 - x^2| + C$

7. $u = x^4 - 1; \; du = 4x^3 \, dx; \; \dfrac{1}{4}\int \dfrac{4x^3\,dx}{x^4 - 1} = (1/4)\ln|x^4 - 1| + C$

9. $u = \cot x; \; du = -\csc^2 x \, du; \; -\int \dfrac{-\csc^2 x\,dx}{\cot x} = -\ln|\cot x| + C$

11. $u = 1 + \tan 3x; \; du = 3\sec^2 3x \, dx; \; \dfrac{1}{3}\int \dfrac{3\sec^2 3x\,dx}{1 + \tan 3x} = \dfrac{1}{3}\ln|1 + \tan 3x| + C$

13. $u = 1 + \csc; x \; du = -\csc x \cot x \, dx; \; -\int \dfrac{-\csc x \cot x\,dx}{1 + \csc x} = -\ln|1 + \csc x| + C$

15. $u = 1 + \sin x; \; du = \cos x \, dx; \; \int \dfrac{\cos x\,dx}{1 + \sin x} = \ln|1 + \sin x| + C$

17. $u = \ln x; \; du = (1/x) \, dx; \; \int \dfrac{dx}{x \ln x} = \ln|\ln x| + C$

19. $(1/2)\displaystyle\int e^{2x} \cdot 2 \, dx = (1/2)e^{2x} + C$

21. $(-1/4)\displaystyle\int e^{-4x}(-4) \, dx = -\dfrac{1}{4e^{4x}} + C$

23. $(1/2)\displaystyle\int 2xe^{x^2} \, dx = (1/2)e^{x^2} + C$

25. $(-1/2) \int e^{-(x^2+9)}(-2x)\,dx = \dfrac{-1}{2e^{x^2+9}} + C$

27. $-\int -\sin x\, e^{\cos x}\, dx = -e^{\cos x} + C$

29. $\dfrac{1}{2}\int_0^2 2xe^{x^2+2}\,dx = (1/2)e^{x^2+2}\ \Big|_0^2 = (e^2/2)(e^4 - 1)$

31. $\dfrac{1}{\ln 4}\int 4^x \ln 4\, dx = \dfrac{4^x}{\ln 4} + C$

33. $u = e^x + 4;\ du = e^x\, dx;\ \displaystyle\int \dfrac{2e^x\,dx}{e^x+4} = 2\ln\left|e^x + 4\right| + C$

35. $\dfrac{1}{2}\displaystyle\int \dfrac{2x\,dx}{x^2+1} = \dfrac{1}{2}\ln\left|x^2 + 1\right|\Big|_0^1 = (1/2)\ln 2$ or 0.347

37. $2\ln\left|2x - 1\right|\Big|_1^5 = 2\ln 9$ or $\ln 81$ or 4.39

39. $2e^{x/2}\ \Big|_0^2 = 2(e - 1)$ or 3.44

41. $\dfrac{1}{3}\displaystyle\int_0^1 3x^2 e^{x^3}\,dx = \dfrac{1}{3}e^{x^3}\ \Big|_0^1 = \dfrac{1}{3}(e - 1)$ or 0.573

43. $-\displaystyle\int_0^{\pi/6} \dfrac{-\cos x\,dx}{1-\sin x} = -\ln\left|1 - \sin x\right|\ \Big|_0^{\pi/6} = -\ln(1/2) = \ln 2$ or 0.693

45. $A = \displaystyle\int_0^1 \dfrac{dx}{1+2x} = (1/2)\ln\left|1 + 2x\right|\Big|_0^1 = (1/2)\ln 3$ or 0.549

47. $A = \displaystyle\int_0^4 e^{2x}\,dx = (1/2)e^{2x}\Big|_0^4 = (1/2)(e^8 - 1)$ or 1490

Section 7.3

1. $(1/5)\displaystyle\int \sin 5x \cdot 5\, dx = (-1/5)\cos 5x + C$

3. $(1/3)\displaystyle\int \cos(3x - 1)\cdot 3\, dx = (1/3)\sin(3x - 1) + C$

5. $(1/2)\displaystyle\int 2x\sin(x^2 + 5)\,dx = (-1/2)\cos(x^2 + 5) + C$

7. $\sin(x^3 - x^2) + C$

9. $(1/5)\displaystyle\int \csc^2 5x \cdot 5\, dx = (-1/5)\cot 5x + C$

Sections 7.2 – 7.3

Solutions to Odd-Numbered Exercises

11. $(1/3) \int \sec 3x \tan 3x \cdot 3 \, dx = (1/3) \sec 3x + C$

13. $(1/4) \int \sec^2 (4x + 3) \cdot 4 \, dx = (1/4) \tan (4x + 3) + C$

15. $(1/2) \int \csc(2x - 3) \cot(2x - 3) \cdot 2 \, dx = (-1/2) \csc(2x - 3) + C$

17. $(1/2) \int 2x \sec^2 (x^2 + 3) \, dx = (1/2) \tan(x^2 + 3) + C$

19. $(1/3) \int 3x^2 \csc(x^3 - 1) \cot(x^3 - 1) \, dx = (-1/3) \csc(x^3 - 1) + C$

21. $(1/4) \int 4 \tan 4x \, dx = (-1/4) \ln \left| \cos 4x \right| + C$

23. $(1/5) \int \sec 5x \cdot 5 \, dx = (1/5) \ln \left| \sec 5x + \tan 5x \right| + C$

25. $\ln \left| \sin e^x \right| + C$　　　　27. $\int (1 + 2 \sec x + \sec^2 x) \, dx = x + 2 \ln \left| \sec x + \tan x \right| + \tan x + C$

29. $\int \dfrac{5 + \sin x}{\cos x} dx = \int (5 \sec x + \tan x) \, dx$

$\qquad = 5 \ln \left| \sec x + \tan x \right| + \ln \left| \sec x \right| + C$ or $5 \ln \left| \sec x + \tan x \right| - \ln \left| \cos x \right| + C$

31. $\dfrac{1}{2} \displaystyle\int_0^{\pi/4} \sin 2x \cdot 2 \, dx = (-1/2) \cos 2x \Big|_0^{\pi/4} = 1/2$　　　　33. $3 \sin(x - \pi/2) \Big|_0^{\pi/2} = 3$

35. $\tan x \Big|_0^{\pi/4} = 1$　　　　37. $(1/2) \sec 2x \Big|_0^{\pi/8} = \dfrac{\sqrt{2}}{2} - \dfrac{1}{2} = \dfrac{\sqrt{2} - 1}{2}$

39. $A = \displaystyle\int_0^{\pi} \sin x \, dx = - \cos x \Big|_0^{\pi} = 2$　　　　41. $A = \displaystyle\int_0^{\pi/4} \sec^2 x \, dx = \tan x \Big|_0^{\pi/4} = 1$

43. $A = \displaystyle\int_0^{\pi/4} \tan x \, dx = - \ln \left| \cos x \right| \Big|_0^{\pi/4} = \dfrac{1}{2} \ln 2$

45. $V = \displaystyle\int_0^{\pi/4} \pi \sec^2 x \, dx = \pi \tan x \Big|_0^{\pi/4} = \pi$

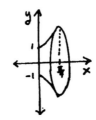

Section 7.3

Section 7.4

1. $\int \sin^3 x \, dx = \int \sin^2 x \sin x \, dx = \int (1 - \cos^2 x) \sin x \, dx = -\cos x + (1/3)\cos^3 x + C$

3. $\int \cos^5 x \, dx = \int \cos^4 x \cos x \, dx = \int (1 - \sin^2 x)^2 \cos x \, dx = \int (1 - 2\sin^2 x + \sin^4 x) \cos x \, dx$
$$= \sin x - (2/3) \sin^3 x + (1/5)\sin^5 x + C$$

5. $\int \sin^2 x \cos \, dx = (1/3) \sin^3 x + C$

7. $\int \cos^{-3} x \sin x \, dx = \dfrac{-1}{2\cos^2 x} + C$

9. $\int \sin^3 x \cos^2 x \, dx = \int \sin x \, (1 - \cos^2 x) \cos^2 x \, dx = \int \sin(\cos^2 x - \cos^4 x) \, dx$
$= (-1/3) \cos^3 x + (1/5)\cos^5 x + C$

11. $\int \sin^2 x \, dx \int (1/2) (1 - \cos 2x) \, dx = x/2 - (1/4) \sin 2x + C$

13. $\int \cos^4 3x \, dx = \int \left\{\dfrac{1 + \cos 6x}{2}\right\}^2 dx = (1/4) \int (1 + 2\cos 6x + \cos^2 6x) \, dx$
$= x/4 + (1/12) \sin 6x + (1/8) \int (1 + \cos 12x) \, dx$
$= x/4 + (1/12) \sin 6x + x/8 + (1/96) \sin 12x + C$
$= 3x/8 + (1/12) \sin 6x + (1/96) \sin 12x + C$

15. $\int \sin^2 x \cos^2 x \, dx = (1/4) \int \sin^2 2x \, dx = (1/8) \int (1 - \cos 4x) \, dx$
$$= x/8 - (1/32) \sin 4x + C$$

17. $\int \sin^2 x \cos^4 x \, dx = \int (\sin^2 x \cos^2 x) \cos^2 x \, dx$
$= (1/8) \int \sin^2 2x(1 + \cos 2x) \, dx$
$= (1/8) \int \sin^2 2x \, dx + (1/8) \int \sin^2 2x \cos 2x \, dx$
$= (1/16) \int (1 - \cos 4x) \, dx + (1/8) \int \sin^2 2x \cos 2x \, dx$
$= x/16 - (1/64) \sin 4x + (1/48) \sin^3 2x + C$

19. $\int \tan^3 x \, dx = \int (\sec^2 x - 1) \tan x \, dx = \int \sec^2 x \tan x \, dx - \int \tan x \, dx = (1/2) \tan^2 x + \ln |\cos x| + C$

21. $\int \cot^4 2x \, dx = \int \cot^2 2x(\csc^2 2x - 1) \, dx = \int \cot^2 2x \csc^2 2x \, dx - \int \cot^2 2x \, dx$
$= \int \cot^2 2x \csc^2 2x \, dx - \int \csc^2 2x \, dx + \int dx = (-1/6)\cot^3 2x + (1/2)\cot 2x + x + C$

Section 7.4

23. $\int \sec^6 x \, dx = \int (1 + \tan^2 x)^2 \sec^2 x \, dx = \int \sec^2 x \, dx + \int 2 \tan^2 x \sec^2 x \, dx + \int \tan^4 x \sec^2 x \, dx$

$\qquad = \tan x + (2/3) \tan^3 x + (1/5) \tan^5 x + C$

25. $\int \tan^4 2x \, dx = \int (\sec^2 2x - 1) \tan^2 2x \, dx = (1/2) \int \sec^2 2x \tan^2 2x \, dx - \int \tan^2 2x \, dx$

$\qquad = (1/6) \tan^3 2x - (1/2) \tan 2x + x + C$

27. $A = \int\limits_0^\pi \sin^2 x \, dx = \dfrac{1}{2} \int\limits_0^\pi (1 - \cos 2x) \, dx = (1/2)[x - (1/2) \sin 2x] \Big|_0^\pi = \dfrac{\pi}{2}$

Section 7.5

1. $u = 3x;$

$du = 3 \, dx; \quad \displaystyle\int \frac{dx}{\sqrt{1 - 9x^2}} = \frac{1}{3} \int \frac{3 \, dx}{\sqrt{1 - 9x^2}} = \frac{1}{3} \text{ Arcsin } 3x + C$

$a = 1$

3. $u = x;$

$du = dx; \quad \displaystyle\int \frac{dx}{\sqrt{9 - x^2}} = \text{Arcsin } \frac{x}{3} + C$

$a = 3$

5. $u = x;$

$du = dx; \quad \displaystyle\int \frac{dx}{\sqrt{x^2 + 25}} = \frac{1}{5} \text{ Arctan } \frac{x}{5} + C$

$a = 5$

7. $u = 3x;$

$du = 3 \, dx; \quad \displaystyle\int \frac{dx}{9x^2 + 4} = \frac{1}{3} \int \frac{3 \, dx}{9x^2 + 4} = \frac{1}{6} \text{ Arctan } \frac{3x}{2} + C$

$a = 2$

9. $u = 5x;$

$du = 5 \, dx; \quad \displaystyle\int \frac{dx}{\sqrt{36 - 25x^2}} = \frac{1}{5} \int \frac{5 \, dx}{\sqrt{36 - 25x^2}} = \frac{1}{5} \text{ Arcsin } \frac{5x}{6} + C$

$a = 6$

11. $u = \sqrt{12}x;$

$du = \sqrt{12} dx; \quad \displaystyle\int \frac{dx}{\sqrt{3 - 12x^2}} = \frac{1}{\sqrt{12}} \int \frac{\sqrt{12} \, dx}{\sqrt{3 - 12x^2}} = \frac{1}{\sqrt{12}} \text{ Arcsin} \frac{\sqrt{12}x}{\sqrt{3}} + C = \frac{1}{2\sqrt{3}} \text{ Arcsin } 2x + C$

$a = \sqrt{3}$

13. $u = x - 1$;

$du = dx$; $\displaystyle\int \frac{dx}{4 + (x-1)^2} = \frac{1}{2} \text{Arctan} \frac{x-1}{2} + C$

$a = 2$

15. $u = x + 3$

$du = dx$; $\displaystyle\int \frac{dx}{(x^2 + 6x + 9) + 16} = \int \frac{dx}{(x+3)^2 + 4^2} = \frac{1}{4} \text{Arctan} \frac{x+3}{4} + C$

$a = 4$

17. $u = e^x$;

$du = e^x \, dx$; $\displaystyle\int \frac{e^x \, dx}{\sqrt{1 - (e^x)^2}} \ \ \text{Arcsin } e^x = C$

$a = 1$

19. $u = \cos x$;

$du = -\sin x \, dx$; $\displaystyle\int \frac{\sin x \, dx}{1 + \cos^2 x} = -\int \frac{-\sin x \, dx}{1 + \cos^2 x} = -\text{Arctan } (\cos x) + C$

$a = 1$

21. $u = x$;

$du = dx$; $\displaystyle\int_0^1 \frac{dx}{1 + x^2} = \text{Arctan } x \ \Big|_0^1 = \pi/4$

$a = 1$

23. $u = 3x$;

$du = 3 \, dx$; $\displaystyle\int_0^1 \frac{dx}{\sqrt{25 - 9x^2}} = \frac{1}{3} \int \frac{3 \, dx}{\sqrt{25 - 9x^2}} = \frac{1}{3} \text{Arcsin} \frac{3x}{5} \Big|_0^1 = \frac{1}{3} \text{Arcsin} \frac{3}{5} = 0.215$

$a = 5$

25. $u = 2x$;

$du = 2 \, dx$; $W = \displaystyle\int_1^2 \frac{100 \, dx}{1 + 4x^2} = 50 \int_1^2 \frac{2 \, dx}{1 + 4x^2} = 50 \, \text{Arctan } 2x \ \Big|_1^2 = 10.9 \text{ N}$

$a = 1$

Section 7.5

1. $\dfrac{8x-29}{(x+2)(x-7)} = \dfrac{A}{x+2} + \dfrac{B}{x-7}$

    Multiply each side by

    L.C.D: $(x+2)(x-7)$.

    $8x-29 = A(x-7) + B(x+2)$

    $8x-29 = (A+B)x - 7A + 2B$

    Thus: $A+B = 8$

    $\qquad -7A + 2B = -29$

    The solution of this system

    of equations is: $A = 5$, $B = 3$.

    $\dfrac{5}{x+2} + \dfrac{3}{x-7}$

3. $\dfrac{-x-18}{2x^2 - 5x - 12} =$

    $\dfrac{-x-18}{(2x+3)(x-4)} = \dfrac{A}{2x+3} + \dfrac{B}{x-4}$

    Multiply each side by

    L.C.D.: $(2x+3)(x-4)$.

    $-x-18 = A(x-4) + B(2x+3)$

    $-x-18 = (A+2B)x - 4A + 3B$

    Thus: $A + 2B = -1$

    $\qquad -4A + 3B = -18$

    The solution of this

    system is: $A = 3$, $B = -2$

    $\dfrac{3}{2x+3} + \dfrac{-2}{x-4}$

5. $\dfrac{61x^2 - 53x - 28}{x(3x-4)(2x+1)} = \dfrac{A}{x} + \dfrac{B}{3x-4} + \dfrac{C}{2x+1}$

    Multiply each side by the L.C.D.: $x(3x-4)(2x+1)$.

    $61x^2 - 53x - 28 = A(3x-4)(2x+1) + Bx(2x+1) + Cx(3x-4)$

    $61x^2 - 53x - 28 = A(6x^2 - 5x - 4) + 2Bx^2 + Bx + 3Cx^2 - 4Cx$

    $61x^2 - 53x - 28 = 6Ax^2 - 5Ax - 4A + 2Bx^2 + Bx + 3Cx^2 - 4Cx$

    Thus: $6A + 2B + 3C = 61$ The solution to this

    $\qquad -5A + B - 4C = -53$ system is:

    $\qquad -4A \qquad\qquad = -28$ $A = 7$, $B = 2$, $C = 5$

    $\dfrac{7}{x} + \dfrac{2}{3x-4} + \dfrac{5}{2x+1}$

7.

$\dfrac{x^2 + 7x - 10}{(x+1)(x+3)^2} = \dfrac{A}{x+1} + \dfrac{B}{x+3} + \dfrac{C}{(x+3)^2}$

Multiply each side by the L.C.D.: $(x+1)(x+3)^2$.

$x^2 + 7x + 10 = A(x+3)^2 + B(x+1)(x+3) + C(x+1)$

$x^2 + 7x + 10 = A(x^2 + 6x + 9) + B(x^2 + 4x + 3) + Cx + C$

$x^2 + 7x - 10 = Ax^2 + 6Ax + 9A + Bx^2 + 4Bx + 3B + Cx + C$

Thus: $A + B = 1$ The solution to this

$\qquad 6A + 4B + C = 7$ system is:

$\qquad 9A + 3B + C = 10$ $A = 1$, $B = 0$, $C = 1$

$\dfrac{1}{x+1} + \dfrac{1}{(x+3)^2}$ Section 7.6

9. $$\frac{48x^2 - 20x - 5}{(4x-1)^3} = \frac{A}{4x-1} + \frac{B}{(4x-1)^2} + \frac{C}{(4x-1)^3}$$

Multiply each side by the L.C.D.: $(4x-1)^3$.

$$48x^2 - 20x - 5 = A(4x-1)^2 + B(4x-1) + C$$

$$48x^2 - 20x - 5 = A(16x^2 - 8x + 1) + 4Bx - B + C$$

$$48x^2 - 20x - 5 = 16Ax^2 - 8Ax + A + 4Bx - B + C$$

Thus:
$$16A = 48$$
$$-8A + 4B = -20$$
$$A - B + C = -5$$

The solution to this system is:

$$A = 3, \ B = 1, \ C = -7$$

$$\frac{3}{4x-1} + \frac{1}{(4x-1)^2} + \frac{-7}{(4x-1)^3}$$

11. $$\frac{11x^2 - 18x + 3}{x(x-1)^2} = \frac{A}{x} + \frac{B}{x-1} + \frac{C}{(x-1)^2}$$

Multiply each side by the L.C.D.: $x(x-1)^2$.

$$11x^2 - 18x + 3 = A(x-1)^2 + Bx(x-1) + Cx$$

$$11x^2 - 18x + 3 = A(x^2 - 2x + 1) + Bx^2 - Bx + Cx$$

$$11x^2 - 18x + 3 = Ax^2 - 2Ax + A + Bx^2 - Bx + Cx$$

Thus:
$$A + B = 11$$
$$-2A - B + C = -18$$
$$A = 3$$

The solution to this system is:

$$A = 3, \ B = 8, \ C = -4$$

$$\frac{3}{x} + \frac{8}{x-1} + \frac{-4}{(x-1)^2}$$

13. $$\frac{-x^2 - 4x + 3}{(x^2+1)(x^2-3)} = \frac{Ax+B}{x^2+1} + \frac{Cx+D}{x^2-3}$$

Multiply each side by the L.C.D.: $(x^2+1)(x^2-3)$.

$$-x^2 - 4x + 3 = (Ax+B)(x^2-3) + (Cx+D)(x^2+1)$$

$$-x^2 - 4x + 3 = Ax^3 + Bx^2 - 3Ax - 3B + Cx^3 + Dx^2 + Cx + D$$

Thus:
$$A + C = 0$$
$$B + D = -1$$
$$-3A + C = -4$$
$$-3B + D = 3$$

The solution to this system is:

$$A = 1, \ B = -1, \ C = -1, \ D = 0$$

$$\frac{x-1}{x^2+1} + \frac{-x}{x^2-3}$$

Section 7.6

15.

$$\frac{4x^3 - 21x - 6}{\left(x^2 + x + 1\right)\left(x^2 - 5\right)} = \frac{Ax + B}{x^2 + x + 1} + \frac{Cx + D}{x^2 - 5}$$

Multiply each side by the L.C.D.: $\left(x^2 + x + 1\right)\left(x^2 - 5\right)$.

$$4x^3 - 21x - 6 = \left(Ax + B\right)\left(x^2 - 5\right) + \left(Cx + D\right)\left(x^2 + x + 1\right)$$

$$4x^3 - 21x - 6 = Ax^3 + Bx^2 - 5Ax - 5B + Cx^3 + Cx^2 + Cx + Dx^2 + Dx + D$$

Thus:

$$A \qquad + C \qquad = 4$$
$$B + C + D = 0$$
$$-5A \qquad + C + D = -21$$
$$-5B \qquad + D = -6$$

The solution to this system is:

$$A = 4,\ B = 1,\ C = 0,\ D = -1$$

$$\frac{4x + 1}{x^2 + x + 1} + \frac{-1}{x^2 - 5}$$

17.

$$\frac{4x^3 - 16x^2 - 93x - 9}{\left(x^2 + 5x + 3\right)(x + 3)(x - 3)} = \frac{A}{x + 3} + \frac{B}{x - 3} + \frac{Cx + D}{x^2 + 5x + 3}$$

Multiply each side by the L.C.D.: $(x + 3)(x - 3)\left(x^2 + 5x + 3\right)$

$$4x^3 - 16x^2 - 93x - 9 = A(x - 3)\left(x^2 + 5x + 3\right)$$
$$+ B(x + 3)\left(x^2 + 5x + 3\right) + \left(Cx + D\right)(x + 3)(x - 3)$$

$$4x^3 - 16x^2 - 93x - 9 = A\left(x^3 + 5x^2 + 3x - 3x^2 - 15x - 9\right)$$
$$+ B\left(x^3 + 5x^2 + 3x + 3x^2 + 15x + 9\right) + \left(Cx + D\right)\left(x^2 - 9\right)$$

$$4x^3 - 16x^2 - 93x - 9 = Ax^3 + 2Ax^2 - 12Ax - 9A +$$
$$Bx^3 + 8Bx^2 + 18Bx + 9B + Cx^3 + Dx^2 - 9Cx - 9D$$

Thus:

$$A + B + C = 4$$
$$2A + 8B \qquad + D = -16$$
$$-12A + 18B - 9C \qquad = -93$$
$$-9A + 9B \qquad\quad -9D = -9$$

The solution to this system is:

$$A = 1,\ B = -2,\ C = 5,\ D = -2$$

$$\frac{1}{x + 3} + \frac{-2}{x - 3} + \frac{5x - 2}{x^2 + 5x + 3}$$

19.

$$\frac{8x^4 - x^3 + 13x^2 - 6x + 5}{x\left(x^2 + 1\right)^2} = \frac{A}{x} + \frac{Bx + C}{x^2 + 1} + \frac{Dx + E}{\left(x^2 + 1\right)^2}$$

Multiply each side by the L.C.D.: $x\left(x^2 + 1\right)^2$.

$$8x^4 - x^3 + 13x^2 - 6x + 5 = A\left(x^2 + 1\right)^2 + \left(Bx + C\right)(x)\left(x^2 + 1\right) + \left(Dx + E\right)x$$

$$8x^4 - x^3 + 13x^2 - 6x + 5 = A\left(x^4 + 2x^2 + 1\right) + Bx^4 + Cx^3 + Bx^2 + Cx + Dx^2 + Ex$$

$$8x^4 - x^3 + 13x^2 - 6x + 5 = Ax^4 + 2Ax^2 + A + Bx^4 + Cx^3 + Bx^2 + Cx + Dx^2 + Ex$$

Thus:

$$A + B \qquad = 8$$
$$C \qquad = -1$$
$$2A + B \qquad + D = 13$$
$$C \qquad + E = -6$$
$$A \qquad = 5$$

The solution to this system is:

$$A = 5,\ B = 3,\ C = -1,\ D = 0,\ E = -5$$

$$\frac{5}{x} + \frac{3x - 1}{x^2 + 1} + \frac{-5}{\left(x^2 + 1\right)^2}$$

Section 7.6

21. $$\frac{x^5 - 2x^4 - 8x^2 + 4x - 8}{x^2\left(x^2 + 2\right)^2} = \frac{A}{x} + \frac{B}{x^2} + \frac{Cx + D}{x^2 + 2} + \frac{Ex + F}{\left(x^2 + 2\right)^2}$$

Multiply each side by the L.C.D.: $x^2\left(x^2 + 2\right)^2$.

$$x^5 - 2x^4 - 8x^2 + 4x - 8 = Ax\left(x^2 + 2\right)^2 + B\left(x^2 + 2\right)^2 + (Cx + D)x^2\left(x^2 + 2\right) + (Ex + F)x^2$$

$$x^5 - 2x^4 - 8x^2 + 4x - 8 = Ax\left(x^4 + 4x^2 + 4\right) + B\left(x^4 + 4x^2 + 4\right) + Cx^5 + Dx^4 + 2Cx^3 + 2Dx^2 + Ex^3 + Fx^2$$

$$x^5 - 2x^4 - 8x^2 + 4x - 8 = Ax^5 + 4Ax^3 + 4Ax + Bx^4 + 4Bx^2 + 4B + Cx^5 + Dx^4 + 2Cx^3 + 2Dx^2 + Ex^3 + Fx^2$$

Thus:

$$\begin{aligned}
A \quad\ + C \quad\quad\quad\quad &= 1 \\
B \quad\ + D \quad\quad &= -2 \\
4A \ + 2C \ + E \quad &= 0 \\
4B \quad\ + 2D \ + F &= -8 \\
4A \quad\quad\quad\quad &= 4 \\
4B \quad\quad\quad\quad &= -8
\end{aligned}$$

The solution to this system is:

$A = 1, \ B = -2, \ C = 0, \ D = 0, \ E = -4, \ F = 0$

$$\frac{1}{x} + \frac{-2}{x^2} + \frac{-4x}{\left(x^2 + 2\right)^2}$$

23. $$\frac{6x^2 + 108x + 54}{x^4 - 81} = \frac{6x^2 + 108x + 54}{\left(x^2 + 9\right)\left(x^2 - 9\right)} = \frac{6x^2 + 108x + 54}{\left(x^2 + 9\right)(x + 3)(x - 3)}$$

$$\frac{6x^2 + 108x + 54}{\left(x^2 + 9\right)(x + 3)(x - 3)} = \frac{A}{x + 3} + \frac{B}{x - 3} + \frac{Cx + D}{x^2 + 9}$$

Multiply each side by the L.C.D.: $(x + 3)(x - 3)\left(x^2 + 9\right)$.

$$6x^2 + 108x + 54 = A(x - 3)\left(x^2 + 9\right) + B(x + 3)\left(x^2 + 9\right) + (Cx + D)\left(x^2 - 9\right)$$

$$6x^2 + 108x + 54 = A\left(x^3 - 3x^2 + 9x - 27\right) + B\left(x^3 + 3x^2 + 9x + 27\right) + Cx^3 + Dx^2 - 9Cx - 9D$$

$$6x^2 + 108x + 54 = Ax^3 - 3Ax^2 + 9Ax - 27A + Bx^3 + 3Bx^2 + 9Bx + 27B + Cx^3 + Dx^2 - 9Cx - 9D$$

Thus:

$$\begin{aligned}
A + B + C &= 0 \\
-3A + 3B \quad\quad\ + D &= 6 \\
9A + 9B \ - 9C \quad &= 108 \\
-27A + 27B \quad - 9D &= 54
\end{aligned}$$

The solution to this system is:

$A = 2, \ B = 4, \ C = -6, \ D = 0$

$$\frac{2}{x + 3} + \frac{4}{x - 3} + \frac{-6x}{x^2 + 9}$$

Section 7.6

25. $\dfrac{x^3}{x^2-1}$ Since the degree of the numerator is greater than the degree of the denominator, divide first.

$$x^2-1\overline{\smash)x^3} \qquad \text{or} \quad x+\frac{x}{x^2-1}=x+\frac{x}{(x-1)(x+1)}$$
$$\underline{x^3-x}$$
$$x$$

$$\frac{x}{(x-1)(x+1)}=\frac{A}{x-1}+\frac{B}{x+1}$$

Multiply each side by the L.C.D.: $(x-1)(x+1)$.

$$x=A(x+1)+B(x-1)=Ax+A+Bx-B$$

Thus: $A+B=1$ The solution to this system is:

$$A-B=0 \qquad A=\frac{1}{2},\ B=\frac{1}{2}$$

$$x+\frac{\frac{1}{2}}{x-1}+\frac{\frac{1}{2}}{x+1}$$

27. $\dfrac{x^3-x^2+8}{x^2-4}$ Since the degree of the numerator is greater than the degree of the denominator, divide first.

$$x^2-4\overline{\smash)x^3-x^2+0x+8} \quad \text{or} \quad x-1+\frac{4x+4}{(x+2)(x-2)}$$
$$\underline{x^3\qquad -4x}$$
$$-x^2+4x+8$$
$$\underline{-x^2\qquad +4}$$
$$4x+4$$

$$\frac{4x+4}{(x+2)(x-2)}=\frac{A}{x+2}+\frac{B}{x-2}$$

Multiply each side by the L.C.D.: $(x+2)(x-2)$.

$$4x+4=A(x-2)+B(x+2)=Ax-2A+Bx+2B$$

Thus: $A+B=4$ The solution to this system is:

$$-2A+2B=4 \qquad A=1,\ B=3$$

$$x-1+\frac{1}{x+2}+\frac{3}{x-2}$$

Section 7.6

29. $\dfrac{3x^4 - 2x^3 - 2x + 5}{x\left(x^2 + 1\right)}$ Since the degree of the numerator is greater than the degree of the denominator, divide first.

$$\begin{array}{r} 3x - 2 \\ x^3 + x \overline{\smash{\big)}\,3x^4 - 2x^3 + 0x^2 - 2x + 5} \\ \underline{3x^4 \qquad\quad + 3x^2} \\ -2x^3 - 3x^2 - 2x + 5 \\ \underline{-2x^3 \qquad\quad - 2x} \\ -3x^2 + 5 \end{array} \quad \text{or} \quad 3x - 2 + \dfrac{-3x^2 + 5}{x^3 + x}$$

$$\dfrac{-3x^2 + 5}{x\left(x^2 + 1\right)} = \dfrac{A}{x} + \dfrac{Bx + C}{x^2 + 1}$$

Multiply each side by the L.C.D.: $x\left(x^2 + 1\right)$.

$$-3x^2 + 5 = A\left(x^2 + 1\right) + \left(Bx + C\right)x$$

$$-3x^2 + 5 = Ax^2 + A + Bx^2 + Cx$$

Thus: $A + B = -3$ The solution to this system is:

$\qquad\quad C = 0$ $A = 5,\ B = -8,\ C = 0$

$\qquad\quad A = 5$

$$3x - 2 + \dfrac{5}{x} + \dfrac{-8x}{x^2 + 1}$$

Section 7.7

1. $\dfrac{1}{1 - x^2} = \dfrac{A}{1 - x} + \dfrac{B}{1 + x}$; $1 = (A - B)x + (A + B)$; $A - B = 0$; $A + B = 1$; $A = 1/2,\ B = 1/2$

$$\int \dfrac{dx}{1 - x^2} = \dfrac{1}{2}\int \dfrac{dx}{1 - x} + \dfrac{1}{2}\int \dfrac{dx}{1 + x} = (-1/2)\ln |x - 1| + (1/2)\ln |x + 1| + C$$

$$= \dfrac{1}{2}\ln \left|\dfrac{x + 1}{x - 1}\right| + C$$

3. $\dfrac{1}{(x + 4)(x - 2)} = \dfrac{A}{x + 4} + \dfrac{B}{x - 2}$; $1 = (A + B)x + (-2A + 4B)$; $A + B = 0$; $-2A + 4B = 1$;

$$A = -1/6,\ B = 1/6$$

$$\int \dfrac{dx}{(x + 4)(x - 2)} = -\dfrac{1}{6}\int \dfrac{dx}{x + 4} + \dfrac{1}{6}\int \dfrac{dx}{x - 2} = -\dfrac{1}{6}\ln |x + 4| + \dfrac{1}{6}\ln |x - 2| + C = \dfrac{1}{6}\ln \left|\dfrac{x - 2}{x + 4}\right| + C$$

5. $\dfrac{1}{(x - 2)(x - 1)} = \dfrac{A}{x - 2} + \dfrac{B}{x - 1}$; $x = A(x - 1) + B(x - 2)$; $x = (A + B)x + (-A - 2B)$;

$$A + B = 1;\ -A - 2B = 0;\ A = 2,\ B = -1$$

$$\int \dfrac{x\,dx}{(x - 2)(x - 1)} = \int \dfrac{2\,dx}{x - 2} - \int \dfrac{dx}{x - 1} = 2\ln |x - 2| - \ln |x - 1| + C = \ln \left|\dfrac{(x - 2)^2}{x - 1}\right| + C$$

Sections 7.6 – 7.7

7. $\dfrac{x+1}{(x+5)(x-1)} = \dfrac{A}{x+5} + \dfrac{B}{x-1}$; $x+1 = A(x-1) + B(x+5)$; $x+1 = (A+B)x + (-A+5B)$;

$$A+B = 1; -A + 5B = 1; A = 2/3, B = 1/3$$

$$\int \dfrac{(x+1)dx}{x^2+4x-5} = \dfrac{2}{3}\int\dfrac{dx}{x+5} + \dfrac{1}{3}\int\dfrac{dx}{x-1} = \dfrac{2}{3}\ln|x+5| + \dfrac{1}{3}\ln|x-1| + C$$

9. $\dfrac{1}{x(x+1)^2} = \dfrac{A}{x} = \dfrac{B}{B+1} + \dfrac{C}{(x+1)^2}$; $1 = A(x+1)^2 + Bx(x+1) + Cx$

$1 = Ax^2 + 2Ax + A + Bx^2 + Bx + Cx$; $1 = (A+B)x^2 + (2A+B+C)x + A$;

$A+B = 0$; $2A+B+C = 0$; $A = 1$; Thus $A = 1, B = -1, C = -1$

$$\int\dfrac{dx}{x(x+1)^2} = \int\dfrac{dx}{x} - \int\dfrac{dx}{x+1} - \int\dfrac{dx}{(x+1)^2}$$

$$= \ln|x| - \ln|x+1| + \dfrac{1}{x+1} + C = \ln\left|\dfrac{x}{x+1}\right| + \dfrac{1}{x+1} + C$$

11. $\dfrac{2x^2+x+3}{x^2(x+3)} = \dfrac{A}{x} + \dfrac{B}{x^2} + \dfrac{C}{x+3}$; $2x^2+x+3 = Ax(x+3) + B(x+3) + Cx^2$

$2x^2+x+3 = Ax^2 + 3Ax + Bx + 3B + Cx^2$;

$2x^2+x+3 = (A+C)x^2 + (3A+B)x + 3B$; $A+C = 2$;

$3A+B = 1$; $3B = 3$; Thus, $A = 0, B = 1, C = 2$

$$\int\dfrac{2x^2+x+3}{x^2(x+3)}dx = \int\dfrac{dx}{x^2} + \int\dfrac{2\,dx}{x+3} = -\dfrac{1}{x} + 2\ln|x+3| + C$$

13. $\dfrac{x^3}{x^2+3x+2} = x-3 + \dfrac{7x+6}{x^2+3x+2}$; $\dfrac{7x+6}{x^2+3x+6} = \dfrac{A}{x+1} + \dfrac{B}{x+2}$;

$7x+6 = A(x+2) + B(x+1)$; $7x+6 = (A+B)x + (2A+B)$;

$A+B = 7, 2A+B = 6, A = -1, B = 8$

$$\int\dfrac{x^3dx}{x^2+3x+2} = \int\left(x-3-\dfrac{1}{x+1}+\dfrac{8}{x+2}\right)dx = \dfrac{x^2}{2} - 3x - \ln|x+1| + 8\ln|x+2| + C$$

$$= \dfrac{x^2}{2} - 3x + \ln\left|\dfrac{(x+2)^8}{x+1}\right| + C$$

15. $\dfrac{x^2-2}{(x^2+1)x} = \dfrac{Ax+B}{x^2+1} + \dfrac{C}{x}$; $x^2-2 = (Ax+B)x + C(x^2+1)$;

$x^2-2 = (A+C)x^2 + Bx + C$; $A+C = 1$; $B = 0$; $C = -2$; Thus $A = 3$

$$\int\dfrac{(x^2-2)dx}{(x^2+1)x} = \int\left\{\dfrac{3x}{x^2+1} - \dfrac{2}{x}\right\}dx = \dfrac{3}{2}\ln|x^2+1| - 2\ln|x| + C$$

Section 7.7

17. $\dfrac{x^3 + 2x^2 - 9}{x^2(x^2 + 9)} = \dfrac{A}{x} + \dfrac{B}{x^2} + \dfrac{Cx + D}{x^2 + 9}$; $x^3 + 2x^2 - 9 = Ax(x^2 + 9) + B(x^2 + 9) + (Cx + D)x^2$;

$x^3 + 2x^2 - 9 = (A + C)x^3 + (B + D)x^2 + 9Ax + 9B$; $A + C = 1$; $B + D = 2$; $9A = 0$; $9B = -9$;

Thus $A = 0$, $B = -1$, $C = 1$, $D = 3$

$$\int \dfrac{x^3 + 2x^2 - 9}{x^2(x^2 + 9)}\, dx = \int \left\{ \dfrac{-1}{x^2} + \dfrac{x + 3}{x^2 + 9} \right\} dx$$

$$= -\int \dfrac{dx}{x^2} + \int \dfrac{x\,dx}{x^2 + 9} \int \dfrac{3\,dx}{x^2 + 9} = \dfrac{1}{x} + \dfrac{1}{2}\, \ln\, |x^2 + 9| + \text{Arctan}\, \dfrac{x}{3} + C$$

19. $\dfrac{x^3}{(x^2 + 1)^2} = \dfrac{Ax + B}{x^2 + 1} + \dfrac{Cx + D}{(x^2 + 1)^2}$; $x^3 = Ax^3 + Bx^2 + (A + C)x + (B + D)$

$A = 1$; $B = 0$; $A + C = 0$; $B + D = 0$; $C = -1, D = 0$

$$\int \dfrac{x^3 dx}{(x^2 + 1)^2} = \int \left\{ \dfrac{x}{x^2 + 1} - \dfrac{x}{(x^2 + 1)^2} \right\} dx = \dfrac{1}{2} \ln|x^2 + 1| + \dfrac{1}{2(x^2 + 1)} + C$$

21. $\dfrac{3}{(1 - x)(1 + x)} = \dfrac{A}{1 - x} + \dfrac{B}{1 + x}$; $3 = A(1 + x) + B(1 - x)$;

$3 = (A - B)x + (A + B)$; $A - B = 0$; $A + B = 3$; $A = 3/2, B = 3/2$

$$\int_2^3 \dfrac{3\,dx}{1 - x^2} = \int_2^3 \left\{ \dfrac{3/2}{1 - x} + \dfrac{3/2}{1 + x} \right\} dx = -\dfrac{3}{2} \ln|1 - x| + \dfrac{3}{2} \ln\, |1 + x| \Big|_2^3$$

$$= -\dfrac{3}{2} \ln 2 + \dfrac{3}{2} \ln 4 - \left(-\dfrac{3}{2} \ln 1 + \dfrac{3}{2} \ln 3 \right) = -\dfrac{3}{2} \ln \dfrac{3}{2} \ \text{or} \ \dfrac{3}{2} \ln \dfrac{2}{3}$$

23. $\dfrac{x}{x^2 + 4x - 5} = \dfrac{A}{x + 5} + \dfrac{B}{x - 1}$; $x = A(x - 1) + B(x + 5)$; $A + B = 1$; $-A + 5B = 0$

Thus $A = 5/6$, $B = 1/6$

$$\int_2^4 \dfrac{x\,dx}{x^2 + 4x - 5} = \int_2^4 \left\{ \dfrac{5/6}{x + 5} + \dfrac{1/6}{x - 1} \right\} dx = \dfrac{5}{6}\, \ln\, |x + 5| + \dfrac{1}{6} \ln|x - 1| \Big|_2^4$$

$$= \dfrac{5}{6} \ln 9 + \dfrac{1}{6} \ln 3 - \left(\dfrac{5}{6} \ln 7 + \dfrac{1}{6} \ln 1 \right) = \dfrac{11}{6} \ln 3 - \dfrac{5}{6}\, \ln 7 \ \text{or} \ 0.393$$

25. $\dfrac{4x}{(x + 3)(x - 1)} = \dfrac{A}{x + 3} + \dfrac{B}{x - 1}$; $4x = A(x - 1) + B(x + 3)$; $A + B = 4$; $-A + 3B = 0$; Thus $A = 3, B = 1$

$$\text{Area} = \int_2^4 \dfrac{4x\,dx}{x^2 + 2x - 3} = \int_2^4 \left\{ \dfrac{3}{x + 3} + \dfrac{1}{x - 1} \right\} dx = 3\ln|x + 3| + \ln|x - 1| \Big|_2^4$$

$$= 3 \ln 7 + \ln 3 - (3 \ln 5 + \ln 1) = 3 \ln \dfrac{7}{5} + \ln 3 = 2.108$$

Section 7.8

1. $u = \ln x$; $dv = dx$

$$\int \ln x\, dx = x \ln x - \int x(1/x)dx = x \ln\, |x| - x + C$$

$du = (1/x)\, dx$; $v = x$

3. $u = x$; $dv = e^x\, dx$

$$\int xe^x\, dx = xe^x - \int e^x dx = xe^x - e^x + C$$

$du = dx$; $v = e^x$

5. $u = \ln x$; $dv = x^{1/2}\, dx$ $\int \sqrt{x} \ln x\, dx = (2/3)x^{3/2} \ln |x|$

$du = (1/x)\, dx$; $v = (2/3)x^{3/2} - \int (2/3)x^{3/2}(1/x)\, dx = (2/3)x^{3/2} \ln |x| - (2/3)\int x^{1/2}\, dx$

$$= (2/3)x^{3/2} \ln |x| - (4/9)x^{3/2} + C$$

7. $\int \ln x^2\, dx = 2\int \ln x\, dx = 2[x \ln |x| - x] + C$ (from Exercise 1)

$$= 2x \ln |x| - 2x + C$$

9. $u = \text{Arccos } x$; $dv = dx$ $\int \text{Arccos } x\, dx = x \text{ Arccos } x - \int x \dfrac{-1}{\sqrt{1 - x^2}}\, dx$

$du = -\dfrac{dx}{\sqrt{1 - x^2}}$; $v = x$

$$= x \text{ Arccos } x + \int \frac{x\, dx}{\sqrt{1 - x^2}} = x \text{ Arccos } x - \frac{1}{2}\int (1 - x^2)^{-1/2}(-2x)\, dx$$

$$= x \text{ Arccos } x - \sqrt{1 - x^2} + C \qquad\qquad u = 1 - x^2$$
$$du = -2x\, dx$$

11. $u = e^x$; $dv = \cos x\, dx$ $u = e^x$; $dv = \sin x\, dx$

    $du = e^x dx$; $v = \sin x$; $du = e^x\, dx$ $v = -\cos x$

$$\int e^x \cos dx = e^x \sin x - \int \sin x \cdot e^x dx = e^x \sin x - [e^x(-\cos x) - \int -\cos x\, e^x dx]$$

Thus $\int e^x \cos x\, dx = e^x \sin x + e^x \cos x - \int e^x \cos x\, dx$

$2\int e^x \cos x\, dx = e^x(\sin x + \cos x)$

$\int e^x \cos x\, dx = (e^x/2)(\sin x + \cos x) + C$

13. $u = x^2$; $dv = \cos x\, dv$ $u = 2x$; $dv = \sin x$

    $du = 2x\, dx$; $v = \sin x$ $du = 2\, dx$; $v = -\cos x$

$$\int x^2 \cos dx = x^2 \sin x - \int 2x \sin x\, dx = x^2 \sin x - 2\int x \sin x\, dx$$

$$= x^2 \sin x - [-2x \cos x - \int -\cos x \cdot 2x\, dx] = x^2 \sin x + 2x \cos x - 2 \sin x + C$$

15. $u = x$; $dv = \sec^2 x\, dx$

    $du = dx$; $v = \tan x$

$\int x \sec^2 x dx = x \tan x - \int \tan x\, dx = x \tan x + \ln |\cos x| + C$

Section 7.8

17. $u = \ln^2 x;$ $dv = dx$
     $du = (2/x) \ln x \, dx;$ $v = x$

$$\int \ln^2 x \, dx = x \ln^2 |x| - \int x(2/x) \, \ln x \, dx = x \ln^2 |x| - 2[x \ln |x| - x] + C$$
$$= x \ln^2 |x| - 2x \ln |x| + 2x + C$$

19. $u = x;$ $dv = \sec x \tan x \, dx$
     $du = dx;$ $v = \sec x$

$$\int x \sec x \tan x \, dx = x \sec x - \int \sec x \, dx = x \sec x - \ln |\sec x + \tan x| + C$$

21. $u = x;$ $dv = e^{3x} \, dx$
     $du = dx;$ $v = (1/3)e^{3x}$

$$\int x e^{3x} dx = (1/3)x e^{3x} - \int (1/3) e^{3x} dx$$
$$= [(1/3)x e^{3x} - (1/9) e^{3x}] \Big|_0^1$$
$$= \frac{1}{9}(2e^3 + 1)$$

23. $u = x; \, dv = (x-1)^{1/2} \, dx$
     $du = dx; \, v = (2/3)(x-1)^{3/2}$

$$\int x\sqrt{x-1} \, dx = (2/3)x(\sqrt{x-1})^3 - (2/3)\int (x-1)^{3/2} \, dx = (2x/3)(x-1)^{3/2} - (4/15)(x-1)^{5/2} + C$$
$$\int_1^2 x\sqrt{x-1} \, dx = [(2x/3)(x-1)^{3/2}) - (4/15)(x-1)^{5/2}] \Big|_1^2 = 16/15$$

25. $u = \ln(x+1);$ $dv = dx$ $\int \ln(x+1) \, dx = x \ln(x+1) - \int x \dfrac{dx}{x+1} = x \ln(x+1) - x + \ln |x+1| + C$

     $du = \dfrac{dx}{x+1};$ $v = x$

$$\int_1^2 \ln(x+1) \, dx = [x \ln(x+1) - x + \ln(x+1)] \Big|_1^2 = \ln \frac{27}{4} - 1$$

27. $u = \ln 2x;$ $dv = dx$
     $du = (1/x) \, dx$ $v = x$

$$A = \int_{1/2}^1 \ln 2x \, dx = x \ln x - \int x(1/x) dx = x \ln 2x - x \Big|_{1/2}^1 = \ln 2 - 1/2$$

29. $u = x;$ $du = e^x \, dx$
     $du = dx;$ $v = e^x$

$$V = 2\pi \int_0^1 x e^x dx = 2\pi[x e^x - \int e^x dx] = 2\pi[x e^x - e^x] \Big|_0^1 = 2\pi$$

Section 7.8

Section 7.9

1. $u = a \tan \theta$
$2x = 3 \tan \theta$
$x = (3/2) \tan \theta$
$dx = (3/2) \sec^2 \theta \, d\theta$

$$\int \frac{dx}{\sqrt{9+4x^2}} = \int \frac{(3/2)\sec^2\theta \, d\theta}{\sqrt{9+4[(3/2)\sec]^2}}$$
$$= \int (1/2)\sec\theta \, d\theta$$
$$= (1/2) \ln \left| \sec\theta + \tan\theta \right| + C$$
$$= (1/2) \ln \left| \sqrt{9+4x^2} + 2x \right| + C$$

3. $u = a \sin \theta$
$3x = 2 \sin \theta$
$x = (2/3) \sin \theta$
$dx = (2/3) \cos\theta \, d\theta$

$$\int \frac{x^2 dx}{\sqrt{4-9x^2}} = \frac{[(2/3)\sin\theta]^2 (2/3)\cos\theta \, d\theta}{\sqrt{4-9[(2/3)\sin\theta]^2}}$$
$$\frac{4}{27} \int \frac{\sin^2\theta \cos\theta \, d\theta}{\cos\theta} = \frac{4}{27} \int \sin^2\theta \, d\theta$$
$$= 2\theta/27 - (1/27)\sin 2\theta + C$$
$$= 2\theta/27 - (2/27)\sin\theta + \cos\theta + C$$
$$= \frac{2}{27} \text{Arcsin} \frac{3x}{2} - \frac{x\sqrt{4-9x^2}}{18} + C$$

5. $\displaystyle\int_0^1 \frac{dx}{\sqrt{9-x^2}} = \text{Arc}\sin\frac{x}{3}\Big|_0^1 = \text{Arcsin}\frac{1}{3}$

7. $u = a \sin \theta$
$x = 2 \sin \theta$
$dx = 2 \cos\theta \, d\theta$

$$\int_0^1 \frac{dx}{\sqrt{(4-x^2)^3}} = \int \frac{2\cos\theta \, d\theta}{\left[\sqrt{(4-2\sin\theta)^2}\right]^3}$$
$$= \int \frac{2\cos\theta \, d\theta}{\left[\sqrt{4\cos^2\theta}\right]^3} = \int \frac{2\cos\theta \, d\theta}{8\cos^3\theta \, d\theta}$$
$$= (1/4)\int \sec^2\theta \, d\theta = (1/4)\tan\theta = \int \frac{1}{4} \frac{x}{\sqrt{4-x^2}}\Big|_0^1 = \frac{1}{4\sqrt{3}}$$

9. $u = a \tan \theta$
$x = 2 \tan \theta$
$dx = 2 \sec^2 \theta \, d\theta$

$$\int \frac{dx}{x\sqrt{x^2+4}} = \int \frac{2\sec^2\theta \, d\theta}{2\tan\theta\sqrt{(2\tan\theta)^2+4}}$$
$$\int \frac{2\sec\theta^2 \, d\theta}{2\tan\theta \, 2\sec\theta} = \frac{1}{2} \int \frac{\sec\theta}{\tan\theta} d\theta$$
$$= (1/2)\int \csc\theta \, d\theta = (1/2)\ln\left| \csc\theta - \cot\theta \right| + C = \frac{1}{2}\ln\left| \frac{\sqrt{x^2+4}-2}{x} \right| + C$$

11. $u = a \sec \theta$
$x = 3 \sec \theta$
$dx = 3 \sec\theta \tan \theta \, d\theta$

$$\int \frac{\sqrt{x^2 - 9}}{x^2} \, dx =$$

$$\int \frac{\sqrt{(3 \sec \theta)^2 - 9}}{(3 \sec \theta)^2} \, 3 \sec \theta \tan \theta \, d\theta$$

$$= \int \frac{3 \tan^2 \theta}{3 \sec \theta} \, d\theta = \frac{\sec^2 \theta - 1}{\sec \theta} \, d\theta = \int (\sec \theta - \cos \theta) d\theta$$

$$= \ln \left| \sec \theta + \tan \theta \right| - \sin \theta + C = \ln \left| x + \sqrt{x^2 - 9} \right| - \frac{\sqrt{x^2 - 9}}{2} + C$$

13. $u = a \sin \theta$
$x = 4 \sin \theta$
$dx = 4 \cos\theta \, d\theta$

$$\int \frac{dx}{x^2 \sqrt{16 - x^2}}$$

$$\int \frac{4 \cos \theta \, d\theta}{(4 \sin \theta)^2 \sqrt{16 - (4 \sin \theta)^2}}$$

$$= \int \frac{4 \cos\theta \, d\theta}{16 \sin^2 \theta \cdot 4 \cos\theta} = (1/16) \int \csc^2 \theta \, d\theta = -\frac{1}{16} \cot \theta + C = -\frac{\sqrt{16 - x^2}}{16x} + C$$

15. $u = \tan \theta$
$x = 3 \tan \theta$
$dx = 3 \sec^2\theta \, d\theta$

$$\int \frac{\sqrt{9 + x^2}}{x} \, dx =$$

$$\int \frac{\sqrt{9 + (3 \tan \theta)^2} \, 3 \sec^2 \theta \, d\theta}{3 \tan \theta} =$$

$$\int \frac{3 \sec\theta \, 3 \sec^2 \, d\theta}{3 \tan \theta} = 3 \int \sec^2 \theta \csc\theta \, d\theta \text{ by parts:}$$

$\quad u = \csc \theta, \qquad\qquad dv = \sec^2\theta \, d\theta;$
$\quad du = -\csc \theta \cot \theta \, d\theta, \qquad v = \tan \theta$

$$= 3[\csc\theta \tan \theta - \int \tan \theta(-\csc\theta \cot \theta)d\theta] = 3 \csc\theta \tan \theta + 3 \int \csc\theta \, d\theta$$

$$= 3 \sec \theta + 3 \ln \left| \csc \theta - \cot \theta \right| + C = 3\frac{\sqrt{9 + x^2}}{3} + \ln \left| \frac{\sqrt{9 + x^2}}{3} - \frac{3}{x} \right| + C$$

$$\text{or } \sqrt{9 + x^2} - 3 \ln \left| \frac{\sqrt{9 + x^2} + 3}{x} \right| + C \text{ if use } \int \csc\theta \, d\theta = -\ln \left| \csc \theta + \cot \theta \right|.$$

17. $u = a \sin \theta$
$x = 5 \sin \theta$
$dx = 5 \cos\theta \, d\theta$

$$\int \frac{dx}{(25 - x^2)^{3/2}} = \frac{5 \cos \theta \, d\theta}{(25 - 25 \sin^2 \theta)^{3/2}}$$

$$\int \frac{5 \cos \theta \, d\theta}{(5 \cos \theta)^3} = (1/25) \int \sec^2 \theta \, d\theta$$

$$= (1/25) \tan \theta + C = \frac{1}{25} \frac{x}{\sqrt{25 - x^2}} + C$$

Section 7.9

19. $u = a \sec \theta$
$x = 3 \sec \theta$
$dx = 3 \sec \theta \tan \theta \, d\theta$

$$\int \frac{dx}{\sqrt{x^2 - 9}} = \int \frac{3 \sec \theta \tan \theta \, d\theta}{\sqrt{(3 \sec \theta)^2 - 9}}$$

$$= \int \frac{3 \sec \theta \tan \theta \, d\theta}{3 \tan \theta} = \int \sec \theta \, d\theta$$

$$\ln \left| \sec \theta + \tan \theta \right| + C = \ln \left| \frac{x}{3} + \frac{\sqrt{x^2 - 9}}{3} \right| + C = \ln \left| x + \sqrt{x^2 - 9} \right| + C$$

21. $u = a \tan \theta$
$3x = 2 \tan \theta$
$dx = (2/3) \sec^2 \theta \, d\theta$

$$\int \frac{x^3 \, dx}{\sqrt{9x^2 + 4}}$$

$$= \int \frac{[(2/3)\tan \theta]^3}{\sqrt{9[(2/3)\tan \theta]^2 + 4}} (2/3) \sec^2 \theta \, d\theta$$

$$= \frac{8}{81} \int \frac{\tan^3 \theta \sec^2 \theta \, d\theta}{\sec \theta} = (8/81) \int \tan^3 \theta \sec \theta \, d\theta$$

$$= (8/81) \int (\sec^2 \theta - 1) \sec \theta \tan \theta \, d\theta = \frac{8}{81} \left(\frac{\sec^3 \theta}{3} - \sec \theta \right) + C$$

$$= \frac{8}{81} \left\{ \frac{(\sqrt{9x^2 + 4})^3}{3 \cdot 8} - \frac{\sqrt{9x^2 + 4}}{2} \right\} + C = \frac{8}{81} \frac{\sqrt{9x^2 + 4}}{2} \left\{ \frac{9x^2 + 4}{12} - 1 \right\} = \frac{9x^2 - 8}{243} \sqrt{9x^2 + 4} + C$$

23. $u = a \sec \theta$
$x - 3 = \sec \theta$
$dx = \sec \theta \tan \theta \, d\theta$

$$\int \frac{dx}{\sqrt{(x-3)^2 - 1}} = \int \frac{\sec \theta \tan \theta \, d\theta}{\sqrt{\sec^2 \theta - 1}}$$

$$= \int \frac{\sec \theta \tan \theta \, d\theta}{\tan \theta} = \int \sec \theta \, d\theta$$

$$= \ln \left| \sec \theta + \tan \theta \right| + C = \ln \left| x - 3 + \sqrt{x^2 - 6x + 8} \right| + C$$

25. $u = a \sec \theta$
$x + 4 = \sec \theta$
$dx = \sec \theta \tan \theta \, d\theta$

$$\int \frac{dx}{(x^2 + 8x + 15)^{3/2}} = \int \frac{dx}{[(x+4)^2 - 1]^{3/2}}$$

$$= \int \frac{\sec \theta \tan \theta \, d\theta}{[\sec^2 \theta - 1]^{3/2}} = \int \frac{\sec \theta \tan \theta \, d\theta}{\tan^3 \theta}$$

$$= \int \sin^{-2} \theta \cos \theta \, d\theta = -\frac{1}{\sin \theta} + C = -\frac{x + 4}{\sqrt{x^2 + 8x + 15}} + C$$

27. $u = a \tan \theta$
$x = 2 \tan \theta$
$dx = 2 \sec^2 \theta \, d\theta$

$$A = \int_0^2 \frac{dx}{\sqrt{x^2 + 4}} = \int \frac{2 \sec^2 \theta \, d\theta}{\sqrt{(2 \tan \theta)^2 + 4}}$$

$$= \int \frac{2 \sec^2 \theta \, d\theta}{2 \sec \theta} = \int \sec \theta \, d\theta$$

$$= \ln \left| \sec \theta + \tan \theta \right| = \ln \left| \frac{\sqrt{x^2 + 4}}{2} + \frac{x}{2} \right| \Big|_0^2 = \ln(\sqrt{2} + 1)$$

Section 7.9

Section 7.10

1. $u = x$, $a = 5$, $b = 1$; Formula 15; $\dfrac{1}{\sqrt{5}} \ln \left| \dfrac{\sqrt{x+5} - \sqrt{5}}{\sqrt{x+5} + \sqrt{5}} \right| + C$

3. $u = x$, $a = 2$, Formula 35; $\ln \left| x + \sqrt{x^2 - 4} \right| + C$

5. $u = x$, $a = 3$, $b = 2$, Formula 13; $\dfrac{-(3-x)\sqrt{3+2x}}{2} + C$

7. $m = 7$, $n = 3$, Formula 75; $-\dfrac{\sin 10x}{2 \cdot 10} + \dfrac{\sin 4x}{2 \cdot 4} + C = \dfrac{1}{4} \left\{ \dfrac{\sin 4x}{2} - \dfrac{\sin 10x}{5} \right\}$

9. $u = x$, $a = 3$, Formula 24; $-\dfrac{x}{2}\sqrt{9 - x^2} + \dfrac{9}{2} \operatorname{Arcsin} \dfrac{x}{3} + C$

11. $a = 1$, $b = 9$, $u = x$, Formula 9; $\dfrac{1}{1 + 9x} + \ln \left| \dfrac{x}{1 + 9x} \right| + C$

13. $u = x$, $a = 5$, Formula 20; $\dfrac{1}{10} \ln \left| \dfrac{x - 5}{x + 5} \right| + C$

15. $u = x$, $a = 2$, Formula 30;

$(1/2)\left[x\sqrt{x^2 + 4} + 4\ln \left| x + \sqrt{x^2 + 4} \right| \right] + C = (x/2)\sqrt{x^2 + 4} + 2\ln \left| x + \sqrt{x^2 + 4} \right| + C$

17. $u = x$, $a = 3$, $b = 4$, Formula 7; $\dfrac{1}{16} \left\{ \ln|3 + 4x| + \dfrac{3}{3 + 4x} \right\} + C$

19. $u = 3x$; $a = 4$; $du = 3\, dx$; Formula 36 $\displaystyle\int \dfrac{dx}{x\sqrt{9x^2 - 16}} = \int \dfrac{3dx}{3x\sqrt{9x^2 - 16}} = \dfrac{1}{4} \operatorname{Arccos} \dfrac{4}{3x} + C$

21. $u = x$, $a = 3$, $n = 4$, Formula 59; $\dfrac{e^{3x}(3\sin 4x - 4\cos 4x)}{25} + C$

23. $u = 2x - 3$; $du = 2\, dx$; Formula 80; $\displaystyle\int (2x - 3)\sin(2x - 3)dx$

$= (1/2) \displaystyle\int (2x - 3)\, \sin(2x - 3) \cdot 2\, dx = (1/2)\sin(2x - 3) - (1/2)(2x - 3)\cos(2x - 3) + C$

25. $u = x$, $n = 4$, Formula 83: $\displaystyle\int \sin^4 x\, dx = (-1/4)\sin^3 x \cos x + (3/4) \int \sin^2 x\, dx =$

$= (-1/4)\sin^3 x \cos x + (3/8) \displaystyle\int (1 - \cos 2x)\, dx$

$= (-1/4)\sin^3 x \cos x + 3x/8 - (3/16)\sin 2x + C$

$= (-1/4)\sin^3 x \cos x + 3x/8 - (3/8)\sin x \cos x + C$

Section 7.10

27. $u = 3x, du = 3\ dx, a = 4$, Formula 33;

$$\int \frac{\sqrt{9x^2 - 16}}{x}\ dx = \int \frac{\sqrt{9x^2 - 16}}{3x}\ 3dx = \sqrt{9x^2 - 16} - 4\ \text{Arccos}\frac{4}{3x} + C$$

Section 7.11

1. $\dfrac{2/4}{2}\ [f(1) + 2f(3/2) + 2f(2) + 2f(5/2) + f(3)] = (1/4)[1 + 4/3 + 1 + 4/5 + 1/3] = 1.117$

3. $\dfrac{1/4}{2}\ [f(0) + 2f(1/4) + 2f(1/2) + 2f(3/4) + f(1)] = (1/8)[1 + 32/17 + 8/5 + 32/25 + 1/2] = 0.783$

5. $\dfrac{1/10}{2}\ [f(0) + 2f(0.1) + 2f(0.2) + 2f(0.3) + 2f(0.4) + 2f(0.5) + 2f(0.6)$

$\qquad\qquad + 2f(0.7) + 2f(0.8) + 2f(0.9) + f(1)]$

$\qquad = (1/20)[2 + 3.995 + 3.980 + 3.955 + 3.919 + 3.873 + 3.816$

$\qquad\quad + 3.747 + 3.666 + 3.572 + 1.732] = 1.913$

7. $\dfrac{1/2}{2}\ [f(0) + 2f(1/2) + 2f(1) + 2f(3/2) + 2f(2) + 2f(5/2) + 2f(3)\ ]$

$\qquad = (1/4)[1 + 2.121 + 2.828 + 4.183 + 6 + 8.155 + 5.292] = 7.395$

9. $\dfrac{\pi/24}{2}\ [f(0) + 2f(\pi/24) + 2f(\pi/12) + 2f(\pi/8) + 2f(\pi/6) + 2f(5\pi/24) + 2f(\pi/4) + 2f(7\pi/24) + f(\pi/3)]$

$\qquad = (\pi/48)[1 + 2 + 1.995 + 1.976 + 1.925 + 1.819 + 1.631 + 1.336 + 0.457] = 0.925$

11. $(2/2)[24 + 2(21) + 2(18) + 2(17) + 2(15) + 2(12) + 2(10) + 9] = 219$ ft-lb

13. $(3/2)[12 + 2(11) + 2(18) + 2(25) + 2(19) + 2(6) + 10] = 270$

15. $\dfrac{1/2}{3}\ [f(0) + 4f(1/2) + 2f(1) + 4f(3/2) + f(2)\ ]$

$\qquad = (1/6)[1 + 3.578 + 1.414 + 2.219 + 0.447] = 1.443$

17. $\dfrac{1/2}{3}\ [f(2) + 4f(5/2) + 2f(3) + 4f(7/2) + 2f(4) + 4f(9/2) + 2f(5) + 4f(11/2) + f(6)]$

$\qquad = (1/6)[0.111 + 0.241 + 0.071 + 0.091 + 0.031 + 0.043 + 0.016 + 0.024 + 0.005] = 0.105$

19. $\dfrac{1/4}{3}\ [f(0) + 4f(1/4) + 2f(1/2) + 4f(3/4) + f(1)]$

$\qquad = (1/12)[1 + 4.258 + 2.568 + 7.020 + 2.718] = 1.464$

21. $\dfrac{\pi/16}{3}\ [f(0) + 4f(\pi/16) + 2f(\pi/8) + 4f(3\pi/16) + f(\pi/4)]$

$\qquad = (\pi/48)[0 + 0.156 + 0.325 + 1.574 + 0.785] = 0.186$

Sections 7.10 − 7.11

23. $\dfrac{\pi/12}{3}$ $[f(0) + 4f(\pi/12) + 2f(\pi/6) + 4f(\pi/4) + 2f(\pi/3) + 4f(5\pi/12) + f(\pi/2)]$

$\qquad = (\pi/36)[1.414 + 5.561 + 2.646 + 4.899 + 2.236 + 4.132 + 1] = 1.910$

25. a) $\dfrac{1/2}{2}$ $[f(0) + 2f(1/2) + 2f(1) + 2f(3/2) + 2f(2) + 2f(5/2) + f(3)]$

$\qquad = (1/4)[3 + 5.916 + 5.657 + 5.196 + 4.742 + 3.317 + 0] = 6.889$

   b) $\dfrac{1/2}{3}$ $[f(0) + 4f(1/2) + 2f(1) + 4f(3/2) + 2f(2) + 4f(5/2) + f(3)]$

$\qquad = (1/6)[3 + 11.832 + 5.657 + 10.392 + 4.472 + 6.633 + 0] = 6.998$

27. a) $\dfrac{1/4}{2}$ $[f(0) + 2f(1/4) + 2f(1/2) + 2f(3/4) + f(1)]$

$\qquad = (1/8)[0 + 0.642 + 1.649 + 3.176 + 2.718] = 1.023$

   b) $\dfrac{1/4}{3}$ $[f(0) + 4f(1/4) + 2f(1/2) + 4f(3/4) + f(1)]$

$\qquad = (1/12)[0 + 1.284 + 1.649 + 6.351 + 2.718] = 1.000$

29. a) $\dfrac{\pi/12}{2}$ $[f(0) + 2f(\pi/12) + 2f(\pi/6) + 2f(\pi/4) + 2f(\pi/3) + 2f(5\pi/12) + f(\pi/2)]$

$\qquad = (\pi/24)[1 + 1.744 + 1.499 + 1.265 + 1.041 + 0.828 + 0.312] = 1.006$

$\qquad \dfrac{\pi/12}{3}$ $[f(0) + 4f(\pi/12) + 2f(\pi/6) + 4f(\pi/4) + 2f(\pi/3) + 4f(5\pi/12) + f(\pi/2)]$

$\qquad = (\pi/36)[1 + 3.488 + 1.499 + 2.529 + 1.041 + 1.655 + 0.312] = 1.006$

31. a) $A = \dfrac{150}{2}[475 + 2(417) + 2(275) + 2(353) + 2(405) + 2(521) + 565] = 373,650 \text{ ft}^2$

$\qquad \approx 374,000 \text{ ft}^2$

   b) $A = \dfrac{150}{3}[475 + 4(417) + 2(275) + 4(353) + 2(405) + 4(521) + 565] = 378,200 \text{ ft}^2$

$\qquad \approx 378,000 \text{ ft}^2$

Section 7.12

1. $A = \dfrac{1}{2}\displaystyle\int_{0}^{2\pi/3} r^2 d\theta = \dfrac{1}{2}\int_{0}^{2\pi/3} 4\,d\theta = 2\theta \,\Big|_{0}^{2\pi/3} = \dfrac{4\pi}{3}$

3. $A = \dfrac{1}{2}\displaystyle\int_{0}^{\pi} 9\cos^2\theta\, d\theta = \dfrac{9}{4}\int(1 + \cos 2\theta)d\theta = \dfrac{9}{4}(\theta + \dfrac{1}{2}\sin 2\theta)\,\Big|_{0}^{\pi} = \dfrac{9\pi}{4}$

5. $A = 2\left\{\dfrac{1}{2}\displaystyle\int_0^{\pi/2} 9\sin 2\theta\, d\theta\right\} = -\dfrac{9}{2}\cos 2\theta\ \Big|_0^{\pi/2} = 9$

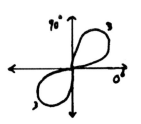

7. $A = 2\left\{\dfrac{1}{2}\displaystyle\int_0^{\pi}(1+\cos\theta)^2\, d\theta\right\} = \displaystyle\int_0^{\pi}(1+2\cos\theta+\cos^2\theta)d\theta$

$\quad = \displaystyle\int_0^{\pi}(3/2+2\cos\theta-(1/2)\cos 2\theta)d\theta$

$\quad = \dfrac{3\theta}{2}+2\sin\theta-\dfrac{1}{4}\sin 2\theta\ \Big|_0^{\pi} = \dfrac{3\pi}{2}$

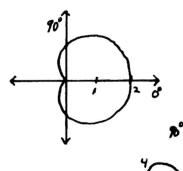

9. $A = 4\left\{\dfrac{1}{2}\displaystyle\int_0^{\pi/2}(4\sin 2\theta)^2\, d\theta\right\} = 16\displaystyle\int_0^{\pi/2}(1-\cos 4\theta)\, d\theta$

$\quad = 16(\theta-(1/4)\sin 4\theta)\ \Big|_0^{\pi/2} = 8\pi$

11. $A = 4\left\{\dfrac{1}{2}\displaystyle\int_0^{\pi/2}16\cos^2\theta\, d\theta\right\} = 32\displaystyle\int_0^{\pi/2}\cos^2\theta\, d\theta$

$\quad = 16\displaystyle\int_0^{\pi/2}(1+\cos 4\theta)\, d\theta = 16(\theta+(1/4)\sin 4\theta\ \Big|_0^{\pi/2} = 8\pi$

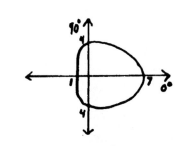

13. $A = 2\left\{\dfrac{1}{2}\displaystyle\int_0^{\pi}(4+3\cos\theta)^2\, d\theta\right\} = \displaystyle\int_0^{\pi}(16+24\cos\theta+9\cos^2\theta)\, d\theta$

$\quad = \displaystyle\int_0^{\pi}[41/2+24\cos\theta+(9/2)\cos 2\theta]\, d\theta$

$\quad = \dfrac{41\theta}{2}+24\sin\theta+\dfrac{9}{4}\sin 2\theta\ \Big|_0^{\pi} = \dfrac{41\pi}{2}$

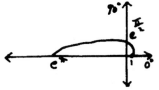

15. $A = \dfrac{1}{2}\displaystyle\int_0^{\pi}e^{2\theta}\, d\theta = \dfrac{1}{4}e^{2\theta}\ \Big|_0^{\pi} = \dfrac{1}{4}(e^{2\pi}-1)$

Section 7.12

17. $A = 2\left\{\dfrac{1}{2}\displaystyle\int_0^{\pi/3}(1-2\cos\theta)^2\,d\theta\right\}$

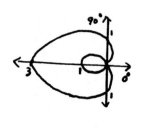

$= \displaystyle\int_0^{\pi/3}(1-4\cos\theta+4\cos^2\theta)\,d\theta$

$= \displaystyle\int_0^{\pi/3}(3-4\cos\theta+2\cos 2\theta\,d\theta = 3\theta-4\sin\theta+\sin 2\theta\ \Big|_0^{\pi/3}$

$= \pi - \dfrac{3\sqrt{3}}{2}$

19. $2\sin\theta+2\cos\theta=0;$
    $\tan\theta=-1;\ \theta=\pi/4,\ -\pi/4$

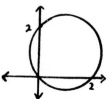

$A = 2\left\{\dfrac{1}{2}\displaystyle\int_{-\pi/4}^{\pi/4}(2\sin\theta+2\cos\theta)^2\,d\theta\right\}$

$= 4\displaystyle\int_{-\pi/4}^{\pi/4}(\sin^2\theta+\cos^2\theta+2\sin\theta\cos\theta)\,d\theta = 4\displaystyle\int_{-\pi/4}^{\pi/4}(1+\sin 2\theta)\,d\theta$

$= 4[\theta-(1/2)\cos 2\theta]\ \Big|_{-\pi/4}^{\pi/4} = 2\pi$

21. $\sin\theta=\sin 2\theta;\ \sin\theta=2\sin\theta\cos\theta$
    $\sin\theta(2\cos\theta-1)=0$
    $\sin\theta=0;\ \cos\theta=1/2$
    $\theta=0,\ \pi,\ \pi/3,\ 5\pi/3$

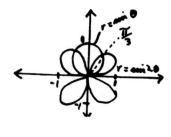

$A = 2\left\{\dfrac{1}{2}\displaystyle\int_0^{\pi/3}\sin^2\theta\,d\theta+\dfrac{1}{2}\displaystyle\int_{\pi/3}^{\pi/2}\sin^2 2\theta\,d\theta\right\}$

$= \dfrac{1}{2}\left\{\displaystyle\int_0^{\pi/3}(1-\cos 2\theta)\,d\theta+\displaystyle\int_{\pi/3}^{\pi/2}(1-\cos 4\theta)\,d\theta\right\}$

$= \dfrac{1}{2}\left\{[\theta-(1/2)\sin 2\theta]\ \Big|_0^{\pi/3}+[\theta-(1/4)\sin 4\theta]\ \Big|_{\pi/3}^{\pi/2}\right\} = \dfrac{\pi}{4}-\dfrac{3\sqrt{3}}{16}$

23. $2\cos 2\theta=1;\ \cos 2\theta=1/2;$
    $2\theta=\pi/3,\ 5\pi/3,\ 7\pi/3,\ 11\pi/3;$
    $\theta=\pi/6,\ 5\pi/6,\ 7\pi/6,\ 11\pi/6$

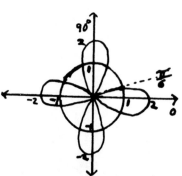

$A = 8\left\{\dfrac{1}{2}\displaystyle\int_0^{\pi/6}[(2\cos 2\theta)^2-1^2]\,d\theta = 4\displaystyle\int_0^{\pi/6}(4\cos^2 2\theta-1)\,d\theta\right.$

$= 4\displaystyle\int_0^{\pi/6}(1+2\cos 4\theta)\,d\theta = 4[\theta+(1/2)\sin 4\theta]\ \Big|_0^{\pi/6} = \dfrac{2\pi}{3}+\sqrt{3}$

Section 7.12

25. $5 \sin \theta = 2 + \sin \theta,$

$4 \sin \theta = 2; \sin \theta = 1/2;$

$\theta = \pi/6, 5\pi/6$

$$A = 2\left\{\frac{1}{2}\int_{-\pi/2}^{0}(2+\sin\theta)^2\,d\theta + \frac{1}{2}\int_{0}^{\pi/6}[(2+\sin\theta)^2 - (5\sin\theta)^2]d\theta\right\}$$

$$= \int_{-\pi/2}^{0}(4+4\sin\theta+\sin^2\theta)^2\,d\theta + \int_{0}^{\pi/6}(4+4\sin\theta-24\sin^2\theta)\,d\theta$$

$$= \int_{-\pi/2}^{0}[9/2+4\sin\theta-(1/2)\cos2\theta]\,d\theta \int_{0}^{\pi/6}(-8+4\sin\theta+12\cos2\theta)\,d\theta$$

$$= \frac{9\theta}{2} - 4\cos\theta - \frac{1}{4}\sin2\theta\ \Big|_{-\pi/2}^{0} + (-80-4\cos\theta+6\sin2\theta)\ \Big|_{0}^{\pi/6} = \frac{11\pi}{12} + \sqrt{3}$$

27. $3 + 3\cos\theta = 3 + 3\sin\theta$

$\tan\theta = 1; \theta = \pi/4, 5\pi/4$

$$A = \frac{1}{2}\int_{\pi}^{3\pi/2}(3+3\cos\theta)^2\,d\theta$$

$$+ \frac{1}{2}\int_{3\pi/2}^{\pi/4}[(3+3\cos\theta)^2 - (3+3\sin\theta)^2]d\theta$$

$$= \frac{9}{2}\int_{\pi}^{3\pi/2}(1+2\cos\theta+\cos^2\theta)\,d\theta + \frac{9}{2}\int_{3\pi/2}^{\pi/4}(2\cos\theta-2\sin\theta+\cos2\theta)\,d\theta$$

$$= \frac{9}{2}\left\{[3\theta/2+2\sin\theta+(1/4)\sin2\theta]\ \Big|_{\pi}^{3\pi/2} + [2\sin\theta+2\cos\theta+(1/2)\sin2\theta]\ \Big|_{3\pi/2}^{\pi/4}\right. = \frac{27\pi}{8} + \frac{9}{4} + 9\sqrt{2}$$

Section 7.13

1. $\displaystyle\int_{1}^{\infty}\frac{2}{x^3}dx$

$= \displaystyle\lim_{t\to\infty}\int_{1}^{t}2x^{-3}dx$

$= \displaystyle\lim_{t\to\infty}\frac{2x^{-2}}{-2}\Big|_{1}^{t}$

$= \displaystyle\lim_{t\to\infty}(-x^2)\Big|_{1}^{t}$

$= \displaystyle\lim_{t\to\infty}\left(-t^{-2} - [-1^{-2}]\right)$

$= \displaystyle\lim_{t\to\infty}\left(-\frac{1}{t^2}+1\right) = 1$

3. $\displaystyle\int_{0}^{\infty}e^{-3x}dx$

$= \displaystyle\lim_{t\to\infty}\int_{0}^{t}\left(-\frac{1}{3}\right)e^{-3x}(-3)dx$

$= \displaystyle\lim_{t\to\infty}\frac{-1}{3}e^{-3x}\Big|_{0}^{t}$

$= \displaystyle\lim_{t\to\infty}\left(\frac{-e^{-3t}}{3} - \left[-\frac{1}{3}e^0\right]\right)$

$= \displaystyle\lim_{t\to\infty}\left(\frac{-1}{3e^{3t}}+\frac{1}{3}\right) = \frac{1}{3}$

Sections 7.12 – 7.13

5. $\int_1^\infty \dfrac{1}{\sqrt{x}}dx$

$= \lim\limits_{t\to\infty}\int_1^t x^{\frac{-1}{2}}dx$

$= \lim\limits_{t\to\infty}\dfrac{-2x^{\frac{1}{2}}}{\frac{1}{2}}\Big|_1^t$

$= \lim\limits_{t\to\infty}-4t^{\frac{1}{2}}-\left(-4\left(1^{\frac{1}{2}}\right)\right)$

$= \lim\limits_{t\to\infty}-4\sqrt{t}+4\sqrt{1}=\infty$

diverges

7. $\int_{-\infty}^0 e^{5x}dx$

$= \lim\limits_{t\to-\infty}\int_t^0 e^{5x}dx$

$= \lim\limits_{t\to-\infty}\dfrac{1}{5}e^{5x}\Big|_t^0$

$= \lim\limits_{t\to-\infty}\dfrac{1}{5}e^0-\dfrac{1}{5}e^{5t}=\dfrac{1}{5}$

9. $\int_0^1 x^{-2/3}dx$

$= \lim\limits_{t\to0^+}\int_t^1 x^{-2/3}dx$

$= \lim\limits_{t\to0^+}3x^{1/3}\Big|_t^1$

$= \lim\limits_{t\to0^+}\left(3\left(1^{1/3}\right)-3\left(t^{1/3}\right)\right)$

$= \lim\limits_{t\to0^+}\left(3-\sqrt[3]{t}\right)=3$

11. $\int_0^5 \dfrac{1}{(x-5)^2}dx$

$= \lim\limits_{t\to5^-}\int_0^t(x-5)^{-2}dx$

$= \lim\limits_{t\to5^-}\left[-1(x-5)^{-1}\right]\Big|_0^t$

$= \lim\limits_{t\to5^-}\left[\dfrac{-1}{t-5}+\dfrac{1}{0-5}\right]=\infty$

diverges

13. $\int_0^\infty xe^{-x^2}dx$

$= \lim\limits_{t\to\infty}\int_0^t xe^{-x^2}dx$

$= \lim\limits_{t\to\infty}\left(-\dfrac{1}{2}\right)e^{-x^2}\Big|_0^1$

$= \lim\limits_{t\to\infty}\left(\dfrac{-1}{2}e^{-t^2}+\dfrac{1}{2}e^0\right)=\dfrac{1}{2}$

15. $\int_{-\infty}^0 -2xe^x dx$

$= \lim\limits_{t\to-\infty}\int_t^0 -2xe^x dx$

let $u=x$ $\qquad dv=2e^x dx$

then $du=dx$ $\qquad v=-2e^x$

$= \lim\limits_{t\to-\infty}\left(-2xe^x\Big|_t^0 - \int_t^0 -2e^x dx\right)$

$= \lim\limits_{t\to-\infty}\left(-0+2te^t+2e^x\Big|_t^0\right)$

$= \lim\limits_{t\to-\infty}\left(2te^t+2e^0-2e^t\right)$

$= \lim\limits_{t\to-\infty}\left(2te^t\right)+2-0$

$= \lim\limits_{t\to-\infty}\dfrac{2t}{e^{-t}}+2$

$= \lim\limits_{t\to-\infty}\dfrac{2}{-e^{-t}}+2$ using l' Hopital's Rule

$= 2$

Section 7.13

166

17. $\displaystyle\int_0^1 -x\ln x\,dx$

$\displaystyle = \lim_{t\to 0^+}\int_t^1 -x\ln x\,dx$

$\displaystyle = \lim_{t\to 0^+}\left(\frac{-1}{2}x^2\ln x\Big|_t^1 - \int_t^1 -\frac{1}{2}x\,dx\right)$

$\displaystyle = \lim_{t\to 0^+}\left(\frac{-1}{2}(1^2)\ln 1 + \frac{1}{2}t^2\ln t + \frac{1}{2}x^2\left(\frac{1}{2}\right)\Big|_t^1\right)$

$\displaystyle = \lim_{t\to 0^+}\left(\frac{-1}{2}t^2\ln t + \frac{1}{4}(1^2) - \frac{1}{4}t^2\right)$

$\displaystyle = \lim_{t\to 0^+}\left(\frac{1}{2}t^2\ln t\right) + \frac{1}{4} - 0$

$\displaystyle = \lim_{t\to 0^+}\left(\frac{\ln t}{2t^{-2}}\right) + \frac{1}{4}$

$\displaystyle = \lim_{t\to 0^+}\frac{\frac{1}{t}}{-4t^{-3}} + \frac{1}{4}$ using l' Hopital's Rule

$\displaystyle = \lim_{t\to 0^+}\frac{t^2}{-4} + \frac{1}{4}$

$\displaystyle = \frac{1}{4}$

19. $\displaystyle\int_1^\infty \frac{4}{x^2+1}\,dx$

$\displaystyle = \lim_{t\to\infty}\int_1^t \frac{4}{x^2+1}\,dx$

$\displaystyle = \lim_{t\to\infty} 4\arctan(x)\Big|_1^t$

$\displaystyle = \lim_{t\to\infty}\left(4\arctan(t) - 4\arctan 1\right)$

$\displaystyle = 4\left(\frac{\pi}{2}\right) - 4\left(\frac{\pi}{4}\right) = 2\pi - \pi = \pi$

Chapter 7 Review

1. $u = 2 + \sin 3x$;

$\displaystyle\frac{1}{3}\int(2+\sin 3x)^{-1/2}\,3\cos 3x\,dx$

$du = 3\cos x\,dx$;

$\displaystyle = \frac{2}{3}\sqrt{2+\sin 3x} + C$

2. $u = 5 + \tan 2x$;

$\displaystyle\frac{1}{2}\int(5+\tan 2x)^3\,2\sec^2 2x\,dx = \frac{1}{8}(5+\tan 2x)^4 + C$

$du = 2\sec^2 2x\,dx$

3. $\displaystyle(1/3)\int\cos 3x\cdot 3\,dx = (1/3)\sin 3x + C$

4. $u = x^2 - 5$;

$\displaystyle\frac{1}{2}\int\frac{2x\,dx}{x^2-5} = \frac{1}{2}\ln\,\left|x^2-5\right| + C$

$du = 2x\,dx$;

Section 7.13 – Chapter 7 Review

5. $u = 3x^2$; $du = 6x\,dx$; $(1/6) \int 6xe^{3x^2}\,dx = (1/6)e^{3x^2} + C$

6. $u = 2x$; $du = 2\,dx$; $a = 3$; $(1/3)$ Arctan $2x/3 + C$

7. $u = x$; $du = dx$; $a = 4$; Arcsin $x/4 + C$

8. $u = 7x + 2$; $du = 7\,dx$; $(1/7) \int \sec^2(7x + 2) \cdot 7\,dx = (1/7) \tan(7x + 2) + C$

9. $u = 3 + 5\tan x$;

$$\frac{1}{5} \int \frac{5\sec^2 x\,dx}{3 + 5\tan x} = (1/5)\ln\,|\,3 + 5\tan x\,| + C$$

$du = 5\sec^2 x\,dx$;

10. $u = x^3 + 4$; $du = 3x^2\,dx$; $(1/3) \int 3x^2 \sin(x^3 + 4)\,dx = (-1/3)\cos(x^3 + 4) + C$

11. $u = 3x$; $du = 3\,dx$; $a = 4$; $\dfrac{1}{4}$ Arctan $\dfrac{3x}{4}\ \Big|_0^1 = \dfrac{1}{4}$ Arctan $\dfrac{3}{4}$

12. $\dfrac{1}{\pi} \displaystyle\int_0^{1/2} \sin \pi x \cdot \pi\,dx = -\dfrac{1}{\pi} \cos \pi x\ \Big|_0^{1/2} = \dfrac{1}{\pi}$

13. $\displaystyle\int_0^{\pi/4} \tan x\,dx = -\ln|\cos x|\ \Big|_0^{\pi/4} = -\ln(1/\sqrt{2})$ or $\ln \sqrt{2}$

14. $u = 2x$;

$du = 2\,dx$; $\qquad \dfrac{1}{2} \displaystyle\int_0^1 \dfrac{2\,dx}{\sqrt{9 - 4x^2}} = \dfrac{1}{2}$ Arcsin $\dfrac{2x}{3}\ \Big|_0^1 = \dfrac{1}{2}$ Arcsin $\dfrac{2}{3}$

$a = 3$

15. $u = 4x$

$du = 4\,dx$; $\qquad \displaystyle\int \dfrac{4\,dx}{4x\sqrt{16x^2 - 9}} = \dfrac{1}{3}$ Arcsec $\dfrac{4x}{3} + C$ or $\dfrac{1}{3}$ Arccos $\dfrac{3}{4x} + C$

$a = 3$

16. $u = \cos 3x$;

$$\int \frac{\tan 3x\,dx}{\sec^4 3x} = -\frac{1}{3} \int \cos^3 3x\,\sin 3x\,(-3)\,dx = \frac{-\cos^4 3x}{12} + C$$

$d = -3\sin 3x\,dx$

17. $u = $ Arctan $3x$;

$$\frac{1}{3} \int \frac{\text{Arctan } 3x(3)\,dx}{1 + 9x^2} = \frac{1}{6} \text{ Arctan}^2 3x + C$$

$du = \dfrac{3}{1 + 9x^2}\,dx$

Chapter 7 Review

18. $u = 5x;$

$$(1/5) \ln |\sec 5x + \tan 5x| + C$$

$du = 5\, dx;$

19. $(1/2) \int 2x \tan x^2\, dx = (-1/2) \ln |\cos x^2| + C$

20. $\int \sin^5 2x \cos^2 2x\, dx = \int \sin 2x (1 - \cos^2 2x)^2 \cos^2 2x\, dx$

$\qquad = \int (\cos^6 2x \sin 2x - 2\cos^4 2x \sin 2x + \cos^2 2x \sin 2x)\, dx$

$\qquad = (-1/2) \int \cos^6 2x(-2\sin 2x) - 2\cos^4 2x(-2\sin 2x) + \cos^2 2x(-2\sin 2x)dx$

$\qquad = (-1/14) \cos^7 2x + (1/5) \cos^5 2x - (1/6) \cos^3 2x + C$

21. $\int \cos^4 3x \sin^2 3x\, dx = \int [(1/2)(1 + \cos 6x)]^2 [(1/2)(1 - \cos 6x)]\, dx$

$\qquad = (1/8) \int (1 + 2\cos 6x + \cos^2 6x)(1 - \cos 6x)\, dx$

$\qquad = (1/8) \int (1 + \cos 6x - \cos^2 6x - \cos^3 6x)\, dx$

$\qquad = (1/8) \int (1/2 - (1/2)\cos 12x + \sin^2 6x \cos 6x)\, dx$

$\qquad = x/16 - (1/192) \sin 12x + (1/144) \sin^3 6x + C$
$\qquad 1 = A(x - 1) + B(x + 3)$

22. $\dfrac{1}{(x+3)(x-1)} = \dfrac{A}{x+3} + \dfrac{B}{x-1};$

$\qquad 1 = (A + B)x + (-A + 3B);$
$\qquad A + B = 0;\ -A + 3B = 1;\ A = -1/4;\ B = 1/4$

$$\int \frac{dx}{x^2 + 2x - 3} = \int \left\{ \frac{-1/4}{x+3} + \frac{1/4}{x-1} \right\} dx = (-1/4) \ln |x + 3| + (1/4) \ln |x - 1| + C$$

$$= \frac{1}{4} \ln \left| \frac{x-1}{x+3} \right| + C$$

23. $u = \cos 4x;\ dv = e^{3x}\, dx$ Then: $u = \sin 4x;\ dv = e^{3x}$
$\qquad du = -4 \sin 4x\, dx;\ v = (1/3)e^{3x}\ du = 4\cos 4x\, dx;\ v = (1/3)e^{3x}$
$\qquad \int e^{3x} \cos 4x\, dx = (1/3)e^{3x} \cos 4x + (4/3) \int e^{3x} \sin 4x\, dx$

$\qquad = (1/3)e^{3x} \cos 4x + (4/3)[(1/3)e^{3x} \sin 4x - \int (1/3)e^{3x} 4\cos 4x\, dx\,]$

$\qquad = (1/3)e^{3x} \cos 4x + (4/9)e^{3x} \sin 4x - (16/9) \int e^{3x} 4\cos 4x\, dx$

Thus $(25/9) \int e^{3x} \cos 4x\, dx = (1/3)e^{3x} \cos 4x + (4/9)e^{3x} \sin 4x$

And, $\int e^{3x} \cos 4x\, dx = \dfrac{3e^{3x} \cos 4x + 4e^{3x} \sin x}{25} + C = (e^{3x}/25)(3 \cos 4x + 4 \sin 4x) + C$

Chapter 7 Review

24. $u = \cos x;$ $\quad -\int -\sin x \ln(\cos x)\, dx = -\cos x \ln \,|\cos x| + \cos x + C = \cos x(1 - \ln\,|\cos x|) + C$

25. $\int \tan^4 x\, dx = \int (\sec^2 x - 1)\tan^2 x\, dx = \int (\sec^2 x \tan^2 x - \tan^2 x)dx$

$\quad = \int (\sec^2 x \tan^2 x - \sec^2 x + 1)\, dx = (1/3)\tan^3 x - \tan x + x + C$

26. $\int \cos^2 5x\, dx = \int [(1/2) + (1/2)\cos 10x]dx = x/2 + (1/20)\sin 10x + C$

27. $\dfrac{x-3}{(3x+1)(2x-1)} = \dfrac{A}{3x+1} + \dfrac{B}{2x-1};$

$\qquad\qquad x - 3 = A(2x - 1) + B(3x + 1)$

$\qquad\qquad x - 3 = (2A + 3B)x + (B - A)$

$\qquad\qquad 2A + 3B = 1;\;\; B - A = -3;\;\; A = 2,\, B = -1$

$\int \dfrac{x-3}{6x^2 - x - 1}\, dx - \int \left\{ \dfrac{2}{3x+1} - \dfrac{1}{2x-1} \right\} dx = \dfrac{2}{3}\ln|3x+1| - \dfrac{1}{2}\ln|2x-1| + C$

28. $u = \ln x;\;\; dv = x^{1/2}\, dx$

$\quad du = (1/x)\, dx;$

$\quad v = (2/3)x^{3/2}$

$\quad \int \sqrt{x}\, \ln x\, dx = (2/3)x^{3/2}\ln x - \int (2/3)x^{3/2}(1/x)dx$

$\quad = (2/3)x^{3/2}\ln|x| - (4/9)x^{3/2} + C = (2/3)x^{3/2}(\ln|x| - 2/3) + C$

29. $u = \cos e^{-x};$ $\quad \int e^{-x}(\cos^2 e^{-x})(\sin e^{-x})\, dx = (1/3)\cos^3 e^{-x} + C$

$\quad du = e^{-x}\sin e^{-x}\, dx$

30. $\dfrac{4}{(x-2)(x+2)} = \dfrac{A}{x-2} + \dfrac{B}{x+2}$

$\quad 4 = A(x+2) + B(x-2);$

$\quad 4 = (A + B) + (2A - 2B);$

$\quad A + B = 0;\;\; 2A - 2B = 4;\;\; A = 1;\;\; B = -1$

$\quad \int \dfrac{4\, dx}{x^2 - 4} = \int \left\{ \dfrac{1}{x-2} - \dfrac{1}{x+2} \right\} dx = \ln|x-2| - \ln|x+2| = \ln\left| \dfrac{x-2}{x+2} \right| + C$

31. $u = x^2;$ $\qquad dv = \sin x\, dx;$ $\qquad u = 2x;$ $\qquad dv = \cos x\, dx;$

$\quad du = 2x\, dx;$ $\qquad v = -\cos x;$ $\qquad du = 2\, dx;$ $\qquad v = \sin x$

$\quad \int x^2 \sin dx = -x^2 \cos x + \int \cos x \cdot 2x\, dx = -x^2 \cos x + 2x\sin x - \int 2\sin x\, dx$

$\quad = -x^2 \cos x + 2x\sin x + 2\cos x + C$

Chapter 7 Review

32. $u = \text{Arcsin } 5x;$

$$\frac{1}{5} \int \frac{5 \text{ Arcsin } 5x \, dx}{\sqrt{1 - 25x^2}} = \frac{1}{10} \text{ Arcsin}^2 5x + C$$

$$du = \frac{5 \, dx}{\sqrt{1 - 25x^2}}$$

33. $u = x;$

$$\int_2^3 \frac{dx}{x\sqrt{x - 1^2}} = \text{Arcsec } x \Big|_2^3 = \text{Arcsec } 3 - \text{Arcsec } 2$$

$du = dx;$

34. $u = x; \quad dv = e^{4x} \, dx;$

$$\int_0^1 xe^{4x} dx = \frac{1}{4} xe^{4x} - \int \frac{1}{4} e^{4x} dx$$

$du = dx;$

$v = (1/4)e^{4x};$

$$= \frac{1}{4} xe^{4x} - \frac{1}{16} e^{4x} \Big|_0^1 = \frac{3e^4 + 1}{16}$$

35. $$\int_0^{\pi/8} 4 \tan 2x \, dx = -2 \ln|\cos 2x| \Big|_0^{\pi/8} = -2 \ln(1/\sqrt{2}) = \ln 2$$

36. $\dfrac{8}{(2 - x)(2 + x)} = \dfrac{1}{2 - x} + \dfrac{B}{2 + x};$

$8 = A(2 + x) + B(2 - x);$

$8 = (A - B)x + (2A + 2B);$

$A - B = 0; \quad 2A + 2B = 8; \quad A = 2; \quad B = 2$

$$\int_0^1 \frac{8 \, dx}{4 - x^2} = \int_0^1 \left\{ \frac{2}{2 - x} + \frac{2}{2 + x} \right\} dx = 2 \ln |2 + x| - 2 \ln |2 - x| = 2 \ln \left| \frac{2 + x}{2 - x} \right| \Big|_0^1 = 2 \ln 3 = \ln 9$$

37. $\dfrac{3x^2 - 11x + 12}{x(x - 2)^2} = \dfrac{A}{x} + \dfrac{B}{x - 2} + \dfrac{C}{(x - 2)^2}$

$3x^2 - 11x + 12 = A(x - 2)^2 + Bx(x - 2) + Cx$

$3x^2 - 11x + 12 = (A + B)x^2 + (-4A - 2B + C)x + 4A$

$A + B = 3; \quad -4A - 2B + C = -11; \quad 4A = 12; \quad A = 3, B = 0 \; C = 1$

$$\int \frac{3x^2 - 11x + 12}{x(x - 2)^2} dx = \int \left\{ \frac{3}{x} + \frac{1}{(x - 2)^2} \right\} dx = 3 \ln |x| - \frac{1}{x - 2} + C$$

Chapter 7 Review

38. $u = a \sin \theta$;

$3x = 4 \sin \theta$;

$$\int \frac{dx}{x\sqrt{16-9x^2}} = \int \frac{(4/3)\cos\theta \, d\theta}{[(4/3)\sin\theta]\sqrt{16(1-\sin^2\theta)}}$$

$dx = (4/3)\cos\theta \, d\theta$

$$= \int \frac{\cos\theta \, d\theta}{\sin\theta \, 4\cos\theta} = \frac{1}{4}\int \csc\theta \, d\theta = \frac{1}{4} \ln \left| \csc\theta - \cot\theta \right| + C$$

$$= \frac{1}{4} \ln \left| \frac{4}{3x} - \frac{\sqrt{16-9x^2}}{3x} \right| + C = \frac{1}{4} \ln \left| \frac{4-\sqrt{16-9x^2}}{3x} \right| + C$$

39. $A = \displaystyle\int_2^4 \frac{1}{x-1}dx = \ln|x-1| \ \Big|_2^4 = \ln 3$

40. $A = \displaystyle\int_1^3 e^{x+2}dx = e^{x+2}\Big|_1^3 = e^5 - e^3 = e^3(e^2-1)$

41. $A = \displaystyle\int_0^1 e^{2x}dx = \frac{1}{2}\int 2e^{2x}dx = \frac{1}{2}e^{2x}\Big|_0^1 = \frac{1}{2}e^2 - \frac{1}{2} = \frac{1}{2}(e^2-1)$

42. $A = \displaystyle\int_0^1 \frac{1}{1+x^2}dx = \text{Arctan } x \ \Big|_0^1 = \text{Arctan } 1 = \frac{\pi}{4}$

43. $A = \displaystyle\int_0^1 \frac{dx}{\sqrt{4-x^2}} = \text{Arcsin}\frac{x}{2}\Big|_0^1 = \frac{\pi}{6}$

44. $A = \displaystyle\int_0^1 x \, dx + \int_1^2 \frac{1}{x}dx = \frac{x^2}{2}\Big|_0^1 + \ln|x| \ \Big|_1^2 = \frac{1}{2} + \ln 2$

45. $A = \displaystyle\int_0^{\pi/4} \sec^2 x \, dx = \tan x \ \Big|_0^{\pi/4} = 1$

46. $A = \displaystyle\int_0^2 \frac{1}{4+x^2}dx = \frac{1}{2}\text{Arctan}\frac{x}{2}\Big|_0^2 = \frac{\pi}{8}$

47. $A = \displaystyle\int_0^{\pi/4} \tan x \, dx = -\ln|\cos x| \ \Big|_0^{\pi/4} = -\ln(1/\sqrt{2}) = \ln\sqrt{2} = \frac{1}{2}\ln 2$

Chapter 7 Review

48. $u = \ln^2 x;$ $dv = dx;$ $u = \ln x;$ $dv = dx$
     $du = 2 \ln x \cdot (1/x)\, dx;$ $v = x;$ $du = (1/x)\, dx;$ $v = x;$

$$V = \pi \int_1^2 \ln^2 x\, dx = \pi \left\{ x \ln^2 x - \int x(2/x) \ln x\, dx \right\} = \pi (x \ln^2 x - 2 \int \ln x\, dx)$$

$$= \pi [x \ln^2 x - 2(x \ln x - \int x(1/x)\, dx)]$$

$$= \pi (x \ln^2 x - 2x \ln x + 2x)\Big|_1^2 = \pi(2 \ln^2 2 - 4 \ln 2 + 2)$$

49. $V = \pi \int_0^1 e^{2x}\, dx = \dfrac{\pi}{2} e^{2x}\Big|_0^1 = \dfrac{\pi}{2}(e^2 - 1)$

50. $u = t;$ $dv = \sin 3t\, dt$
     $du = dt;$ $v = (-1/3) \cos 3t$
     $q = \int 4t \sin 3t\, dt = 4[(-1/3)t \cos 3t + \int (1/3) \cot 3t\, dt = (-4/3)t \cos 3t + (4/9) \sin 3t + C$

51. $\dfrac{5t + 1}{(t + 2)(t - 1)} = \dfrac{A}{t + 2} + \dfrac{B}{t - 1};$
     $5t + 1 = A(t - 1) + B(t + 2)$
     $5t + 1 = (A + B)t + (2B - A)$
     $A + B = 5;\ -A + 2B = 1;\ A = 3, B = 2$
     $q = \int \dfrac{5t + 1}{t^2 + t - 2}\, dt = \int \left\{ \dfrac{3}{t + 2} + \dfrac{2}{t - 1} \right\} dt = 3 \ln|t + 2| + 2 \ln|t - 1| + C = \ln\left| \dfrac{(t + 2)^3}{(t - 1)^2} \right| + C$

52. $u = x^2$
     $du = 2x\, dx;$ $F = \dfrac{1}{2} \int_1^2 2x e^{x^2}\, dx = \dfrac{1}{2} e^{x^2}\Big|_1^2 = \dfrac{1}{2}(e^4 - e) = \dfrac{e}{2}(e^3 - 1)$

53. Entry 5. $u = x, a = 5, b = 3$

$$\int \dfrac{dx}{x(5 + 3x)} = \dfrac{1}{5} \ln\left| \dfrac{x}{5 + 3x} \right| + C$$

54. Entry 29. $u = x, a = 4.$

$$\int \dfrac{\sqrt{16 - x^2}}{x^2}\, dx = \dfrac{-\sqrt{16 - x^2}}{x} - \arcsin \dfrac{x}{4} + C$$

55. Entry 47. $u = x, a = 3, b = 6, c = 1$

$$\int \dfrac{dx}{\sqrt{3 + 6x + x^2}} = \ln\left| 2x + 6 + 2\sqrt{3 + 6x + x^2} \right| + C$$

Chapter 7 Review

56. Entry 78. $u = x, a = 1, n = 2$

$$\int e^x \sin 2x \, dx = \frac{e^x}{5}(\sin 2x - 2\cos 2x) + C$$

57. Entry 33. $u = 2x, du = 2dx, a = 3$

$$\int \frac{\sqrt{4x^2 - 9}}{x} \, dx = \int \frac{\sqrt{(2x)^2 - 9}}{2x} \cdot 2dx$$

$$= \sqrt{4x^2 - 9} - 3\arccos\frac{3}{2x} + C$$

$$= \sqrt{4x^2 - 9} - 3\operatorname{arcsec}\frac{2x}{3} + C$$

58. Entry 73. $u = 3x \qquad du = 3\,dx, \qquad n = 4$

$$\int \cos^4 3x \sin 3x \, dx = \frac{1}{3}\int \cos^4 3x \sin 3x \, 3dx$$

$$= \frac{1}{3}\left(\frac{-\cos^5 3x}{5}\right) + C = -\frac{1}{15}\cos^5 3x + C$$

59. Entry 17. $u = x, a = 9, b = 4$

$$\int \frac{\sqrt{9 + 4x}}{x} \, dx = 2\sqrt{9 + 4x} + 9\int \frac{dx}{x\sqrt{9 + 4x}} + C$$

Now use entry 15. $u = x, a = 9, b = 4$

$$= 2\sqrt{9 + 4x} + 9\left\{\frac{1}{3}\ln\left|\frac{\sqrt{9 + 4x} - 3}{\sqrt{9 + 4x} + 3}\right|\right\} + C$$

$$= 2\sqrt{9 + 4x} + 3\ln\left|\frac{\sqrt{9 + 4x} - 3}{\sqrt{9 + 4x} + 3}\right| + C$$

60. Entry 85 (three times). $u = x, n = 6$ for first time.

$$\int \tan^6 x \, dx = \frac{\tan^5 x}{5} - \int \tan^4 x \, dx + C$$

$$= \frac{1}{5}\tan^5 x - \left[\frac{1}{3}\tan^3 x \, dx - \int \tan^2 x \, dx\right] + C$$

$$= \frac{1}{5}\tan^5 x - \frac{1}{3}\tan^3 x + \tan x - \int dx + C$$

$$= \frac{1}{5}\tan^5 x - \frac{1}{3}\tan^3 x + \tan x - x + C$$

61. $\dfrac{1/2}{2}[f(1) + 2f(3/2) + 2f(2) + 2f(5/2) + 2f(3) + 2f(7/2) + f(4)]$

$= (1/4)[1 + 1 + 2/3 + 1/2 + 2/5 + 1/3 + 1/7] = 1.011$

Chapter 7 Review

62. $\dfrac{1/2}{2}[f(1) + 2\,f(3/2) + 2\,f(2) + 2\,f(5/2) + f(3)]$

$\quad = (1/4)[0.1 + 0.178 + 0.154 + 0.131 + 0.056] = 0.155$

63. $\dfrac{1/2}{2}[f(0) + 2\,f(1/2) + 2\,f(1) + 2\,f(3/2) + 2\,f(2) + 2\,f(5/2) + 2f(3) + 2\,f(7/2) + f(4)]$

$\quad = (1/4)[4 + 7.937 + 7.746 + 7.416 + 6.928 + 6.245 + 5.292 + 3.873 + 0] = 12.395$

64. $\dfrac{1/2}{2}[f(1) + 2\,f(3/2) + 2\,f(2) + 2\,f(5/2) + 2\,f(3) + 2\,f(7/2) + f(4)]$

$\quad = (1/4)[2.571 + 5.013 + 4.820 + 4.547 + 4.160 + 3.583 + 1.260] = 6.489$

65. $(2/2)[2.3 + 2(2.8) + 2(3.4) + 2.1] = 22.2$

66. $(0.2/2)[0.9 + 2(0.7) + 2(0.8) + 2(1.1) + 2(1.3) + 0.9] = 0.96$

67. $\dfrac{1/2}{2}[f(0) + 4\,f(1/2) + 2\,f(1) + 4\,f(3/2) + 2\,f(2) + 4\,f(5/2) + 2\,f(3) + 4\,f(7/2) + f(4)]$

$\quad = (1/6)[4 + 16.125 + 8.246 + 17.088 + 8.944 + 18.868 + 10 + 21.260 + 5.657] = 18.365$

68. $(2/3)[f(0) + 4\,f(2) + 2\,f(4) + 4\,f(6) + 2\,f(8) + 4\,f(10) + f(12)]$

$\quad = (2/3)[4 + 3.2 + 0.471 + 0.432 + 0.123 + 0.158 + 0.028] = 5.608$

69. $\dfrac{\pi/8}{3}[f(0) + 4\,f(\pi/8) + 2\,f(\pi/4) + 4\,f(3\pi/8) + f(\pi/2)]$

$\quad = (\pi/24)(1/2 + 1.679 + 0.739 + 1.368 + 1.3) = 0.605$

70. $(1/3)[f(0) + 4\,f(1) + 2\,f(2) + 4\,f(3) + 2\,f(4) + 4\,f(5) + 2\,f(6) + 4\,f(7) + f(8)]$

$\quad = (1/3)[0 + 2.828 + 5.333 + 20.410 + 15.876 + 45.544 + 29.326 + 73.973 + 22.605] = 71.632$

71. $A = 4\left\{\dfrac{1}{2}\displaystyle\int_0^{\pi/4} 4\cos 2\theta\, d\theta\right\}$

$\quad = 8\displaystyle\int_0^{\pi/4}\cos 2\theta\, d\theta$

$\quad = 4\sin 2\theta\ \Big|_0^{\pi/4} = 4$

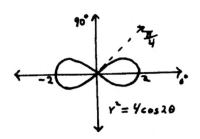

$r^2 = 4\cos 2\theta$

72. $A = 2\left\{\dfrac{1}{2}\displaystyle\int_{-\pi/2}^{\pi/2}(1-\sin\theta)^2\, d\theta\right\}$

$\quad = \displaystyle\int_{-\pi/2}^{\pi/2}(1 - 2\sin\theta + \sin^2\theta)\, d\theta$

$\quad = \displaystyle\int_{-\pi/2}^{\pi/2}[3/2 - 2\sin\theta - (1/2)\cos 2\theta]\, d\theta$

$\quad = 3\theta/2 + 2\cos\theta - (1/4)\sin 2\theta\ \Big|_{-\pi/2}^{\pi/2} = 3\pi/2$

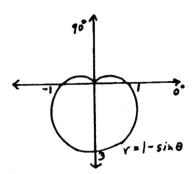

$r = 1 - \sin\theta$

Chapter 7 Review

73. $A = 2\left\{\dfrac{1}{2}\displaystyle\int_{-\pi/2}^{\pi/2}(3 - 2\sin\theta)^2\,d\theta\right\}$

$\quad = \displaystyle\int_{-\pi/2}^{\pi/2}(9 - 12\sin\theta + 4\sin^2\theta)\,d\theta$

$\quad = \displaystyle\int_{-\pi/2}^{\pi/2}(11 - 12\sin\theta - 2\cos 2\theta)\,d\theta$

$\quad = 11\theta + 12\cos\theta - \sin 2\theta\ \Big|_{-\pi/2}^{\pi/2} = 11\pi$

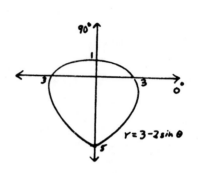

$r = 3 - 2\sin\theta$

74. $A = 6\left\{\dfrac{1}{2}\displaystyle\int_0^{\pi/6}[(1/2)\cos 3\theta]^2\,d\theta\right\}$

$\quad = \dfrac{3}{4}\displaystyle\int_0^{\pi/6}\cos^2 3\theta\,d\theta$

$\quad = \dfrac{3}{8}\displaystyle\int_0^{\pi/6}(1 + \cos 6\theta)\,d\theta = \dfrac{3}{8}[\theta + (1/6)\sin 6\theta]\ \Big|_0^{\pi/6} = \dfrac{\pi}{16}$

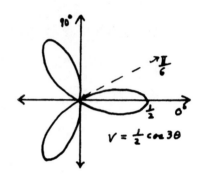

$r = \tfrac{1}{2}\cos 3\theta$

75. $A = \dfrac{1}{2}\displaystyle\int_0^{\pi}\{[(1/2)(\theta + \pi)]^2 - \theta^2\}\,d\theta = \dfrac{1}{2}\displaystyle\int_0^{\pi}\left\{-\dfrac{3}{4}\theta^2 + \dfrac{1}{2}\pi\theta + \dfrac{\pi^2}{4}\right\}\,d\theta$

$\quad = \dfrac{1}{2}\left\{-\dfrac{1}{4}\theta^3 + \dfrac{1}{4}\pi\theta^2 + \dfrac{\pi^2}{4}\theta\right\}\Big|_0^{\pi} = \dfrac{\pi^3}{8}$

76. $1 = 2\cos\theta,\ \cos\theta = 1/2$;

$\quad \theta = \pi/3,\ 5\pi/3$

$\quad A = 2\left\{\dfrac{1}{2}\displaystyle\int_0^{\pi/3}[(2\cos\theta)^2 - 1^2]\,d\theta\right\} = \displaystyle\int_0^{\pi/3}(4\cos^2\theta - 1)\,d\theta$

$\quad = \displaystyle\int_0^{\pi/3}(1 + 2\cos 2\theta)\,d\theta = (\theta + \sin 2\theta)\ \Big|_0^{\pi/3} = \dfrac{\pi}{3} = \dfrac{\sqrt{3}}{2}$

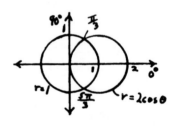

$r = 1$

$r = 2\cos\theta$

77. $A = 2\left\{\dfrac{1}{2}\displaystyle\int_0^{\pi/3}(2 - 4\cos\theta)^2\,d\theta\right\} = 4\displaystyle\int_0^{\pi/3}(1 - 4\cos\theta + 4\cos^2\theta)\,d\theta$

$\quad = 4\displaystyle\int_0^{\pi/3}(3 - 4\cos\theta + 2\cos 2\theta)\,d\theta = 4(3\theta - 4\sin\theta + \sin 2\theta)\ \Big|_0^{\pi/3}$

$\quad = 4\pi - 6\sqrt{3}$

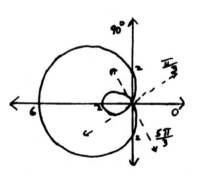

Chapter 7 Review

Solutions to Odd-Numbered Exercises

78. $A = 2\left\{\dfrac{1}{2}\displaystyle\int_{-\pi/2}^{0}[1^2 - (1+\sin\theta)^2]\,d\theta\right\} = \displaystyle\int_{-\pi/2}^{0}(-2\sin\theta - \sin^2\theta)\,d\theta$

$\qquad = \displaystyle\int_{-\pi/2}^{0}[-2\sin\theta - (1/2) + (1/2)\cos 2\theta]\,d\theta$

$\qquad = [2\cos\theta - \theta/2 + (1/4)\sin 2\theta]\ \Big|_{-\pi/2}^{0} = 2 - \dfrac{\pi}{4} = \dfrac{8-\pi}{4}$

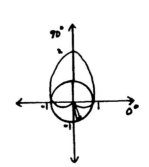

79. $r = \sqrt{8\sin 2\theta} = 2;$

$\quad \sin 2\theta = 4/8 = 1/2$

$\quad 2\theta = \pi/6,\ 5\pi/6,\ 13\pi/6,\ 17\pi/6$

$\quad \theta = \pi/12,\ 5\pi/12,\ 13\pi/12,\ 17\pi/12$

$\quad A = 4\left\{\dfrac{1}{2}\displaystyle\int_{\pi/4}^{5\pi/12}(8\sin 2\theta - 2^2)\,d\theta\right\} = 8\displaystyle\int_{\pi/4}^{5\pi/12}(2\sin 2\theta - 1)\,d\theta$

$\quad = 8(-\cos 2\theta - \theta)\ \Big|_{\pi/4}^{5\pi/12} = 4\sqrt{3} - 4\pi/3$

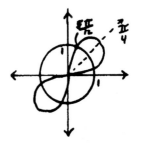

80. $\cos 2\theta = \sin 2\theta;\ \tan 2\theta = 1$

$\quad 2\theta = \pi/4,\ 5\pi/4,\ \dots$

$\quad \theta = \pi/8,\ 5\pi/8,\ \dots$

$\quad A = 2\left\{\dfrac{1}{2}\displaystyle\int_{0}^{\pi/8}\sin 2\theta\,d\theta + \dfrac{1}{2}\displaystyle\int_{\pi/8}^{\pi/4}\cos 2\theta\,d\theta\right\} = -\dfrac{1}{2}\cos 2\theta\ \Big|_{0}^{\pi/8} + \dfrac{1}{2}\sin 2\theta\ \Big|_{\pi/8}^{\pi/4} = 1 - \dfrac{\sqrt{2}}{2}$

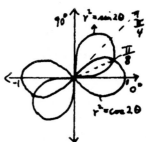

81. $\displaystyle\int_{1}^{\infty}\dfrac{1}{x^{4/3}}\,dx =$

$\quad \displaystyle\lim_{t\to\infty}\int_{1}^{t}x^{-4/3}\,dx =$

$\quad \displaystyle\lim_{t\to\infty}\left(-3x^{-1/3}\right)\Big|_{1}^{t} =$

$\quad \displaystyle\lim_{t\to\infty}\left(-3t^{-1/3} + 3(1)^{-1/3}\right) =$

$\quad \displaystyle\lim_{t\to\infty}\left(\dfrac{-3}{t^{1/3}} + 3\right) = 3$

82. $\displaystyle\int_{1}^{\infty}\dfrac{1}{x^{2/3}}\,dx =$

$\quad \displaystyle\lim_{t\to\infty}\int_{1}^{t}x^{-2/3}\,dx =$

$\quad \displaystyle\lim_{t\to\infty}3x^{1/3}\Big|_{1}^{t} =$

$\quad \displaystyle\lim_{t\to\infty}\left(3t^{1/3} - 3\right) = \infty$

$\quad diverges$

Chapter 7 Review

83. $\int_{-\infty}^{0} e^{x/2} dx =$

$\displaystyle\lim_{t \to -\infty} \int_{t}^{0} 2e^{x/2} \left(\frac{1}{2}\right) dx =$

$\displaystyle\lim_{t \to -\infty} \left(2e^{x/2}\right)\Big|_{t}^{0} =$

$\displaystyle\lim_{t \to -\infty} \left(2 - 2e^{t/2}\right) = 2$

85. $\int_{3}^{4} \frac{1}{\sqrt{x-3}} dx =$

$\displaystyle\lim_{t \to 3^{+}} \int_{t}^{4} (x-3)^{-1/2} dx =$

$\displaystyle\lim_{t \to 3^{+}} \left(2(x-3)^{1/2}\right)\Big|_{t}^{4} =$

$\displaystyle\lim_{t \to 3^{+}} \left(2(1) - 2(t-3)^{1/2}\right) = 2$

86. $\int_{1}^{2} \frac{1}{x \ln x} dx$

$\displaystyle\lim_{t \to 1^{+}} \int_{t}^{2} (\ln x)^{-1} \left(\frac{1}{x}\right) dx$

$\displaystyle\lim_{t \to 1^{+}} \left(\ln(\ln x)\right)\Big|_{t}^{2}$

$\displaystyle\lim_{t \to 1^{+}} \left(\ln \ln 2 - \ln(\ln t)\right)$

$\ln \ln 2 - (-\infty) = \infty$

diverges

84. $\int_{-\infty}^{0} xe^{x} dx$

$= \displaystyle\lim_{t \to -\infty} \int_{t}^{0} xe^{x} dx$

let $u = x \qquad dv = e^{x} dx$
then $du = dx \qquad v = e^{x}$

$= \displaystyle\lim_{t \to -\infty} \left(xe^{x}\Big|_{t}^{0} - \int_{t}^{0} e^{x} dx\right)$

$= \displaystyle\lim_{t \to -\infty} \left(0 - te^{t} - (e^{0} - e^{t})\right)$

$= \displaystyle\lim_{t \to -\infty} \left(-te^{t} - 1 + e^{t}\right)$

$= \displaystyle\lim_{t \to -\infty} \left(-te^{t}\right) - 1$

$= \displaystyle\lim_{t \to -\infty} \left(\frac{-t}{e^{-t}}\right) - 1$ using 1' Hopital's Rule

$= \displaystyle\lim_{t \to -\infty} \frac{-1}{-e^{-t}} - 1$

$= -1$

CHAPTER 8

Section 8.1

1.

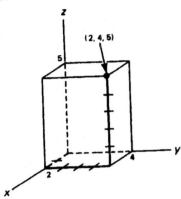

3.

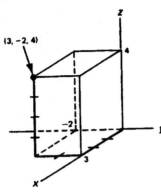

5.

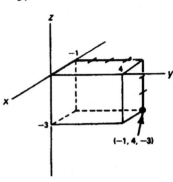

7. Plane

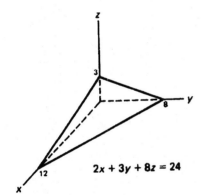

$2x + 3y + 8z = 24$

9. Plane

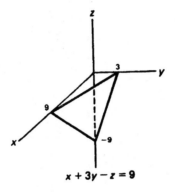

$x + 3y - z = 9$

11. Plane

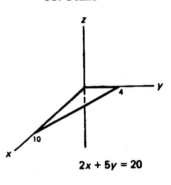

$2x + 5y = 20$

13. Plane

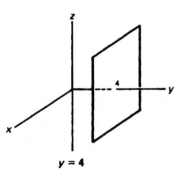

$y = 4$

15. Sphere

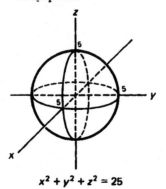

$x^2 + y^2 + z^2 = 25$

17. Sphere

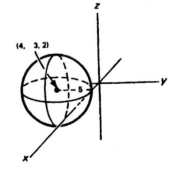

$x^2 + y^2 + z^2 - 8x + 6y - 4z + 4 = 0$

Section 8.1

19. Cylindrical surface

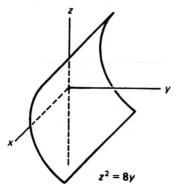

$$z^2 = 8y$$

21. Cylindrical surface

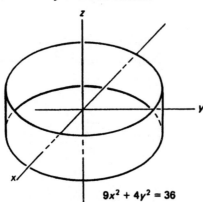

$$9x^2 + 4y^2 = 36$$

23. Ellipsoid

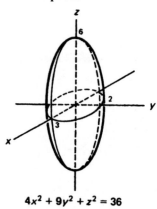

$$4x^2 + 9y^2 + z^2 = 36$$

25. Elliptic paraboloid

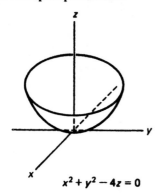

$$x^2 + y^2 - 4z = 0$$

27. Hyperboloid of two sheets

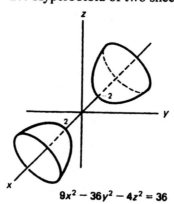

$$9x^2 - 36y^2 - 4z^2 = 36$$

29. Cylindrical surface

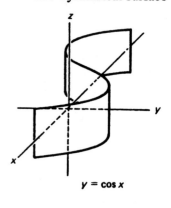

$$y = \cos x$$

31. Hyperbolic paraboloid

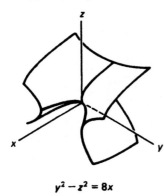

$$y^2 - z^2 = 8x$$

33. Elliptic cone

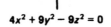

$$4x^2 + 9y^2 - 9z^2 = 0$$

35. Hyperboloid of one sheet

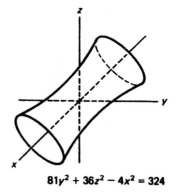

$$81y^2 + 36z^2 - 4x^2 = 324$$

37. $(x - 3)^2 + (y + 2)^2 + (z - 4)^2 = 36$

39. $(x - 3)^2 + (y + 2)^2 + (z - 4)^2 = 16$

Section 8.1

41. Consider two points $P_1(x_1, y_1, z_1)$ and $P_2(x_2, y_2, z_2)$ as shown in the figure at the right. These points determine a rectangular box with P_1 and P_2 as opposite vertices and with edges parallel to the coordinate axes. Triangles P_1RQ and P_1QP_2 are right triangles. By the Pythagorean theorem:

$d^2 = |P_1Q|^2 + |QP_2|^2$ and $|P_1Q|^2 = |P_1R|^2 + |RQ|^2$

Thus $d^2 = |P_1R|^2 + |RQ|^2 + |QP_2|^2$

$\quad = (x_2 - x_1)^2 + (y_2 - y_1)^2 + (z_2 - z_1)^2$

And, $d = \sqrt{(x_2 - x_1)^2 + (y_2 - y_1)^2 + (z_2 - z_1)^2}$

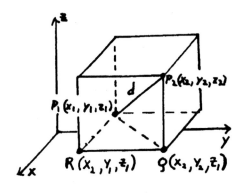

43. $d = \sqrt{(3-5)^2 + (1-(-3))^2 + (-2-2)^2} = \sqrt{36} = 6$

45. $d_1 = \sqrt{(5-2)^2 + (6-8)^2 + (3-4)^2} = \sqrt{14}$

$\quad d_2 = \sqrt{(5-3)^2 + (6-5)^2 + (3-6)^2} = \sqrt{14}$

$\quad d_3 = \sqrt{(2-3)^2 + (8-5)^2 + (4-6)^2} = \sqrt{14}$

Thus the triangle is equilateral.

Section 8.2

1. a) $\partial z/\partial x = 12x^2y^2$ b) $\partial z/\partial y = 8x^3y$

3. a) $\partial z/\partial x = 12xy^4 + 2y^2$ b) $\partial z/\partial y = 24x^2y^3 = 4xy$

5. a) $\partial z/\partial x = (1/2)(x^2 + y^2)^{-1/2}(2x) = \dfrac{x}{\sqrt{x^2 + y^2}}$ b) $\dfrac{\partial z}{\partial y} = \dfrac{y}{\sqrt{x^2 + y^2}}$

7. a) $\dfrac{\partial z}{\partial x} = \dfrac{(2xy)(2x) - (x^2 - y^2)(2y)}{(2xy)^2} = \dfrac{4x^2y - 2x^2y + 2y^3}{4x^2y^2} = \dfrac{2y(x^2 + y^2)}{4x^2y^2} = \dfrac{x^2 + y^2}{2x^2y}$

   b) $\dfrac{\partial z}{\partial y} = \dfrac{(2xy)(-2y) - (x^2 - y^2)(2x)}{(2xy)^2} = \dfrac{-4xy^2 - 2x^3 + 2xy^2}{4x^2y^2} = \dfrac{2x(-y^2 - x^2)}{4x^2y^2} = \dfrac{-y^2 - x^2}{2xy^2}$

9. a) $\partial z/\partial x = ay$ b) $\partial z/\partial x = ax$

11. a) $\dfrac{\partial z}{\partial x} = \dfrac{y}{x} \cdot \dfrac{1}{y} = \dfrac{1}{x}$ b) $\dfrac{\partial z}{\partial y} = \dfrac{y}{x}(-xy^{-2}) = -\dfrac{1}{y}$

13. a) $\partial z/\partial x = \sec^2(x - y)$ b) $\partial z/\partial y = -\sec^2(x - y)$

15. a) $\partial z/\partial x = e^{3x}y \cos xy + 3e^{3x} \sin xy = e^{3x}(y \cos xy + 3 \sin xy)$
    b) $\partial z/\partial y = e^{3x}x \cos xy$

Sections 8.1 – 8.2

17. a) $\dfrac{\partial z}{\partial x} = \dfrac{xy\,e^{xy}\sin y - e^{xy}\sin y}{(x\sin y)^2} = \dfrac{e^{xy}\sin y(xy-1)}{x^2\sin^2 y} = \dfrac{e^{xy}(xy-1)}{x^2\sin y}$

   b) $\dfrac{\partial z}{\partial y} = \dfrac{x^2\,e^{xy}\sin y - e^{xy}x\cos y}{(x\sin y)^2} = \dfrac{x\,e^{xy}(x\sin y - \cos y)}{x^2\sin^2 y} = \dfrac{e^{xy}(x\sin y - \cos y)}{x\sin^2 y}$

19. a) $\partial z/\partial x = -2\sin x\sin y$ b) $\partial z/\partial y = 2\cos x\cos y$

21. a) $\partial z/\partial x = xy^2\sec^2 xy + y\tan xy$ b) $\partial z/\partial y = x^2 y\sec^2 xy + x\tan xy$

23. $\partial P/\partial I = 2IR$ 25. $\dfrac{\partial I}{\partial R} = \dfrac{-E}{(R+r)^2}$ 27. $\dfrac{\partial z}{\partial R} = \dfrac{R}{\sqrt{R^2 + X_L^2}}$

29. $\partial e/\partial t = 2\pi f E\cos 2\pi f t$ 31. $\partial E/\partial I_2 = R_2 + R_3$

33. $\dfrac{\partial I}{\partial t} = \dfrac{-E}{R^2 C}\,e^{-t/RC}$

35. $\dfrac{\partial q}{\partial C} = \dfrac{Et}{RC}\,e^{-t/RC} + E\,e^{-t/RC} = E\,e^{-t/RC}\left[\dfrac{t}{RC} + 1\right]$

37. $\dfrac{\partial \phi}{\partial R} = \dfrac{1}{(X_L/R^2)+1}\cdot\dfrac{-X_L}{R^2} = \dfrac{R^2}{X_L^2 + R^2}\cdot\dfrac{-X_L}{R^2} = \dfrac{-X_L}{X_L^2 + R^2}$

39. a) $\partial z/\partial x = 18x\Big|_{x=1} = 18$ b) $\partial z/\partial y = 8y\Big|_{y=-2} = -16$

41. a) $\dfrac{\partial z}{\partial x} = \dfrac{25x}{\sqrt{25x^2 + 36y^2 + 164}}\Bigg| = \dfrac{25}{15} = \dfrac{5}{3}$

                                $(1,-1,15)$

   b) $\dfrac{\partial z}{\partial y} = \dfrac{36y}{\sqrt{25x^2 + 36y^2 + 164}}\Bigg| = -\dfrac{36}{15} = -\dfrac{12}{5}$

                                $(1,-1,15)$

43. $\dfrac{\partial P}{\partial V}\cdot\dfrac{\partial V}{\partial T}\cdot\dfrac{\partial T}{\partial P} = \dfrac{-nRT}{V^2}\cdot\dfrac{nR}{P}\cdot\dfrac{V}{nR} = \dfrac{-nRT}{VP} = -\dfrac{V}{V} = -1.$

   Note: $V = \dfrac{nRT}{P}$

45. $\partial V/\partial r = 16\pi r\Big|_{r=6} = 96\pi\,\text{cm}^3$

Section 8.2

47. $s^2 \, \partial w/\partial s + t \, \partial w/\partial t = s^2[(-t \, e^{1/s}/s^2) \sec^2 te^{1/s}] + t(2t + e^{1/s}\sec^2 te^{1/s})$
$\qquad = -t \, e^{1/s} \sec^2 te^{1/s} + 2t^2 + t \, e^{1/s} \sec^2 te^{1/s} = 2t^2$

Section 8.3

1. $dz = (6x + 4y) \, dx + (4x + 3y^2) \, dy$

3. $dz = 2x \cos y \, dx - x^2 \sin y \, dy$

5. $dz = \dfrac{(xy)(1) - (x - y)y}{(xy)^2} \, dx + \dfrac{(xy)(-1) - (x - y)x}{(xy)^2} \, dy = \dfrac{1}{x^2} \, dx - \dfrac{1}{y^2} \, dy$

7. $dz = \dfrac{1}{\sqrt{1 + xy}} \cdot \dfrac{1}{2}(1 + xy)^{-1/2}(y)dx + \dfrac{1}{\sqrt{1 + xy}} \cdot \dfrac{1}{2}(1 + xy)^{-1/2}(x)dy = \dfrac{y}{2(1 + xy)} \, dx + \dfrac{x}{2(1 + xy)} \, dy$

9. $dV = lw \, dh + lh \, dw + wh \, dl = (26)(26)(0.15) + (26)(12)(0.15) + (26)(12)(0.15) = 195 \text{ cm}^3$

11. $dR = \dfrac{R_2^2}{(R_1 + R_2)^2} \, dR_1 + \dfrac{R_1^2}{(R_1 + R_2)^2} \, dR_2 = \dfrac{(600)^2}{(400 + 600)^2}(25) + \dfrac{(400)^2(50)}{(400 + 600)^2} = 17\Omega$

13. $dV = (2/3)\pi rh \, dr + (1/3)\pi r^2 \, dh = (2/3)\pi(12.00)(21.00)(-0.15) + (1/3)\pi(12.00)^2(0.10) = -64.1 \text{ cm}^3$

15. $f(x, y) = z = 9 + 6x - 8y - 3x^2 - 2y^2$
$\qquad f_x(x, y) = 6 - 6x; \quad f_y(x, y) = -8 - 4y$
$\qquad$ Set each expression equal to zero and solve the resulting system of equations.
$\qquad\qquad 6 - 6x = 0$
$\qquad\qquad -8 - 4y = 0$
$\qquad$ Thus $x = 1$ and $y = -2$. Substitute these values in $z = f(x, y)$ to find
$\qquad z = 9 + 6(1) - 8(-2) - 3(1)^2 - 2(-2)^2 = 20$.
$\qquad$ Next determine if $(1, -2, 20)$ is a relative maximum or a relative minimum.
$\qquad f_{xx}(x, y) = -6; \, f_{yy}(x, y) = -4; \, f_{xy}(x, y) = f_{yx}(x, y) = 0$.
$\qquad D(x, y) = f_{xx}(x, y) \, f_{yy}(x, y) - [f_{xy}(x, y)]^2$
$\qquad D(1, -2) = -6(-4) - 0 = 24$
$\qquad$ Since $D(1, -2) = 24 > 0$ and $f_{xx}(1, 2) = -6 < 0$, then $(1, -2, 20)$ is a relative maximum point.

17. $f(x, y) = z = \dfrac{1}{x} + \dfrac{1}{y} + xy$
$\qquad f_x(x, y) = -x^{-2} + y; \, f_y(x, y) = -y^{-2} + x$
$\qquad$ Set each expression equal to zero and solve the resulting system of equations.
$\qquad\qquad \dfrac{-1}{x^2} + y = 0$
$\qquad\qquad -\dfrac{1}{y^2} + x = 0$
$\qquad x = 0$ or 1, but $f(x, y)$ is not defined as $x = 0$.

Sections 8.2 − 8.3

So $x = 1$ and $y = 1$. Thus $z = \dfrac{1}{1} + \dfrac{1}{1} + (1) = 3$.

Next determine if $(1, 1, 3)$ is a relative maximum or a relative minimum.

$f_{xx}(x, y) = 2x^{-3}; \ f_{yy}(x, y) = 2y^{-3}$

$f_{xy}(x, y) = f_{yx}(x, y) = 1$

$D(x, y) = f_{xx}(x, y)f_{yy}(x, y) - [f_{xy}(x, y)]^2$

$D(1, 1) = 2(2) - 1(1) = 3$

Since $D(1, 1) = 3 > 0$ and $f_{xx}(1, 1) = 2 > 0$,

$(1, 1, 3)$ is a relative minimum point.

19. $f(x, y) = z = x^2 - y^2 - 2x - 4y - 4$

$f_x(x, y) = 2x - 2; \ f_y(x, y) = -2y - 4$

Set each expression equal to zero and solve the resulting system of equations.

$2x - 2 = 0$

$-2y - 4 = 0$

Thus $x = 1$ and $y = -2$. $z = 1^2 - (-2)^2 - 2(1) - 4(-2) - 4 = -1$.

Next determine if $(1, -2, -1)$ is a relative maximum or a relative minimum.

$f_{xx}(x, y) = 2; \ f_{yy}(x, y) = -2; f_{xy}(x, y) = 0$

$D(x, y) = f_{xx}(x, y)f_{yy}(x, y) - [f_{xy}(x, y)]^2$

$D(1, -2) = 2(-2) - 0^2 = -4$.

Since $D(1, -2) = -4 < 0$, $(1, -2, -1)$ is a saddle point.

21. $f(x, y) = z = x^2 - y^2 - 6x + 4y$

$f_x(x, y) = 2x - 6$

$f_y(x, y) = -2y + 4$

Set each expression equal to zero and solve the resulting system of equations.

$2x - 6 = 0$

$-2y + 4 = 0$

Thus, $x = 3$ and $y = 2$. Substitute these values in $z = f(x, y)$ to find

$z = 3^2 - 2^2 - 6(3) + 4(2) = -5$.

Next determine if $(3, 2, -5)$ is a relative maximum or minimum.

$f_{xx}(x, y) = 2; f_{yy}(x, y) = -2, f_{xy}(x, y) = f_{xy}(x, y) = 0$

$D(x, y) = f_{xx}(x, y)\, f_{yy}(x, y) - [f_{xy}(x, y)]^2$

$\qquad = 2(-2) - 0 = -4$

Since $D(3, 2) = -4 < 0$, $(3, 2, -5)$ is a saddle point.

23. $f(x, y) = z = 4x^3 + y^2 - 12x^2 - 36x - 2y$

$f_x(x, y) = 12x^2 - 24x - 36; \ f_y(x, y) = 2y - 2$

Set each expression equal to zero and solve the resulting system of equations.

$12x^2 - 24x - 36 = 0$

$\qquad 2y - 2 = 0$

Thus $x = 3$ or -1 and $y = 1$

For $x = 3$ and $y = 1$

$\qquad z = 4(3)^3 + 1^2 - 12(3)^2 - 36(3) - 2(1) = -109$.

Section 8.3

For $x = -1$ and $y = 1$,

$z = 4(-1) + 1^2 - 12(1) - 36(-1) - 2 = 19.$

Thus $(3, 1, -109)$ and $(-1, 1, 19)$ are critical points.

$f_{xx}(x, y) = 24x - 24; f_{yy}(x, y) = 2, f_{xy}(x, y) = 0$

$D(x, y) = f_{xx}(x, y) f_{yy}(x, y) - [f_{xy}(x, y)]^2$

$D(x, y) = 2(24x - 24) = 48x - 48$

$D(3, 1) = 48(3) - 48 = 96 > 0$

$f_{xx}(3, 1) = 24(3) - 24 = 48 > 0$, thus $(3, 1, -109)$ is a relative minimum point.

$D(-1, 1) = 48(-1) - 48 = -96 < 0$, thus $(-1, 1, 19)$ is a saddle point.

25. $500 = \ell wh$ and $\ell w + 2\ell h + 2wh = M; \; h = 500/\ell w;$

$M = \ell w + 2\ell (500/\ell w) + 2\ell (500/\ell w) = \ell w + 1000/\ell + 1000/\ell$

$\partial M/\partial \ell = w - 1000/\ell^2 = 0; \; \partial M/\partial w = \ell - 1000/w^2 = 0;$

$w\ell^2 = 1000; \; \ell w^2 = 1000; \; \ell = 10 \text{ cm}, w = 10 \text{ cm}, h = 5 \text{ cm}$

27. $x + y + z = 30; \; M = xyz = xy(30 - x - y) = 30xy - x^2y - xy^2$

$\partial M/\partial x = 30y - 2xy - y^2 = y(30 - 2x - y) = 0$

$\partial M/\partial y = 30x - x^2 - 2xy = x(30 - x - 2y) = 0$

$x = 0, y = 0$ gives a min.

$2x + y = 30; \; x + 2y = 30; \; x = 10, y = 10, z = 10$

Section 8.4

1. $\displaystyle\int_0^2 \int_0^{3x} (x + y)\, dy\, dx = \int_0^2 (xy + y^2/2) \Big|_0^{3x} dx = \int_0^2 (3x^2 + 9x^2/2)\, dx = \int_0^2 15x^2/2\, dx = 5x^3/2 \Big|_0^2 = 20$

3. $\displaystyle\int_0^1 \int_y^{2y} (3x^2 + xy)\, dx\, dy = \int_0^1 (x^3 + (1/2)x^2 y) \Big|_y^{2y} dy = \int_0^1 17y^3/2\, dy = 17y^4/8 \Big|_0^1 = 17/8$

5. $\displaystyle\int_{-1}^1 \int_x^{x^2} (4xy + 9x^2 + 6y)\, dy\, dx = \int_{-1}^1 (2xy^2 + 9x^2 y + 3y^2) \Big|_x^{x^2} dx$

$= \displaystyle\int_{-1}^1 (2x^5 + 12x^4 - 11x^3 - 3x^2)\, dx = (x^6/3 + 12x^5/5 - 11x^4/4 - x^3) \Big|_{-1}^1 = \dfrac{14}{5}$

7. $\displaystyle\int_0^1 \int_0^{\sqrt{1-y^2}} (x + y)\, dx\, dy = \int_0^1 (x^2/2 + xy) \Big|_0^{\sqrt{1-y^2}} dy = \int_0^1 \left[\left(\dfrac{1 - y^2}{2} + y\sqrt{1 - y^2} \right) \right] dy$

$= [y/2 - y^3/6 - (1/3)(1 - y^2)^{3/2}] \Big|_0^1 = \dfrac{2}{3}$

9. $\displaystyle\int_0^1 \int_0^x e^{x+y}\, dy\, dx = \int_0^1 e^{x+y} \Big|_0^x dx = \int_0^1 (e^{2x} - e^x)\, dx = (1/2)e^{2x} - e^x \Big|_0^1 = e^2/2 - e + 1/2$

11. $\displaystyle\int_0^3\int_0^{4y}\sqrt{y^2+16}\,dx\,dy = \int_0^3 x\sqrt{y^2+16}\ \Big|_0^{4y}\,dy = \int_0^3 4y\sqrt{y^2+16}\,dy = 2[(2/3)(y^2+16)^{3/2}]\ \Big|_0^3$

$= (4/3)(25)^{3/2} - (4/3)(16)^{3/2} = 244/3$

13. $\displaystyle\int_0^{\pi/2}\int_0^x \cos x\sin y\,dy\,dx\ \int_0^{\pi/2}(-\cos x\cos y)\ \Big|_0^x dx = \int_0^{\pi/2}(-\cos^2 x + \cos x)\,dx$

$= \displaystyle\int_0^{\pi/2}(-1/2-(1/2)\cos 2x + \cos x)\,dx = (-x/2 - (1/4)\sin 2x + \sin x)\ \Big|_0^{\pi/2} = 1 - \pi/4$

15. Region bounded in xy-plane by $x=0$, $y=0$, $y=12-3x$.

$V = \displaystyle\int_0^4\int_0^{12-3x}(2-x/2-y/6)\,dy\,dx = \int_0^4 (2y - xy/2 - y^2/12)\ \Big|_0^{12-3x} dx$

$= \displaystyle\int_0^4 (12-6x+3x^2/4)dx = (12x-3x^2+x^3/4)\ \Big|_0^4 = 16$

17. Region bounded in xy-plane by $x=2$, $y=x$, $y=0$.

$V = \displaystyle\int_0^2\int_0^x xy\,dy\,dx = \int_0^2 (1/2)xy^2\Big|_0^x dx = \int_0^2 x^3/2\,dx = x^4/8\ \Big|_0^2 = 2$

19. $V = 2\displaystyle\int_0^2\int_0^{\sqrt{4-x^2}}(2x+3y)\,dy\,dx = \int_0^2 (4xy+3y)^2\ \Big|_0^{\sqrt{4-x^2}} dx$

$= \displaystyle\int_0^2 [4x\sqrt{4-x^2}+3(4-x^2)]dx = [(-4/3)(4-x^2)^{3/2}+12x-x^3]\ \Big|_0^2 = 16$

Chapter 8 Review

1. Plane
2. Elliptic paraboloid
3. Hyperboloid of one sheet

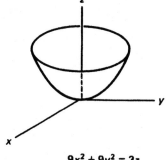

$3x + 6y + 4z = 36$

$9x^2 + 9y^2 = 3z$

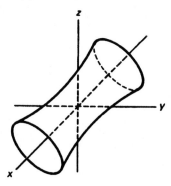

$36y^2 + 9z^2 - 16x^2 = 144$

Section 8.4 – Chapter 8 Review

Solutions to Odd-Numbered Exercises

4. Ellipsoid

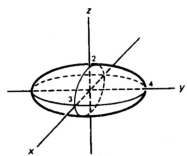

$$16x^2 + 9y^2 + 36z^2 = 144$$

5. Cylindrical surface

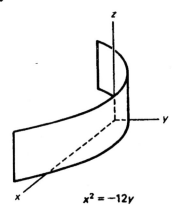

$$x^2 = -12y$$

6. Hyperbolic paraboloid

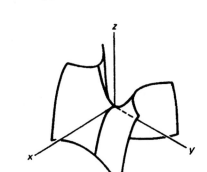

$$9x^2 - 9y^2 = 3z$$

7. Hyperboloid of two sheets

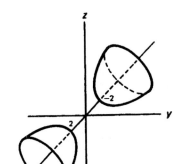

$$9y^2 - 36x^2 - 16z^2 = 144$$

8. Elliptic cone

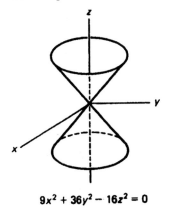

$$9x^2 + 36y^2 - 16z^2 = 0$$

9. $(x+3)^2 + (y-2)^2 + (z-1)^2 = 9$

10. $d = \sqrt{(1-4)^2 + (2+2)^2 + (9-4)^2} = 5\sqrt{2}$

11. a) $\partial z/\partial x = 3x^2 + 6xy$
    b) $\partial z/\partial y = 3x^2 + 4y$

12. a) $\partial z/\partial x = 6x\,e^{2y}$
    b) $\partial z/\partial y = 6x^2\,e^{2y}$

13. a) $\dfrac{\partial z}{\partial x} = \dfrac{1}{3x^2 y}(6xy) = \dfrac{2}{x}$
    b) $\dfrac{\partial z}{\partial y} = \dfrac{1}{3x^2 y}(3x^2) = \dfrac{1}{y}$

14. a) $\partial z/\partial x = 3\sin 3y \cos 3x$
    b) $\partial z/\partial y = 3\sin 3x \cos 3y$

15. a) $\dfrac{\partial z}{\partial x} = \dfrac{(x\ln y)2xy\,e^{x^2} - y\,e^{x^2}\ln y}{(x\ln y)^2} = \dfrac{y\,e^{x^2}\ln y(2x^2 - 1)}{x^2\ln^2 y} = \dfrac{y\,e^{x^2}(2x^2 - 1)}{x^2\ln y}$

    b) $\dfrac{\partial z}{\partial y} = \dfrac{(x\ln y)\,e^{x^2} - y\,e^{x^2}(x/y)}{(x\ln y)^2} = \dfrac{x\,e^{x^2}(\ln y - 1)}{x^2\ln^2 y} = \dfrac{e^{x^2}(\ln y - 1)}{x\ln^2 y}$

Chapter 8 Review

16. a) $\dfrac{\partial z}{\partial x} = \dfrac{y \sin x (e^x \sin y) - e^x \sin y (y \cos x)}{y^2 \sin^2 x} = \dfrac{e^x \sin y (\sin x - \cos x)}{y \sin^2 x}$

b) $\dfrac{\partial z}{\partial y} = \dfrac{y \sin x (e^x \cos y) - e^x \sin y (\sin x)}{y^2 \sin^2 x} = \dfrac{e^x (y \cos y - \sin y)}{y^2 \sin x}$

17. $\partial v / \partial w = (1/2)(p/w)^{-1/2}(-p/w^2) = -p^{1/2}/(2w^{3/2})$

18. $\partial U / \partial I = 2[(1/2)LI] = LI$

19. $\dfrac{\partial Z}{\partial X_C} = X_C / \sqrt{R^2 + X_C^2}$

20. $\dfrac{\partial V}{\partial R} = \dfrac{(R+r)E - RE(1)}{(R+r)^2} = \dfrac{rE}{(R+r)^2}$

21. a) $\partial z / \partial x = y + 6x - 4 \Big|_{(2,\,0,\,4)} = 8$

b) $\partial z / \partial y = 2y + x \Big|_{(2,\,0,\,4)} = 2$

22. $dz = \dfrac{1}{\sqrt{x^2 + xy}} \cdot \dfrac{1}{2}(x^2 + xy)^{-1/2}(2x + y)\, dx + \dfrac{1}{\sqrt{x^2 + xy}} \cdot \dfrac{1}{2}(x^2 + xy)^{-1/2} x\, dy$

$= \dfrac{2x + y}{2(x^2 + xy)}\, dx + \dfrac{x}{2(x^2 + xy)}\, dy$

23. $dz = \dfrac{(x+y)(1) - (x-y)(1)}{(x+y)^2}\, dx + \dfrac{(x+y)(-1) - (x-y)(1)}{(x+y)^2}\, dy = \dfrac{2y}{(x+y)^2}\, dx - \dfrac{2x}{(x+y)^2}\, dy$

24. $dV = (\partial V / \partial r)\, dr + (\partial V / \partial h)\, dh = 2\pi r h\, dr + \pi r^2\, dh$ (2) [(2) for top and bottom]

$= 2\pi(2)(10)(0.003) + \pi(2)^2(0.003)(2) = 0.452\ \text{m}^3 = 452\ \text{L}$

Note: $1\ \text{m}^3 = 1000\ \text{L}$

25. $di = -50\, e^{-t/50R} \cdot \dfrac{tR^{-2}}{50}\, dr + (-50)e^{-t/50R} \cdot \dfrac{-1}{50R}\, dt$

$= (e^{-t/50R}/R)[dt - (t/R)\, dr] = \dfrac{e^{(-6)/[50 \cdot 150]}}{150}\ [0.5 - (6/150)(10)] = 0.000666\ A$

26. $f(x, y) = z = x^2 + 2xy - y^2 - 14x - 6y + 8$

$f_x(x, y) = 2x + 2y - 14;\ f_y(x, y) = 2x - 2y - 6$

Set each expression equal to zero and solve the resulting system of equations.

$2x + 2y - 14 = 0$

$2x - 2y - 6 = 0$

Thus $x = 5$ and $y = 2$

$z = 5^2 + 2(5)(2) - 2^2 - 14(5) - 6(2) + 8 = -33.$

Chapter 8 Review

So (5, 2, –33) is a critical point.

$f_{xx}(x, y) = 2$; $f_{yy}(x, y) = -2$; $f_{xy}(x, y) = 2$

$D(x, y) = f_{xx}(x, y)\, f_{yy}(x, y) - [f_{xy}(x, y)]^2$

$D(5, 2) = 2(-2) - (2)^2 = -8 < 0$, thus

(5, 2, –33) is a saddle point.

27. $f(x, y) = z = y^2 - xy - x^2 + 4x - 3y - 6$

$f_x(x, y) = -y - 2x + 4$; $f_y(x, y) = 2y - x - 3$

Set each expression equal to zero and solve the resulting system of equations.

$-y - 2x + 4 = 0$

$2y - x - 3 = 0$

Thus $x = 1$ and $y = 2$.

$z = 2^2 - 1(2) - (1)^2 + 4(1) - 3(2) - 6 = -7$

(1, 2, –7) is a critical point.

$f_{xx}(x, y) = -2$; $f_{yy}(x, y) = 2$; $f_{xy}(x, y) = -1$

$D(x, y) = f_{xx}(x, y)\, f_{yy}(x, y) - [f_{xy}(x, y)]^2$

$D(1, 2) = -2(2) - (-1)^2 = -5 < 0$, thus

(1, 2, –7) is a saddle point.

28. $f(x, y) = z = -x^2 + xy - y^2 + 4x - 8y + 9$

$f_x(x, y) = -2x + y + 4$; $f_y(x, y) = x - 2y - 8$

Set each expression equal to zero and solve the resulting system of equations.

$-2x + y + 4 = 0$

$x - 2y - 8 = 0$

Thus $x = 0$ and $y = -4$. Then

$z = 0 + 0 - (-4)^2 + 0 - 8(-4) + 9 = 25$.

(0, –4, 25) is a critical point.

$f_{xx}(x, y) = -2$; $f_{yy}(x, y) = -2$; $f_{xy}(x, y) = 1$

$D(x, y) = f_{xx}(x, y)\, f_{yy}(x, y) - [f_{xy}(x, y)]^2$

$D(0, -4) = -2(-2) - (1)^2 = 3 > 0$

$f_{xx}(0, -4) = -2 < 0$, thus (0, –4, 25) is a relative maximum point.

29. $V = 8 = \ell\, wh$ and $C = 2(\ell\, w) + 2(\ell\, h) + 1(2wh)$

$h = 8/\ell\, w$ $\qquad\qquad C = 2\ell\, w + 16/w + 16/\ell$

$\partial C / \partial \ell = 2w - 16/\ell^2 = 0$; $w\ell^2 = 8$

$\partial C / \partial w = 2\ell - 16/w^2 = 0$; $\ell w^2 = 8$ Thus $\ell = 2$, $w = 2$, $h = 2$

Chapter 8 Review

30. $\displaystyle\int_0^1\int_0^{x^2}(x-2y)\,dy\,dx = \int_0^1(xy-y^2)\Big|_0^{x^2}dx = \int_0^1(x^3-x^4)\,dx = \frac{x^4}{4}-\frac{x^5}{5}\Big|_0^1 = \frac{1}{20}$

31. $\displaystyle\int_0^2\int_y^{4y}(4xy+6x^2-9y^2)\,dx\,dy = \int_0^2(2x^2y+2x^3-9xy^2)\Big|_y^{4y} = \int_0^2 129y^3\,dy = (129/4)y^4\Big|_0^2 = 516$

32. $\displaystyle\int_1^3\int_0^{\ln y}y\,e^x\,dx\,dy = \int_1^3 y\,e^x\Big|_0^{\ln y}dy = \int_1^3(y\,e^{\ln y}-y)\,dy$

$\displaystyle = \int_1^3(y^2-y)\,dy = y^3/3-y^2/2\Big|_1^3 = 14/3$

33. $\displaystyle\int_0^{\pi/4}\int_0^x\sec^2 y\,dy\,dx = \int_0^{\pi/4}\tan y\Big|_0^x dx = \int_0^{\pi/4}\tan x\,dx = \ln|\sec x|\Big|_0^{\pi/4} = \ln\sqrt{2} = (1/2)\ln 2$

34. The region is bounded in the xy-plane by $x=0$, $y=0$, $y=b(1-x/a)$.

$$V = \int_0^a\int_0^{b(1-x/a)}c(1-x/a-y/b)\,dy\,dx$$

First, find the first integral:

$$\int_0^{b(1-x/a)}c(1-x/a-y/b)\,dy = c[(1-x/a)y-y^2/2b]\Big|_0^{b(1-x/a)} = \frac{cy}{2b}[2b(1-x/a)-y]\Big|_0^{b(1-x/a)}$$

$$= \frac{c}{2b}\cdot b(1-x/a)[2b(1-x/a)-b(1-x/a)] = (c/2)(1-x/a)\cdot b(1-x/a) = (bc/2)(1-x/a)^2.$$

Now integrate *wrt* x:

$$V = \int_0^a(bc/2)(1-x/a)^2\,dx = (bc/2)(-a)(1/3)(1-x/a)^3\Big|_0^a = (-abc/6)(1-x/a)^3\Big|_0^a = (1/6)abc$$

35. $\displaystyle V = \int_0^1\int_0^{\sqrt{1-x^2}}2x\,dy\,dx = \int_0^1 2xy\Big|_0^{\sqrt{1-x^2}}dx = \int_0^1 2x\sqrt{1-x^2}\,dx = (-2/3)(1-x^2)^{3/2}\Big|_0^1 = 2/3$

36. $\displaystyle V = \int_0^1\int_0^{1-x^2}(1-y-x^2)\,dy\,dx = \int_0^1(y-y^2/2-x^2 y)\Big|_0^{1-x^2}dx$

$\displaystyle = \int_0^1(1/2-x^2+x^4/2)\,dx = (x/2-x^3/3+x^5/10)\Big|_0^1 = 4/15$

Chapter 8 Review

CHAPTER 9

Section 9.1

1. $2, 5, 8, ..., n = 6$
$d = 5 - 2 = 3$
$\ell = a + (n-1)d$
$\ell = 2 + (6-1)(3)$
$\ell = 2 + 5(3) = 17$

3. $3, 4.5, 6, ..., n = 15$
$d = 4.5 - 3 = 1.5$
$\ell = a + (n-1)d$
$\ell = 3 + (15-1)(1.5)$
$\ell = 3 + (14)(1.5) = 24$

5. $4, -5, -14, ..., n = 12$
$d = -5 - 4 = -9$
$\ell = a + (n-1)d$
$\ell = 4 + (12-1)(-9)$
$\ell = 4 + (11)(-9) = -95$

7. Corresponds to problem 1.
$n = 6, a = 2, \ell = 17$
$S_n = \frac{n}{2}(a + \ell)$
$S_6 = \frac{6}{2}(2 + 17)$
$S_6 = 3(19) = 57$

9. Corresponds to problem 3.
$n = 15, a = 3, \ell = 24$
$S_n = \frac{n}{2}(a + \ell)$
$S_{15} = \frac{15}{2}(3 + 24)$
$S_{15} = 7.5(27) = 202.5$

11. Corresponds to problem 5.
$n = 12, a = 4, \ell = -95$
$S_n = \frac{n}{2}(a + \ell)$
$S_{12} = \frac{12}{2}(4 + (-95))$
$S_{12} = 6(-91) = -546$

13. $a = 2, d = -3$
$2, 2 + (-3), 2 + 2(-3),$
$2 + 3(-3), 2 + 4(-3)$
ie $2, -1, -4, -7, -10$

15. $a = 5, d = \frac{2}{3}$
$5, 5 + \frac{2}{3}, 5 + 2\left(\frac{2}{3}\right),$
$5 + 3\left(\frac{2}{3}\right), 5 + 4\left(\frac{2}{3}\right)$
i.e., $5, 5\frac{2}{3}, 6\frac{1}{3}, 7, 7\frac{2}{3}$

17. $n = 10, \ell = 12, S_{10} = 80$
$S_n = \frac{n}{2}(a + \ell)$
$S_{10} = \frac{10}{2}(a + 12)$
$80 = 5a + 60$
$20 = 5a$
$4 = a$

19. $a = 1, d = 2, n = 1000$
$\ell = a + (n-1)d$
$\ell = 1 + (1000-1)(2)$
$\ell = 1999$
$S_n = \frac{n}{2}(a + \ell)$
$S_{1000} = \frac{1000}{2}(1 + 1999)$
$S_{1000} = 1,000,000$

21. $a = \$24,000, d = \$800, n = 10$
$\ell = a + (n-1)d$
$\ell = 24,000 + (10-1)(800)$
$\ell = 24,000 + 9(800)$
$\ell = \$31,200$

Section 9.1

Section 9.2

1. $20, \dfrac{20}{3}, \dfrac{20}{9}, \ldots, n = 8$

   $\ell = ar^{n-1}, \; a = 20, \; r = \dfrac{1}{3}$

   $\ell = 20\left(\dfrac{1}{3}\right)^{8-1} = 20\left(\dfrac{1}{3}\right)^{7}$

   $\ell = \dfrac{20}{2187}$

3. $\sqrt{2}, 2, 2\sqrt{2}, \ldots, n = 6$

   $a = \sqrt{2}, \; r = \sqrt{2}$

   $\ell = ar^{n-1}$

   $\ell = \sqrt{2}\left(\sqrt{2}\right)^{6-1}$

   $\ell = \sqrt{2}\left(\sqrt{2}\right)^{5} = 8$

5. $8, -4, 2, \ldots, n = 10$

   $a = 8, \; r = \dfrac{-1}{2}$

   $\ell = ar^{n-1}$

   $\ell = 8\left(\dfrac{-1}{2}\right)^{10-1} = 8\left(\dfrac{-1}{2}\right)^{9}$

   $\ell = \dfrac{-1}{64}$

7. Corresponds to problem 1.

   $a = 20, \; r = \dfrac{1}{3}, \; n = 8$

   $S_n = \dfrac{a(1-r^n)}{1-r}$

   $S_8 = \dfrac{20\left(1-\left(\dfrac{1}{3}\right)^{8}\right)}{1-\dfrac{1}{3}}$

   $S_8 = \dfrac{20\left(1-\dfrac{1}{6561}\right)}{\dfrac{2}{3}}$

   $S_8 = 20\left(\dfrac{3}{2}\right)\left(\dfrac{6560}{6561}\right)$

   $S_8 = 30\left(\dfrac{6560}{6561}\right)$

   $S_8 = \dfrac{65,600}{2187}$

9. Corresponds to problem 3.

   $a = \sqrt{2}, \; r = \sqrt{2}, \; n = 6$

   $S_n = \dfrac{a(1-r^n)}{1-r}$

   $S_6 = \dfrac{\sqrt{2}\left(1-\left(\sqrt{2}\right)^{6}\right)}{1-\sqrt{2}}$

   $S_6 = \dfrac{\sqrt{2}(1-8)}{1-\sqrt{2}}$

   $S_6 = \dfrac{-7\sqrt{2}}{\left(1-\sqrt{2}\right)} \cdot \dfrac{\left(1+\sqrt{2}\right)}{\left(1+\sqrt{2}\right)}$

   $S_6 = \dfrac{-7\sqrt{2}\left(1+\sqrt{2}\right)}{1-2}$

   $S_6 = 7\sqrt{2} + 7(2)$

   $S_6 = 7\sqrt{2} + 14$

11. Corresponds to problem 5.

   $a = 8, \; r = \dfrac{-1}{2}, \; n = 10$

   $S_n = \dfrac{a(1-r^n)}{1-r}$

   $S_{10} = \dfrac{8\left(1-\left(\dfrac{-1}{2}\right)^{10}\right)}{1-\left(\dfrac{-1}{2}\right)}$

   $S_{10} = \dfrac{8\left(1-\dfrac{1}{1024}\right)}{\dfrac{3}{2}}$

   $S_{10} = 8\left(\dfrac{2}{3}\right)\left(\dfrac{1023}{1024}\right)$

   $S_{10} = \dfrac{341}{64}$

13. $a = 3, \; r = \dfrac{1}{2}$

   $3, 3\left(\dfrac{1}{2}\right), 3\left(\dfrac{1}{2}\right)^{2},$

   $3\left(\dfrac{1}{2}\right)^{3}, 3\left(\dfrac{1}{2}\right)^{4}$

   i.e.: $3, \dfrac{3}{2}, \dfrac{3}{4}, \dfrac{3}{8}, \dfrac{3}{16}$

Section 9.2

15. $a = 5,\ r = \dfrac{-1}{4}$

$$5, 5\left(\frac{-1}{4}\right), 5\left(\frac{-1}{4}\right)^2,$$

$$5\left(\frac{-1}{4}\right)^3, 5\left(\frac{-1}{4}\right)^4$$

i.e.: $5, \dfrac{-5}{4}, \dfrac{5}{16}, \dfrac{-5}{64}, \dfrac{5}{256}$

17. $a = -4,\ r = 3$

$$-4,\ -4(3),\ -4(3)^2,$$

$$-4(3)^3,\ -4(3)^4$$

i.e.: $-4, -12, -36, -108, -324$

19. $a = 6,\ \ell = \dfrac{3}{4},\ n = 4$

$\ell = ar^{n-1}$

$\dfrac{3}{4} = 6r^3$

$\dfrac{1}{8} = r^3$

$r = \dfrac{1}{2}$

21. The amount at the end of the first

year is $\$1000(1.0575) = \$1057.50$, so $a = \$1057.50$

$r = 1.0575$ and $n = 10$.

$$S_n = \frac{a(1-r^n)}{1-r}$$

$$S_{10} = \frac{1057.50(1-1.0575^{10})}{1-1.0575}$$

$$S_{10} = \$13,776$$

23. Bounce one results in the second height, thus, bounce five results in the sixth height.

$r = \dfrac{1}{2},\ n = 6,\ a = 12\,\text{ft}$

$\ell = ar^{n-1} = 12\left(\dfrac{1}{2}\right)^{6-1} = 12\left(\dfrac{1}{2}\right)^5 = \dfrac{12}{32} = \dfrac{3}{8}\,\text{ft}$

25. After minute 8 is the beginning of minute 9.

$a = 90°C,\ r = \dfrac{4}{5},\ n = 9$

$\ell = ar^{n-1} = 90\left(\dfrac{4}{5}\right)^{9-1} = 90\left(\dfrac{4}{5}\right)^8 = 15.1°C$

27. $4 + \dfrac{4}{7} + \dfrac{4}{49} + \ldots + 4\left(\dfrac{1}{7}\right)^{n-1} + \ldots$

$a = 4,\ r = \dfrac{1}{7},\ S = \dfrac{a}{1-r}.$

Thus $S = \dfrac{4}{1-\dfrac{1}{7}} = \dfrac{4}{\dfrac{6}{7}} = \dfrac{14}{3}$

29. $3 - \dfrac{3}{8} + \dfrac{3}{64} - \ldots + 3\left(\dfrac{-1}{8}\right)^{n-1} + \ldots$

$a = 3,\ r = \dfrac{-1}{8},\ S = \dfrac{a}{1-r}.$

Thus $S = \dfrac{3}{1-\left(\dfrac{-1}{8}\right)} = \dfrac{3}{\dfrac{9}{8}} = \dfrac{8}{3}$

Section 9.2

31. $4 + 12 + 36 + \ldots + 4(3)^{n-1} + \ldots$

Since $r = 3$, there is no sum.

33. $5 + 1 + 0.2 + 0.04 + \ldots + 5(0.2)^{n-1} + \ldots$

$a = 5, r = 0.2$

$$S = \frac{a}{1-r} = \frac{5}{1-0.2} = 6.25$$

35. $0.333\ldots$

$0.33\overline{3} = 0.3 + 0.03 + 0.003\ldots$

Then $a = 0.3$ and $r = 0.1$

$$S = \frac{a}{1-r} = \frac{0.3}{1-0.1} = \frac{0.3}{0.9} = \frac{1}{3}$$

37. $0.01\overline{212}$

$0.01\overline{212} = 0.012 + 0.00012 + 0.0000012\ldots$

$a = 0.012, r = 0.01$

$$S = \frac{a}{1-r} = \frac{0.012}{(1-0.01)} = \frac{0.012}{0.99} = \frac{12}{990} = \frac{2}{165}$$

39. $0.8\overline{666}$

$0.8\overline{666} = 0.8 + 0.06 + 0.006 + \ldots$

For $0.06 + 0.006 + \ldots, a = 0.06, r = 0.1,$

$$S = \frac{a}{1-r} = \frac{0.06}{1-0.1} = \frac{0.06}{0.9} = \frac{6}{90} = \frac{1}{15}$$

Thus $0.8\overline{666} = 0.8 + \dfrac{1}{15} = \dfrac{8}{10} + \dfrac{1}{15} = \dfrac{12}{15} + \dfrac{1}{15} = \dfrac{13}{15}$

Section 9.3

Each problem in this section will be done two ways – the first, using the formulas illustrated in the text; the second method will use the notation $_nC_r$, which is found on both the TI-83 and TI-89 \ calculators. For the TI-83, it is under MATH, PRB. Enter n before the symbol $_nC_r$. After $_nC_r$, enter r. For the TI-89, it is under CATALOG alpha n. After getting the symbol $_nC_r$, enter n,r). For $(a+b)^n$, the kth term is $_nC_{k-1}(a)^{n-(k-1)}(b)^{k-1}$.

The formula we are using is in section 9.3.

1. $(3x+y)^3 = (3x)^3 + 3(3x)^2(y)^1 + \dfrac{3(2)}{2!}(3x)^1 y^2 + \dfrac{3(2)(1)}{3!}(3x)^0 y^3$

$(3x+y)^3 = 27x^3 + 27x^2y + 9xy^2 + y^3$

Method 2

$(3x+y)^3 = {_3C_0}(3x)^3 y^0 + {_3C_1}(3x)^2 y^1 + {_3C_2}(3x)^1 y^2 + {_3C_3}(3x)^0 y^3$

$(3x+y)^3 = 1(27x^3) + 3(9x^2)y + 3(3x)y^2 + 1(1)y^3$

$(3x+y)^3 = 27x^3 + 27x^2y + 9xy^2 + y^3$

3.

$$(a-2)^5 = a^5 + 5a^4(-2)^1 + \frac{5(4)}{2!}a^3(-2)^2 + \frac{5(4)(3)}{3!}a^2(-2)^3$$

$$+ \frac{5(4)(3)(2)}{4!}a^1(-2)^4 + \frac{5(4)(3)(2)(1)}{5!}a^0(-2)^5$$

$$(a-2)^5 = a^5 - 10a^4 + 40a^3 - 80a^2 + 80a - 32$$

Method 2

$$(a-2)^5 = {}_5C_0(a)^5(-2)^0 + {}_5C_1(a^4)(-2)^1 + {}_5C_2a^3(-2)^2 + {}_5C_3a^2(-2)^3$$

$$+ {}_5C_4a^1(-2)^4 + {}_5C_5a^0(-2)^5$$

$$(a-2)^5 = a^5 + 5a^4(-2) + 10a^3(-2)^2 + 10a^2(-2)^3$$

$$+ 5a(-2)^4 + 1(1)(-2)^5$$

$$(a-2)^5 = a^5 - 10a^4 + 40a^3 - 80a^2 + 80a - 32$$

5.

$$(2x-1)^4 = (2x)^4 + 4(2x)^3(-1)^1 + \frac{4(3)}{2!}(2x)^2(-1)^2$$

$$+ \frac{4(3)(2)}{3!}(2x)^1(-1)^3 + \frac{4(3)(2)(1)}{4!}(2x)^0(-1)^4$$

$$(2x-1)^4 = 16x^4 - 32x^3 + 24x^2 - 8x + 1$$

Method 2

$$(2x-1)^4 = {}_4C_0(2x)^4 + {}_4C_1(2x)^3(-1)^1 + {}_4C_2(2x)^2 + (-1)^2$$

$$+ {}_4C_3(2x)^1(-1)^3 + {}_4C_4(2x)^0(-1)^4$$

$$(2x-1)^4 = 1(2x)^4 + 4(2x)^3(-1)^1 + 6(2x)^2(-1)^2 + 4(2x)(-1)^3 + 1(2x)^0(-1)^4$$

$$(2x-1)^4 = 16x^4 - 32x^3 + 24x^2 - 8x + 1$$

Section 9.3

7. $$(2a+3b)^6 = (2a)^6 + 6(2a)^5(3b)^1 + \frac{6\cdot5}{2!}(2a)^4(3b)^2$$

$$+\frac{6\cdot5\cdot4}{3!}(2a)^3(3b)^3 + \frac{6\cdot5\cdot4\cdot3}{4!}(2a)^2(3b)^4$$

$$+\frac{6\cdot5\cdot4\cdot3\cdot2}{5!}(2a)^1(3b)^5 + \frac{6\cdot5\cdot4\cdot3\cdot2\cdot1}{6!}(2a)^0(3b)^6$$

$$(2a+3b)^6 = 64a^6 + 576a^5b + 2160a^4b^2 + 4320a^3b^3$$

$$+4860a^2b^4 + 2916ab^5 + 729b^6$$

Method 2

$$(2a+3b)^6 = {_6}C_0(2a)^6(3b)^0 + {_6}C_1(2a)^5(3b)^1 + {_6}C_2(2a)^4(3b)^2$$

$$+ {_6}C_3(2a)^3(3b)^3 + {_6}C_4(2a)^2(3b)^4 + {_6}C_5(2a)^1(3b)^5 + {_6}C_6(3b)^6$$

$$(2a+3b)^6 = 1(64a^6) + 6(32)(a^5)(3)b + 15(16a^4)(9b^2)$$

$$+ 20(2^3)a^3(3^3)b^3 + 15(2^2)a^2(3^4)b^4 + 6(2a)(3^5b^5) + 1(3^6)b^6$$

$$(2a+3b)^6 = 64a^6 + 576a^5b + 2160a^4b^2 + 4320a^3b^3 + 4860a^2b^4 + 2916ab^5 + 729b^6$$

9. $$\left(\frac{2}{3}x - 2\right)^5 = \left(\frac{2}{3}x\right)^5 + 5\left(\frac{2}{3}x\right)^4(-2)^1 + \frac{5\cdot4}{2!}\left(\frac{2}{3}x\right)^3(-2)^2$$

$$+\frac{5\cdot4\cdot3}{3!}\left(\frac{2}{3}x\right)^2(-2)^3 + \frac{5\cdot4\cdot3\cdot2}{4!}\left(\frac{2}{3}x\right)(-2)^4$$

$$+\frac{5\cdot4\cdot3\cdot2\cdot1}{5!}\left(\frac{2}{3}x\right)^0(-2)^5$$

$$\left(\frac{2}{3}x - 2\right)^5 = \frac{32}{243}x^5 - \frac{160}{81}x^4 + \frac{320}{27}x^3 - \frac{320}{9}x^2 + \frac{160}{3}x - 32$$

Method 2

$$\left(\frac{2}{3}x - 2\right)^5 = {_5}C_0\left(\frac{2}{3}x\right)^5 + {_5}C_1\left(\frac{2}{3}x\right)^4(-2)^1 + {_5}C_2\left(\frac{2}{3}x\right)^3(-2)^2$$

$$+ {_5}C_3\left(\frac{2}{3}x\right)^2(-2)^3 + {_5}C_4\left(\frac{2}{3}x\right)(-2)^4 + {_5}C_5\left(\frac{2}{3}x\right)^0(-2)^5$$

$$\left(\frac{2}{3}x - 2\right)^5 = \left(\frac{2}{3}\right)^5 x^5 + 5\left(\frac{2}{3}\right)^4 x^4(-2) + 10\left(\frac{2}{3}\right)^3 x^3(-2)^2$$

$$+ 10\left(\frac{2}{3}\right)^2 x^2(-2)^3 + 5\left(\frac{2}{3}\right)x(-2)^4 + 1(-2)^5$$

$$\left(\frac{2}{3}x - 2\right)^5 = \frac{32}{243}x^5 - \frac{160}{81}x^4 + \frac{320}{27}x^3 - \frac{320}{9}x^2 + \frac{160}{3}x - 32$$

Section 9.3

11.
$$\left(a^{\frac{1}{2}}+3b^2\right)^4 = \left(a^{\frac{1}{2}}\right)^4 + 4\left(a^{\frac{1}{2}}\right)^3\left(3b^2\right)^1 + \frac{4\cdot3}{2!}\left(a^{\frac{1}{2}}\right)^2\left(3b^2\right)^2$$

$$+ \frac{4\cdot3\cdot2}{3!}\left(a^{\frac{1}{2}}\right)\left(3b^2\right)^3 + \frac{4\cdot3\cdot2\cdot1}{4!}\left(a^{\frac{1}{2}}\right)^0\left(3b^2\right)^4$$

$$\left(a^{\frac{1}{2}}+3b^2\right)^4 = a^2 + 12a^{\frac{3}{2}}b^2 + 54ab^4 + 108a^{\frac{1}{2}}b^6 + 81b^8$$

Method 2

$$\left(a^{\frac{1}{2}}+3b^2\right)^4 = {_4}C_0\left(a^{\frac{1}{2}}\right)^4\left(3b^2\right)^0 + {_4}C_1\left(a^{\frac{1}{2}}\right)^3\left(3b^2\right)^1$$

$$+ {_4}C_2\left(a^{\frac{1}{2}}\right)^2\left(3b^2\right)^2 + {_4}C_3\left(a^{\frac{1}{2}}\right)\left(3b^2\right)^3 + {_4}C_4\left(a^{\frac{1}{2}}\right)^0\left(3b^2\right)^4$$

$$\left(a^{\frac{1}{2}}+3b^2\right)^4 = \left(a^{\frac{1}{2}}\right)^4\left(3b^2\right)^0 + 4\left(a^{\frac{1}{2}}\right)^3\left(3b^2\right) + 6\left(a^{\frac{1}{2}}\right)^2\left(3^2\right)\left(b^4\right)$$

$$+ 4\left(a^{\frac{1}{2}}\right)\left(3^3b^6\right) + 1\left(3^4\right)b^8$$

$$\left(a^{\frac{1}{2}}+3b^2\right)^4 = a^2 + 12a^{\frac{3}{2}}b^2 + 54ab^4 + 108a^{\frac{1}{2}}b^6 + 81b^8$$

13.
$$\left(\frac{x}{y}-\frac{2}{z}\right)^4 = \left(\frac{x}{y}\right)^4 + 4\left(\frac{x}{y}\right)^3\left(\frac{-2}{z}\right)^1 + \frac{4\cdot3}{2!}\left(\frac{x}{y}\right)^2\left(\frac{-2}{z}\right)^2$$

$$+ \frac{4\cdot3\cdot2}{3!}\left(\frac{x}{y}\right)\left(\frac{-2}{z}\right)^3 + \frac{4\cdot3\cdot2\cdot1}{4!}\left(\frac{x}{y}\right)^0\left(\frac{-2}{z}\right)^4$$

$$\left(\frac{x}{y}-\frac{2}{z}\right)^4 = \frac{x^4}{y^4} - \frac{8x^3}{y^3z} + \frac{24x^2}{y^2z^2} - \frac{32x}{yz^3} + \frac{16}{z^4}$$

Method 2

$$\left(\frac{x}{y}-\frac{2}{z}\right)^4 = {_4}C_0\left(\frac{x}{y}\right)^4\left(\frac{-2}{z}\right)^0 + {_4}C_1\left(\frac{x}{y}\right)^3\left(\frac{-2}{z}\right) + {_4}C_2\left(\frac{x}{y}\right)^2\left(\frac{-2}{z}\right)^2$$

$$+ {_4}C_3\left(\frac{x}{y}\right)\left(\frac{-2}{z}\right)^3 + {_4}C_4\left(\frac{x}{y}\right)^0\left(\frac{-2}{z}\right)^4$$

$$\left(\frac{x}{y}-\frac{2}{z}\right)^4 = \frac{x^4}{y^4} + 4\left(\frac{-2x^3}{y^3z}\right) + 6\frac{x^2}{y^2}\left(\frac{4}{z^2}\right)$$

$$+ 4\left(\frac{x}{y}\right)\left(\frac{-8}{z^3}\right) + 1\left(\frac{16}{z^4}\right)$$

$$\left(\frac{x}{y}-\frac{2}{z}\right)^4 = \frac{x^4}{y^4} - \frac{8x^3}{y^3z} + \frac{24x^2}{y^2z^2} - \frac{32x}{yz^3} + \frac{16}{z^4}$$

Section 9.3

15. $(x-y)^9$; 6th term. $k=6, n=9, a=x, 6=-y$

$$\text{kth term} = \frac{n(n-1)(n-2)...(n-[k-2])}{(k-1)!}a^{n-(k-1)}b^{k-1}$$

$$\text{6th term} = \frac{9(8)(7)(6)(5)}{(6-1)!}x^{9-5}(-y)^5$$

6th term $= -126x^4y^5$

Method 2

$$\text{kth term} =_n C_{k-1}a^{n-(k-1)}b^{k-1}$$

$$\text{6th term} =_9 C_5 (x)^4 (-y)^5$$

6th term $= -126x^4y^5$

17. $(2x-y)^{13}$; 9th term. $k=9, n=13, a=2x, b=-y$

$$\text{kth term} = \frac{n(n-1)(n-2)...(n-[k-2])}{(k-1)!}a^{n-(k-1)}b^{k-1}$$

$$\text{9th term} = \frac{13\cdot12\cdot11\cdot10\cdot9\cdot8\cdot7\cdot6}{(9-1)!}(2x)^5(-y)^8$$

9th term $= 1287(2^5)x^5(-y)^8 = 41,184x^5y^8$

Method 2

$$\text{kth term} =_n C_{k-1}a^{n-(k-1)}b^{k-1}$$

9th term $=_{13} C_8 (2x)^5 (-y)^8 = 1287(2^5)x^5(-y)^8 = 41,184x^5y^8$

19. $(2x+y^2)^7$; 5th term. $n=7, k=5, a=2x, b=y^2$

$$\text{kth term} = \frac{n(n-1)(n-2)...(n-[k-2])}{(k-1)!}a^{n-(k-1)}b^{k-1}$$

$$\text{5th term} = \frac{7\cdot6\cdot5\cdot4}{4!}(2x)^3(y^2)^4$$

5th term $= 35(2^3)x^3(y^8) = 280x^3y^8$

Method 2

$$\text{kth term} =_n C_{k-1}a^{n-(k-1)}b^{k-1}$$

5th term $=_7 C_4 (2x)^3 (y^2)^4 = 35(2^3)x^3y^8 = 280x^3y^8$

Section 9.3

21. $(3x+2y)^6$; middle term. There are 7 terms. The middle is the 4th. $n=6$, $k=4$, $a=3x$, $b=2y$

$$k\text{th term} = \frac{n(n-1)(n-2)...(n-[k-2])}{(k-1)!}a^{n-(k-1)}b^{k-1}$$

$$4\text{th term} = \frac{6\cdot 5\cdot 4}{(4-1)!}(3x)^3(2y)^3 = 20(3)^3 x^3(2)^3 y^3$$

$$4\text{th term} = 4320x^3 y^3$$

Method 2

$$k\text{th term} =_n C_{k-1}a^{n-(k-1)}b^{k-1}$$

$$4\text{th term} =_6 C_3(3x)^3(2y)^3 = 20(3)^3 x^3(2)^3 y^3$$

$$4\text{th term} = 4320x^3 y^3$$

23. $(2x-1)^{10}$; term containing x^5. If x is to the 5th power, then (-1) is also, since the sum of the exponents is n, i.e., 10. In the kth term, b's exponent is $k-1$. So, k must be 6. We are seeking the 6th term.

$n=10$, $k=6$, $a=2x$, $b=-1$

$$k\text{th term} = \frac{n(n-1)(n-2)...(n-[k-2])}{(k-1)!}a^{n-(k-1)}b^{k-1}$$

$$6\text{th term} = \frac{10\cdot 9\cdot 8\cdot 7\cdot 6}{(6-1)!}(2x)^5(-1)^5 = 252(2^5)x^5(-1)$$

$$6\text{th term} = -8064x^5$$

Method 2

$$k\text{th term} =_n C_{k-1}a^{n-(k-1)}b^{k-1}$$

$$6\text{th term} =_{10} C_5(2x)^5(-1)^5 = 252(2^5)x^5(-1)$$

$$6\text{th term} = -8064x^5$$

Section 9.3

1. $3, 7, 11, 15, \ldots, n = 12$. Arithmetic: $d = 7 - 3 = 4$.
   $\ell = a + (n - 1)d$; $\ell = 3 + (12 - 1)(4) = 47$

2. $4, 2, 1, \dfrac{1}{2}, \ldots, n = 7$. Geometric: $r = \dfrac{1}{2}$

   $\ell = ar^{n-1}$, $\ell = 4\left(\dfrac{1}{2}\right)^{6} = \dfrac{1}{16}$

3. $\sqrt{3}, -3, 3\sqrt{3}, -9, \ldots, n = 8$. Geometric: $r = -\sqrt{3}$

   $l = ar^{n-1}$; $\ell = \sqrt{3}\left(-\sqrt{3}\right)^{7} = -81$

4. $4, -2, -8, -14, \ldots, n = 12$. Arithmetic : $d = -6$
   $\ell = a + (n - 1)d$; $\ell = 4 + (12 - 1)(-6) = -62$

5. $6, 2, \dfrac{2}{3}, \dfrac{2}{9}, \ldots, n = 6$. Geometric: $r = \dfrac{1}{3}$

   $\ell = ar^{n-1}$; $\ell = 6\left(\dfrac{1}{3}\right)^{5} = \dfrac{2}{81}$

6. $5, 15, 25, 35, \ldots, n = 10$. Arithmetic: $d = 10$
   $\ell = a + (n - 1)d$; $\ell = 5 + (10 - 1)(10) = 95$

7. Corresponds to problem 1.

   $S_n = \dfrac{n}{2}(a + \ell)$

   $S_{12} = \dfrac{12}{2}(3 + 47) = 300$

8. Corresponds to problem 2.

   $S_n = \dfrac{a\left(1 - r^n\right)}{1 - r}$

   $S_{12} = \dfrac{4\left(1 - \left(\dfrac{1}{2}\right)^{7}\right)}{1 - \dfrac{1}{2}} = \dfrac{127}{16}$

Chapter 9 Review

9. Corresponds to problem 3.

$$S_n = \frac{a(1-r^n)}{1-r}$$

$$S_8 = \frac{\sqrt{3}\left(1-\left(-\sqrt{3}\right)^8\right)}{1-\left(-\sqrt{3}\right)}$$

$$S_8 = \frac{\sqrt{3}(1-81)}{1+\sqrt{3}} = \frac{-80\sqrt{3}}{1+\sqrt{3}}$$

10. Corresponds to problem 4.

$$S_n = \frac{n}{2}(a+\ell)$$

$$S_{12} = \frac{12}{2}\left(4+(-62)\right) = -348$$

11. Corresponds to problem 5.

$$S_n = \frac{a(1-r^n)}{1-r}$$

$$S_6 = \frac{6\left(1-\left(\frac{1}{3}\right)^6\right)}{1-\frac{1}{3}} = \frac{728}{81}$$

12. Corresponds to problem 6.

$$S_n = \frac{n}{2}(a+\ell)$$

$$S_{10} = \frac{10}{2}(5+95) = 500$$

13. $2, 4, 6, \ldots, n = 1000$. Arithmetic: $d = 2$.

$$\ell = a + (n-1)d$$

$$\ell = 2 + (1000-1)(2) = 2000$$

$$S_n = \frac{n}{2}(a+\ell)$$

$$S_{1000} = \frac{1000}{2}(2+2000) = 1,001,000$$

14. The amount at the end of the first year is $\$500(1.06) = \$530$, so $a = \$530$.

$r = 1.06$, and $n = 5$.

$$S_n = \frac{a(1-r^n)}{1-r}$$

$$S_5 = \frac{\$530\left(1-1.06^5\right)}{1-1.06} = \$2988$$

15. $3 + 6 + 12 + \ldots; r = 2$.

Since $r \geq 1$, no sum.

16. $5 + \dfrac{5}{7} + \dfrac{5}{49} + \ldots; r = \dfrac{1}{7}$.

$$S = \frac{a}{1-r} = \frac{5}{1-\frac{1}{7}} = \frac{35}{6}$$

17. $2 - \dfrac{2}{3} + \dfrac{2}{9} - \ldots; r = \dfrac{-1}{3}$.

$$S = \frac{a}{1-r} = \frac{2}{1+\frac{1}{3}} = \frac{3}{2}$$

18. $3 + \dfrac{9}{2} + \dfrac{27}{4} + \ldots; r = \dfrac{3}{2}$.

Since $r \geq 1$, no sum.

Chapter 9 Review

19. $0.454545... = 0.45 + 0.0045 + 0.000045 + ...;$

$r = 0.01, \ S = \dfrac{a}{1-r} = \dfrac{0.45}{1-0.01} = \dfrac{5}{11}$

20. $0.9212121... = 0.9 + 0.021 + 0.00021 + ...$

For: $0.021 + 0.00021 + ..., \ r = 0.01;$

$S = \dfrac{a}{1-r} = \dfrac{0.021}{1-0.01} = \dfrac{7}{330}$

So the whole sum is $0.9 + \dfrac{7}{330} = \dfrac{152}{165}$

21. $(a-b)^6 = a^6 + 6(a)^5(-b)^1 + \dfrac{6 \cdot 5}{2!}(a)^4(-b)^2$

$\qquad + \dfrac{6 \cdot 5 \cdot 4}{3!}(a)^3(-b)^3 + \dfrac{6 \cdot 5 \cdot 4 \cdot 3}{4!}(a)^2(-b)^4$

$\qquad + \dfrac{6 \cdot 5 \cdot 4 \cdot 3 \cdot 2}{5!}(a)^1(-b)^5 + \dfrac{6 \cdot 5 \cdot 4 \cdot 3 \cdot 2 \cdot 1}{6!}(-b)^6$

$(a-b)^6 = a^6 - 6a^5b + 15a^4b^2 - 20a^3b^3 + 15a^2b^4$

$\qquad - 6ab^5 + b^6$

Method 2

$(a-b)^6 = {_6}C_0 a^6(-b)^0 + {_6}C_1 a^5(-b) + {_6}C_2 a^4(-b)^2 + {_6}C_3 a^3(-b)^3$

$\qquad + {_6}C_4 a^2(-b)^4 + {_6}C_5 a(-b)^5 + {_6}C_6 a^0(-b)^6$

$(a-b)^6 = a^6 - 6a^5b + 15a^4b^2 - 20a^3b^3 + 15a^2b^4 - 6ab^5 + b^6$

22. $(2x^2 - 1)^5 = (2x^2)^5 + 5(2x^2)^4(-1)^1 + \dfrac{5 \cdot 4}{2!}(2x^2)^3(-1)^2$

$\qquad + \dfrac{5 \cdot 4 \cdot 3}{3!}(2x^2)^2(-1)^3 + \dfrac{5 \cdot 4 \cdot 3 \cdot 2}{4!}(2x^2)^1(-1)^4 + \dfrac{5 \cdot 4 \cdot 3 \cdot 2 \cdot 1}{5!}(-1)^5$

$(2x^2 - 1)^5 = 32x^{10} - 80x^8 + 80x^6 - 40x^4 + 10x^2 - 1$

Method 2

$(2x^2 - 1)^5 = {_5}C_0 (2x^2)^5(-1)^0 + {_5}C_1 (2x^2)^4(-1)^1 + {_5}C_2 (2x^2)^3(-1)^2$

$\qquad + {_5}C_3 (2x^2)^2(-1)^3 + {_5}C_4 (2x^2)^1(-1)^4 + {_5}C_5 (2x^2)^0(-1)^5$

$(2x^2 - 1)^5 = 2^5 x^{10} - 5(2^4)x^8 + 10(2^3)x^6 - 10(2^2)x^4 + 5(2x^2) - 1$

$(2x^2 - 1)^5 = 32x^{10} - 80x^8 + 80x^6 - 40x^4 + 10x^2 - 1$

Chapter 9 Review

Solutions to Odd-Numbered Exercises

23. $$(2x+3y)^4 = (2x)^4 + 4(2x)^3(3y) + \frac{4 \cdot 3}{2!}(2x)^2(3y)^2$$
$$+ \frac{4 \cdot 3 \cdot 2}{3!}(2x)(3y)^3 + \frac{4 \cdot 3 \cdot 2 \cdot 1}{4!}(2x)^0(3y)^4$$
$$(2x+3y)^4 = 16x^4 + 96x^3y + 216x^2y^2 + 216xy^3 + 81y^4$$

Method 2

$$(2x+3y)^4 = {}_4C_0(2x)^4(3y)^0 + {}_4C_1(2x)^3(3y)^1 + {}_4C_2(2x)^2(3y)^2$$
$$+ {}_4C_3(2x)^1(3y)^3 + {}_4C_4(2x)^0(3y)^4$$
$$(2x+3y)^4 = 2^4x^4 + 4(2^3)x^3(3y) + 6(2^2)x^2(3^2)y^2$$
$$+ 4(2x)(3^3)y^3 + 1(3^4)y^4$$
$$(2x+3y)^4 = 16x^4 + 96x^3y + 216x^2y^2 + 216xy^3 + 81y^4$$

24. $$(1+x)^8 = 1^8 + 8(1)^7(x) + \frac{8 \cdot 7}{2!}(1)^6(x)^2 + \frac{8 \cdot 7 \cdot 6}{3!}(1)^5(x)^3$$
$$+ \frac{8 \cdot 7 \cdot 6 \cdot 5}{4!}(1)^4x^4 + \frac{8 \cdot 7 \cdot 6 \cdot 5 \cdot 4}{5!}(1)^3(x)^5$$
$$+ \frac{8 \cdot 7 \cdot 6 \cdot 5 \cdot 4 \cdot 3}{6!}(1)^2(x^6) + \frac{8 \cdot 7 \cdot 6 \cdot 5 \cdot 4 \cdot 3 \cdot 2}{7!}(1)(x^7)$$
$$+ \frac{8 \cdot 7 \cdot 6 \cdot 5 \cdot 4 \cdot 3 \cdot 2 \cdot 1}{8!}(1)^0(x^8)$$
$$(1+x)^8 = 1 + 8x + 28x^2 + 56x^3 + 70x^4 + 56x^5 + 28x^6 + 8x^7 + x^8$$

Method 2

$$(1+x)^8 = {}_8C_0(1)^8(x)^0 + {}_8C_1(1)^7(x)^1 + {}_8C_2(1)^6(x)^2$$
$$+ {}_8C_3(1)^5(x)^3 + {}_8C_4(1)^4(x)^4 + {}_8C_5(1)^3(x)^5$$
$$+ {}_8C_6(1)^2(x)^6 + {}_8C_7(1)^2(x)^7 + {}_8C_8(1)^0(x)^8$$
$$(1+x)^8 = 1 + 8x + 28x^2 + 56x^3 + 70x^4 + 56x^5 + 28x^6 + 8x^7 + x^8$$

25. $(1-3x)^5$; 3rd term

$$k\text{th term} = \frac{n(n-1)(n-2)...(n-[k-2])}{(k-1)!}a^{n-(k-1)}b^{k-1}$$

$$3\text{rd term} = \frac{5 \cdot 4}{(3-1)!}(1)^3(-3x)^2 = 90x^2$$

Method 2

$$k\text{th term} = {}_nC_{k-1}a^{n-(k-1)}b^{k-1}$$

$$3\text{rd term} = {}_5C_2(1)^3(-3x)^2 = 10(-3)^2x^2 = 90x^2$$

Chapter 9 Review

26. $(a+4b)^6$; 4th term

$$k\text{th term} = \frac{n(n-1)(n-2)\dots(n-[k-2])}{(k-1)!}a^{n-(k-1)}b^{k-1}$$

$$4\text{th term} = \frac{6\cdot5\cdot4}{3!}a^3(4b)^3 = 1280a^3b^3$$

Method 2

$$k\text{th term} =_n C_{k-1}a^{n-(k-1)}b^{k-1}$$

$$4\text{th term} =_6 C_3 a^3(4b)^3 = 20(4^3)a^3b^3 = 1280a^3b^3$$

27. $(x+2b^2)^{10}$; middle term. There are 11 terms; the middle one is the 6th term.

$$k\text{th term} = \frac{n(n-1)(n-2)\dots(n-[k-2])}{(k-1)!}a^{n-(k-1)}b^{k-1}$$

$$6\text{th term} = \frac{10\cdot9\cdot8\cdot7\cdot6}{(6-1)!}(x)^5(2b^2)^5 = 8064x^5b^{10}$$

Method 2

$$k\text{th term} =_n C_{k-1}a^{n-(k-1)}b^{k-1}$$

$$6\text{th term} =_{10} C_5 x^5(2b^2)^5 = 252(2^5)x^5b^{10} = 8064x^5b^{10}$$

28. $(3x^2-1)^{12}$; term containing x^{16}. This term will contain $(3x^2)^8$; i.e., $n-(k-1)=8$. Since $n=12$, $k=5$. We are seeking the 5th term.

$$k\text{th term} = \frac{n(n-1)(n-2)\dots(n-[k-2])}{(k-1)!}a^{n-(k-1)}b^{k-1}$$

$$5\text{th term} = \frac{12\cdot11\cdot10\cdot9}{(5-1)!}(3x^2)^8(-1)^4$$

$$5\text{th term} = 495(3^8)x^{16} = 3,247,695x^{16}$$

Method 2

$$k\text{th term} =_n C_{k-1}a^{n-(k-1)}b^{k-1}$$

$$5\text{th term} =_{12} C_4 (3x^2)^8(-1)^4 = 495(3^8)x^{16} = 3,247,695x^{16}$$

Chapter 9 Review

CHAPTER 10

1. $5 + 9 + 13 + 17 + 21 + 25$

3. $10 + 17 + 26 + 37 + 50 + 65$

5. $\dfrac{1}{2} + \dfrac{4}{3} + \dfrac{9}{4} + \dfrac{16}{5} + \cdots + \dfrac{n^2}{n+1}$

7. $-1 + \dfrac{1}{4} - \dfrac{1}{9} + \dfrac{1}{16} - \dfrac{1}{25} + \cdots$

9. $\displaystyle\sum_{n=1}^{12} n$

11. $\displaystyle\sum_{n=1}^{50} 2n$

13. $\displaystyle\sum_{n=1}^{n} (2n-1)$

15. $\displaystyle\sum_{n=3}^{n} (n^2+1)$

17. Diverges because $\displaystyle\lim_{n\to\infty} S_n = \infty$

19. Diverges because $\displaystyle\lim_{n\to\infty} \frac{2n}{n-1} = \lim_{n\to\infty} \frac{2}{1-1/n} = 2$

21. Diverges (p-series with $p = 1/4$)

23. Converges (p-series with $p = 2$)

25. Compare $\displaystyle\sum 1/n^2$ and $\displaystyle\sum 1/(n+1)^2$. Since $\displaystyle\sum 1/n^2$ converges and $\dfrac{1}{(n+1)^2} < \dfrac{1}{n^2}$ for $n \geq 1$,

by the comparison test $\displaystyle\sum 1/(n+1)^2$ also converges.

27. Compare $\displaystyle\sum 1/n^2$ and $\displaystyle\sum \frac{1}{n(n+1)}$. Since $\displaystyle\sum 1/n^2$ converges and

$\displaystyle\lim_{n\to\infty} \frac{1/n^2}{1/(n^2+n)} = \lim_{n\to\infty} \frac{n^2+n}{n^2} = \lim_{n\to\infty}(1+1/n) = 1$, both series have the same order of magnitude and

by the limit comparison test both series converge.

29. Compare $\displaystyle\sum 1/n$ and $\displaystyle\sum 1/(2n)$. Since $\displaystyle\sum 1/n$ diverges and

$\displaystyle\lim_{n\to\infty} \frac{1/n}{1/(2n)} = \lim_{n\to\infty} \frac{2n}{n} = 2$, both series have the same order of magnitude and by the limit

comparison test both series diverge.

31. Compare $\displaystyle\sum 1/(2n-1)^2$ and $\displaystyle\sum 1/n^2$. Since $\displaystyle\sum 1/n^2$ converges and $\dfrac{1}{(2n-1)^2} \leq \dfrac{1}{n^2}$ for $n \geq 1$,

then by the comparison test $\displaystyle\sum \frac{1}{(2n-1)^2}$ converges.

33. Compare $\sum 1/\sqrt{n^2+1}$ and $\sum 1/n = \sum 1/\sqrt{n^2}$. Since $\sum 1/n$ diverges and

$$\lim_{n\to\infty}\frac{1/\sqrt{n^2+1}}{1/\sqrt{n^2}} = \lim_{n\to\infty}\sqrt{\frac{n^2}{n^2+1}} = \lim_{n\to\infty}\sqrt{\frac{1}{1+1/n^2}} = 1,$$ both series have the same order of

magnitude and by the limit comparison test both diverge.

35. Compare $\sum\dfrac{1}{\sqrt{n}(n+1)}$ and $\sum\dfrac{1}{n^{3/2}}$. Since $\sum\dfrac{1}{n^{3/2}}$ is a convergent p-series $\dfrac{1}{\sqrt{n}(n+1)} < \dfrac{1}{n^{3/2}}$

for $n \geq 1$, by the comparison test the given series also converges.

37. Compare $\sum\dfrac{1}{2^n+2n}$ and $\sum\left(\dfrac{1}{2}\right)^n$. Since $\sum\left(\dfrac{1}{2}\right)^n$ is a geometric convergent series and

$\dfrac{1}{2^n+2n} < \dfrac{1}{2^n}$ for $n > 1$, the given series converges by the comparison test.

39. Compare $\sum 1/\ln n$ and $\sum 1/n$. Since $0 < \ln n < n$ for $n > 1$, $1/\ln n > 1/n$ for $n \geq 2$. Since

$\sum 1/n$ diverges, by the comparison test the given series also diverges.

41. Compare $\sum\dfrac{1+\sin n\pi}{n^2}$ and $\sum 2/n^2$. Since $\sum 2/n^2$ is a convergent p-series and

$\dfrac{1+\sin n\pi}{n^2} < \dfrac{2}{n^2}$, by the comparison test the given series also converges.

43. Compare $\sum\dfrac{1}{\sqrt{n(n+1)}}$ and $\sum 1/n$. Since $\sum 1/n$ is a divergent p-series and

$$\lim_{n\to\infty}\frac{1/\sqrt{n^2+n}}{1/n} = \lim_{n\to\infty}\sqrt{\frac{n^2}{n^2+n}} = 1,$$ both series have the same order of magnitude and by the

limit comparison test both series diverge.

Section 10.2

1. $r = \lim_{n\to\infty}\dfrac{\dfrac{n+2}{(n+1)3^{n+1}}}{\dfrac{n+1}{n\cdot 3^n}} = \lim_{n\to\infty}\dfrac{n\,3^n(n+2)}{(n+1)^2\,3^{n+1}} = \lim_{n\to\infty}\dfrac{n(n+2)}{3(n+1)^2} = \lim_{n\to\infty}\dfrac{1(1+2/n)}{3(1+1/n)^2} = \dfrac{1}{3} < 1$. The given series

converges by the ratio test.

3. $r = \lim_{n\to\infty}\dfrac{1/(n+1)!}{1/n!} = \lim_{n\to\infty} n!/(n+1)! = \lim_{n\to\infty}\dfrac{1}{n+1} = 0 < 1$.
The given series converges by the ratio test.

Sections 10.1 – 10.2

5. $r = \lim\limits_{n\to\infty} \dfrac{\dfrac{(n+1)^2}{(n+1)!}}{\dfrac{n^2}{n!}} = \lim\limits_{n\to\infty} \dfrac{(n+1)^2\, n!}{(n+1)!\, n^2} = \lim\limits_{n\to\infty} \dfrac{(n+1)}{n^2} = \lim\limits_{n\to\infty}(1/n + 1/n^2) = 0.$ The given series converges

by the ratio test.

7. $r = \lim\limits_{n\to\infty} \dfrac{\dfrac{3^{n+1}}{(n+1)2^{n+1}}}{\dfrac{3^n}{n\,2^n}} = \lim\limits_{n\to\infty} \dfrac{n\,2^n\,3^{n+1}}{(n+1)2^{n+1}3^n} = \lim\limits_{n\to\infty} \dfrac{3n}{2(n+1)} = \dfrac{3}{2} > 1.$

The given series diverges by the ratio test.

9. $r = \lim\limits_{n\to\infty} \dfrac{\dfrac{2n+5}{2^{n+1}}}{\dfrac{2n+3}{2^n}} = \lim\limits_{n\to\infty} \dfrac{(2n+5)2^n}{(2n+3)2^{n+1}} = \lim\limits_{n\to\infty} \dfrac{2n+5}{2(2n+3)} = \lim\limits_{n\to\infty} \dfrac{2+5/n}{4+6/n} = \dfrac{1}{2} < 1.$ The given series

converges by the ratio test.

11. $\displaystyle\int_1^\infty \dfrac{dx}{2x+1} = \lim\limits_{b\to\infty}\int_1^b \dfrac{dx}{2x+1} = \lim\limits_{b\to\infty} \dfrac{1}{2}\ln|2x+1|\ \Big|_1^b = \lim\limits_{b\to\infty} \dfrac{1}{2}[\ln|2b+1| - \ln 3] = \infty.$ The integral does

not exist and the given series diverges by the integral test.

13. $\displaystyle\int_2^\infty \dfrac{dx}{x\ln x^{1/2}} = \lim\limits_{b\to\infty}\int_2^b \dfrac{dx}{x\ln x^{1/2}} = \lim\limits_{b\to\infty} 2\sqrt{\ln x}\ \Big|_2^b = \lim\limits_{b\to\infty} 2(\sqrt{\ln b} - \sqrt{\ln 2}) = \infty.$ The integral does not

exist and the series diverges by the integral test.

15. $\displaystyle\int_1^\infty \dfrac{dx}{2x-1} = \lim\limits_{b\to\infty}\int_1^b \dfrac{dx}{2x-1} = \lim\limits_{b\to\infty} \dfrac{1}{2}\ln|2x-1|\ \Big|_1^b = \lim\limits_{b\to\infty} \dfrac{1}{2}[\ln|2b-1|] = \infty.$ The integral does not exist

and the series diverges by the integral test.

17. $\displaystyle\int_1^\infty \dfrac{x\,dx}{x^2+1} = \lim\limits_{b\to\infty}\int_1^b \dfrac{x\,dx}{x^2+1} = \lim\limits_{b\to\infty} \dfrac{1}{2}\ln|x^2+1|\ \Big|_1^b = \lim\limits_{b\to\infty} \dfrac{1}{2}[\ln|b^2+1| - \ln 2] = \infty.$ The integral does not

exist and the series diverges by the integral test.

19. $\displaystyle\int_1^\infty (x^2/e^x)\,dx = \lim\limits_{b\to\infty}\int_1^b x^2 e^{-x}\,dx = \lim\limits_{b\to\infty} -e^{-x}(x^2+2x+2)\ \Big|_1^b = \lim\limits_{b\to\infty} -\dfrac{b^2+2b+2}{e^b} + \dfrac{5}{e} = \dfrac{5}{e}.$ The

integral exists and the series converges by the integral test.

Section 10.2

Section 10.3

1. Converges conditionally: given alternating series converges because
$$a_{n+1} = \frac{1}{2n+3} < \frac{1}{2n+1} = a_n \text{ and } \lim_{n \to \infty} \frac{1}{2n+1} = 0; \text{ but } \sum \frac{1}{2n+1} \text{ diverges.}$$

3. Converges absolutely: $\sum \frac{1}{(2n)^2}$ converges by limit comparison test with $\sum \frac{1}{n^2}$, which converges.

5. Diverges: $\lim\limits_{n \to \infty} \dfrac{2n}{2n-1} = \lim\limits_{n \to \infty} \dfrac{2}{2-1/n} = 1$

7. Converges conditionally: given alternating series converges because
$$a_{n+1} = \frac{1}{\ln(n+1)} < \frac{1}{\ln n} = a_n \text{ and } \lim_{n \to \infty} \frac{1}{\ln n} = 0; \text{ but } \sum \frac{1}{\ln n} \text{ diverges because } \frac{1}{\ln n} > \frac{1}{n} \text{ and }$$
$$\sum \frac{1}{n} \text{ diverges.}$$

9. Converges absolutely: $\sum (n^2/2^n)$ converges (See Exercise 6, Section 10.2).

11. Diverges: $\lim\limits_{n \to \infty} \dfrac{n^2}{n^2+1} = \lim\limits_{n \to \infty} \dfrac{1}{(1+1/n^2)} = 1$

13. Diverges: $\lim\limits_{n \to \infty} \dfrac{n!}{3^n} = \lim\limits_{n \to \infty} \dfrac{n}{3} \cdot \dfrac{n-1}{3} \cdots \dfrac{2}{3} \cdot \dfrac{1}{3} > \lim\limits_{n \to \infty} \dfrac{n}{3} = \infty$

15. Converges conditionally: given alternating series converges because
$$\lim_{n \to \infty} \frac{2n+1}{n^2} = \lim_{n \to \infty} (2/n + 1/n^2) = 0 \text{ and}$$
$$a_{n+1} = \frac{2n+3}{(n+1)^2} < \frac{2n+1}{n^2} = a_n$$
$$\text{Since } \frac{2n+1}{n^2} > \frac{2n}{n^2} = \frac{2}{n} \text{ and } \sum 2/n \text{ diverges, } \sum \frac{2n+1}{n^2} \text{ diverges.}$$

17. Diverges: $\lim\limits_{n \to \infty} \dfrac{n}{\ln n} = \lim\limits_{n \to \infty} \dfrac{1}{1/n}$ by l'Hopital $= \lim\limits_{n \to \infty} n = \infty$

19. Converges absolutely: $|\cos n| \le 1$ so the given series converges by the comparison test with the convergent p-series $\sum 1/n^2$.

Section 10.4

1. This geometric series converges for $\left|\dfrac{x}{2}\right| < 1$ or $-2 < x < 2$.

3. $\displaystyle \lim_{n\to\infty}\left|\frac{a_{n+1}}{a_n}\right| = \lim_{n\to\infty}\left|\frac{(4n+4)!(x/2)^{n+1}}{(4n)!(x/2)^n}\right| = \lim_{n\to\infty}\left|(4n+4)(4n+3)(4n+2)(4n+1)\frac{x}{2}\right| = \infty$

   Thus the series converges only for $x = 0$.

5. $\displaystyle \lim_{n\to\infty}\left|\frac{(4x)^{n+1}(2n)!}{(2n+2)!(4x)^n}\right| = \lim_{n\to\infty}\frac{4x}{(2n+2)(2n+1)} = 0$

   Thus the interval of convergence is $-\infty < x < \infty$.

7. $\displaystyle \lim_{n\to\infty}\left|\frac{\dfrac{x^{n+1}}{(n+2)(n+3)}}{\dfrac{x^n}{(n+1)(n+2)}}\right| = |x|\lim_{n\to\infty}\frac{n+1}{n+3} = |x| < 1$ or $-1 < x < 1$

   For $x = 1$, the series $\displaystyle\sum\frac{-1}{(n+1)(n+2)}$ converges. For $x = -1$, the series $\displaystyle\sum\frac{1}{(n+1)(n+2)}$

   converges. Thus the interval of convergence is $-1 \le x \le 1$.

9. $\displaystyle \lim_{n\to\infty}\left|\frac{(n+1)x^{n+1}}{(n+2)^2}\cdot\frac{(n+1)^2}{n\,x^n}\right| = |x|\lim_{n\to\infty}\frac{(n+1)^3}{n(n+2)^2} = |x| < 1$ or $-1 < x < 1$. For $x = 1$, the series

   $\displaystyle\sum\frac{n}{(n+1)^2}$ diverges with comparison with $\displaystyle\sum 1/n$. For $x = -1$, the series $\displaystyle\sum\frac{n(-1)^n}{(n+1)^2}$

   converges by the alternating series test. Thus the interval of convergence is $-1 \le x < 1$.

11. $\displaystyle \lim_{n\to\infty}\left|\frac{2^{n+1}x^{n+1}}{3^{n+1}}\cdot\frac{3^n}{2^n\,x^n}\right| = |x|\lim_{n\to\infty}\frac{2}{3} = \frac{2}{3}\,|x| < 1$ or $-1 < \frac{2}{3}x < 1$ or $-3/2 < x < 3/2$.

    For $x = \pm 3/2$, each series clearly diverges.

13. $\displaystyle \lim_{n\to\infty}\left|\frac{x^{n+1}}{(n+1)2^{n+1}}\cdot\frac{n\,2^n}{x^n}\right| = \left|\frac{x}{2}\right|\lim_{n\to\infty}\frac{n}{n+1} = \left|\frac{x}{2}\right| < 1$ or $-2 < x < 2$

    For $x = 2$, the alternating series $\displaystyle\sum\frac{(-1)^{n+1}}{n}$ converges. For $x = -2$, the series

    $\displaystyle\sum\frac{(-1)^{2n+1}}{n} = \sum\frac{-1}{n}$ diverges. Thus the interval of convergence is $-2 < x \le 2$.

15. $\displaystyle \lim_{n\to\infty}\left|\frac{(x-2)^{n+1}}{\sqrt{n+1}}\cdot\frac{\sqrt{n}}{(x-2)^n}\right| = |x-2|\lim_{n\to\infty}\frac{\sqrt{n}}{\sqrt{n+1}} = |x-2| < 1$ or $1 < x < 3$. For $x = 3$, the alternating

    series converges. For $x = 1$, the series $\displaystyle\sum 1/\sqrt{n}$ is a divergent p-series. Thus the interval of

    convergence is $1 < x \le 3$.

Section 10.4

17. $\lim_{n\to\infty} \left| \dfrac{x^{n+1}}{(n+1)^2} \cdot \dfrac{n^2}{x^n} \right| = |x| \lim_{n\to\infty} \dfrac{n^2}{(n+1)^2} = |x| < 1$ or $-1 < x < 1$. For $x = \pm 1$, we have a convergent p-

series. Thus the interval of convergence is $-1 \le x \le 1$.

19. $\lim_{n\to\infty} \left| \dfrac{x^{2n+2}}{(n+1)!} \cdot \dfrac{n!}{x^{2n}} \right| = x^2 \lim_{n\to\infty} \dfrac{1}{n+1} = 0$. Thus the interval of convergence is $-\infty < x < \infty$.

21. $\lim_{n\to\infty} \left| \dfrac{2^{n+1} x^{n+2}}{(n+1)3^{n+2}} \cdot \dfrac{n\,3^{n+1}}{2^n\, x^{n+1}} \right| = \left| \dfrac{2x}{3} \right| \lim_{n\to\infty} \dfrac{n}{n+1} = \left| \dfrac{2}{3} x \right| < 1$ or $-3/2 < x < 3/2$. For $x = 3/2$, the series

$\sum 1/(2n)$ diverges. For $x = -3/2$, the series $\sum \dfrac{(-1)^{n+1}}{2n}$ converges. Thus the interval of

convergence is $-3/2 \le x < 3/2$.

23. $\lim_{n\to\infty} \left| \dfrac{(2x-5)^{n+1}}{(n+1)^2} \cdot \dfrac{n^2}{(2x-5)^n} \right| = |2x-5| \lim_{n\to\infty} \dfrac{n^2}{(n+1)^2} = |2x-5| < 1$ or $2 < x < 3$. For $x = 2$, the

series $\sum \dfrac{(-1)^n}{n^2}$ converges. For $x = 3$, the series $\sum 1/n^2$ converges. Thus the interval of

convergence is $2 \le x \le 3$.

Section 10.5

1. $f(x) = \sin x;\ f(0) = 0$
   $f'(x) = \cos x;\ f'(0) = 1$
   $f''(x) = -\sin x;\ f''(0) = 0$
   $f'''(x) = -\cos x;\ f'''(0) = -1$
   $f^{(4)}(x) = \sin x;\ f^{(4)}(0) = 0$

   $\sin x = x - \dfrac{x^3}{3!} + \dfrac{x^5}{5!} - \cdots$

3. $f(x) = e^{-x};\ f(0) = 1$
   $f'(x) = -e^{-x};\ f'(0) = -1$
   $f''(x) = e^{-x};\ f''(0) = 1$
   $f'''(x) = -e^{-x};\ f'''(0) = -1$
   $f^{(4)}(x) = e^{-x};\ f^{(4)}(0) = 1$

   $e^{-x} = 1 - x + \dfrac{x^2}{2!} - \dfrac{x^3}{3!} + \dfrac{x^4}{4!} - \cdots$

5. $f(x) = \ln(1 + x);\ f(0) = 0$

   $f'(x) = \dfrac{1}{1+x};\ f'(0) = 1$

   $f''(x) = \dfrac{-1}{(1+x)^2};\ f''(0) = -1$

   $f'''(x) = \dfrac{2!}{(1+x)^3};\ f'''(0) = 2$

   $f^{(4)}(x) = \dfrac{-3!}{(1+x)^4};\ f^{(4)}(0) = -3!$

   $\ln(1 + x) = x - \dfrac{x^2}{2!} + 2!\dfrac{x^3}{3!} - 3!\dfrac{x^4}{4!} + \cdots$

   $= x - \dfrac{1}{2} x^2 + \dfrac{1}{3} x^3 - \dfrac{1}{4} x^4 + \cdots$

7. $f(x) = \cos 2x;\ f(0) = 1$
   $f'(x) = -2 \sin 2x;\ f'(0) = 0$
   $f''(x) = -4 \cos 2x;\ f''(0) = -4$
   $f'''(x) = 8 \sin 2x;\ f'''(0) = 0$
   $f^{(4)}(x) = 16 \cos 2x;\ f^{(4)}(0) = 16$

   $\cos 2x = 1 - 2x^2 + \dfrac{16}{4!} x^4 - \cdots$

Sections 10.4 – 10.5

9. $f(x) = xe^x; f(0) = 0$
    $f'(x) = e^x(x + 1); f'(0) = 1$
    $f''(x) = e^x(x + 2); f''(0) = 2$
    $f'''(x) = e^x(x + 3); f'''(0) = 3$
    $f^{(4)}(x) = e^x(x + 4); f^{(4)}(0) = 4$

    $$x\,e^x = x + x^2 + \frac{x^3}{2!} + \frac{x^4}{3!} + \cdots$$

11. $f(x) = (4 - x)^{1/2}; f(0) = 2$
    $f'(x) = (-1/2)(4 - x)^{-1/2}; f'(0) = -1/4$
    $f''(x) = (-1/4)(4 - x)^{-3/2}; f''(0) = -1/32$
    $f'''(x) = (-3/8)(4 - x)^{-5/2}; f'''(0) = \dfrac{-3}{256}$

    $$\sqrt{4 - x} = 2 - \frac{x}{4} - \frac{x^2}{32(2!)} - \frac{3x^3}{256(3!)} - \cdots$$

13. $f(x) = \sin(x - \pi/2) = \sin[-(\pi/2 - x)] = -\sin(\pi/2 - x) = -\cos x;\ f(0) = -1$
    $f'(x) = \sin x; f'(0) = 0$
    $f''(x) = \cos x; f''(0) = 1$
    $f'''(x) = -\sin x; f'''(0) = 0$
    $f^{(4)}(x) = -\cos x; f^{(4)}(0) = -1$ Thus $\sin(x - \pi/2) = -1 + x^2/2! - x^4/4! + \cdots$

15. $f(x) = 1(1 - x)^2;\ f(0) = 1$
    $f'(x) = 2(1 - x)^{-3}; f'(0) = 2$
    $f''(x) = 6(1 - x)^{-4}; f''(0) = 6$
    $f'''(x) = 24(1 - x)^{-5}; f'''(0) = 24$
    $f^{(4)}(x) = 120(1 - x)^{-6}; f^{(4)}(0) = 120$
    $1/(1 - x)^2 = 1 + 2x + 6x^2/2! + 24x^3/3! + 120x^4/4! + \cdots$

17. $f(x) = (1 + x)^5;\ f(0) = 1$
    $f'(x) = 5(1 + x)^4; f'(0) = 5$
    $f''(x) = 20(1 + x)^3; f''(0) = 20$
    $f'''(x) = 60(1 + x)^2; f'''(0) = 60$
    $f^{(4)}(x) = 120(1 + x); f^{(4)}(0) = 120$
    $f^{(5)}(x) = 120; f^{(5)}(0) = 120$
    $(1 + x)^5 = 1 + 5x + 10x^2 + 10x^3 + 5x^4 + x^5$ (Sum is finite.)

19. $f(x) = e^{-x}\sin x;\ f(0) = 0$
    $f'(x) = e^{-x}(\cos x - \sin x); f'(0) = 1$
    $f''(x) = -2e^{-x}\cos x; f''(0) = -2$
    $f'''(x) = 2e^{-x}(\sin x + \cos x); f'''(0) = 2$
    $f^{(4)}(x) = -4e^{-x}\sin x; f^{(4)}(0) = 0$
    $f^{(5)}(x) = 4e^{-x}(\sin x - \cos x);\ f^{(5)}(0) = -4$
    Thus $e^{-x}\sin x = x - x^2 + x^3/3 - x^5/30 + \cdots$

Section 10.6

1. Substitute $-x$ for x in equation A.

    $$f(x) = e^{-x} = 1 - x + \frac{x^2}{2!} + \frac{x^3}{3!} + \frac{x^4}{4!} - \cdots$$

3. Substitute x^2 for x in equation A.

    $$f(x) = e^{x^2} = 1 + x^2 + \frac{x^4}{2!} + \frac{x^6}{3!} + \cdots$$

Sections 10.5 – 10.6

5. Substitute $-x$ for x in equation D.

$$f(x) = \ln(1-x) = -x - \frac{x^2}{2} - \frac{x^3}{3} - \frac{x^4}{4} - \cdots$$

7. Substitute $5x^2$ for x in equation C.

$$f(x) = \cos 5x^2 = 1 - \frac{25x^4}{2!} + \frac{625x^8}{3!} - \cdots$$

9. Substitute x^3 for x in equation D.

$$f(x) = \sin x^3 = x^3 - \frac{x^9}{3!} + \frac{x^{15}}{5!} - \cdots$$

11. $f(x) = x\,e^x = x\left(1 + x + \frac{x^2}{2!} + \frac{x^3}{3!} + \cdots\right) = x + x^2 + \frac{x^3}{2!} + \frac{x^4}{3!} + \cdots$

13. $\dfrac{\cos x - 1}{x} = \dfrac{[1 - x^2/2! + x^4/4! - x^6/6! + \cdots] - 1}{x} = \dfrac{-x}{2!} + \dfrac{x^3}{4!} - \dfrac{x^5}{6!} + \cdots$

15. $\displaystyle\int_0^1 e^{-x^2}\,dx \int_0^1 (1 - x^2 + x^4/2! - x^6/3! + \cdots)\,dx$

$$= (x - x^3/3 + x^5/10 - x^7/42)\,\Big|_0^1 = 1 - 1/3 + 1/10 - 1/42 - (0) = 0.743$$

17. Let $u = x - 1$; $\dfrac{e^{x-1}}{x-1} = \dfrac{e^u}{u} = \dfrac{1 + u + u^2/2! + u^3/3! + \cdots}{u} = \dfrac{1}{u} + 1 + \dfrac{u}{2!} + \dfrac{u^2}{3!} + \cdots$

$$\int_2^3 \frac{e^{x-1}}{x-1} = \int_2^3\left[\frac{1}{x-1} + 1 + \frac{x-1}{2!}\right]dx = \ln(x-1) + x + \frac{(x-1)^2}{4}\,\Bigg|_2^3 = \ln 2 + \frac{7}{4}$$

19. $\displaystyle\int_0^1 \sin\sqrt{x}\,dx \int_0^1\left(x^{1/2} - \frac{x^{3/2}}{3!} + \frac{x^{5/2}}{5!}\right)dx = (2/3)x^{3/2} - (1/15)x^{5/2} + (1/420)x^{7/2}\Big|_0^1 = 0.602$

21. $\sinh x = (1/2)(e^x - e^{-x}) = (1/2)\left(2x + \dfrac{2x^3}{3!} + \dfrac{2x^5}{5!} + \cdots\right) = x + \dfrac{x^3}{3!} + \dfrac{x^5}{5!} + \cdots$

$$e^x = 1 + x + \frac{x^2}{2!} + \frac{x^3}{3!} + \frac{x^4}{4!} + \cdots$$

$$e^{-x} = 1 - x + \frac{x^2}{2!} + \frac{x^3}{3!} + \frac{x^4}{4!} - \cdots$$

$$e^x - e^{-x} = 2x + \frac{2x^3}{3!} + \frac{2x^5}{5!} + \cdots$$

Section 10.6

23. $q = \int\limits_0^{0.5} \sin t^2 \, dt \int\limits_0^{0.5}\left(t^2 - \dfrac{t^6}{3!} + \dfrac{t^{10}}{5!} - \dfrac{t^{14}}{7!}\right) dt = \dfrac{t^3}{3} - \dfrac{t^7}{3!(7)} + \dfrac{t^{11}}{5!(11)} - \dfrac{t^{15}}{7!(15)}\Bigg|_0^{0.5} = 0.041481\,C$

25. $e^{jx} = 1 + jx + \dfrac{(jx)^2}{2!} + \dfrac{(jx)^3}{3!} + \cdots$

$e^{-jx} = 1 - jx + \dfrac{(-jx)^2}{2!} + \dfrac{(-jx)^3}{3!} + \cdots$

_____

$e^{jx} = e^{-jx} = 2jx \qquad\qquad - \dfrac{2jx}{3!} + \cdots$

$\dfrac{e^{jx} - e^{-jx}}{2j} = x - \dfrac{x}{3!} + \dfrac{x^5}{5!} - \cdots = \sin x$

Section 10.7
1. $f(x) = \cos x; = f(\pi/2) = 0$

$f'(x) = -\sin x; f'(\pi/2) = -1$

$f''(x) = -\cos x; f''(\pi/2) = 0$

$f'''(x) = \sin x; f'''(\pi/2) = 1$

$f^{(4)}(x) = \cos x; f^{(4)}(\pi/2) = 0$

$f^{(5)}(x) = -\sin x; f^{(5)}(\pi/2) = -1$

Thus $f(x) = -(x - \pi/2) + \dfrac{(x - \pi/2)^3}{3!} - \dfrac{(x - \pi/2)^5}{5!} + \cdots$

3. $f(x) = e^x = f'(x) = f''(x) = f'''(x) = \cdots$

$f(2) = e^2 = f'(2) = f''(2) = f'''(2) = \cdots$

Thus $f(x) = e^2 + e^2(x - 2) + \dfrac{e^2}{2!}(x - 2)^2 + \dfrac{e^2}{3!}(x - 2)^3 + \cdots$

$\qquad = e^2\left[1 + (x - 2) + \dfrac{(x - 2)^2}{2!} + \dfrac{(x - 2)^3}{3!} + \cdots\right]$

5. $f(x) = x^{1/2}; f(9) = 3$

$f'(x) = (1/2)x^{-1/2}; f'(9) = 1/6$

$f''(x) = (-1/4)x^{-3/2}; f''(9) = -1/108$

$f'''(x) = (3/8)x^{-5/2}; f'''(9) = 1/648$

Thus $f(x) = 3 + \dfrac{x - 9}{6} - \dfrac{(x - 9)^2}{2!(108)} + \dfrac{(x - 9)^3}{3!(648)} + \cdots$

7. $f(x) = x^{-1}; f(2) = 1/2$

$f'(x) = -x^{-2}; f'(2) = -1/4$

$f''(x) = 2x^{-3}; f''(2) = 1/4$

$f'''(x) = -6x^{-4}; f'''(2) = -3/8$

$f(x) = \dfrac{1}{2} - \dfrac{1}{4}(x-2) + \dfrac{(x-2)^2}{4(2!)} - \dfrac{3(x-2)^3}{8(3!)} + \cdots$

9. $f(x) = \ln x; f(1) = 0$

$f'(x) = 1/x; f'(1) = 1$

$f''(x) = -x^{-2}; f''(1) = -1$

$f'''(x) = 2x^{-3}; f'''(1) = 2$

$f^{(4)}(x) = -6x^{-4}; f^{(4)}(1) = -6$

Thus $f(x) = (x-1) - \dfrac{(x-1)^2}{2!} + \dfrac{2(x-1)^3}{3!} - \dfrac{6(x-1)^4}{4!} + \cdots$

11. $f(x) = x^{-1/2}; f(1) = 1$

$f'(x) = (-1/2)x^{-3/2}; f'(1) = -1/2$

$f''(x) = (3/4)x^{-5/2}; f''(1) = 3/4$

$f'''(x) = (-15/8)x^{-7/2}; f'''(1) = -15/8$

$f^{(4)}(x) = (105/16)x^{-9/2}; f^{(4)}(1) = 105/16$

Thus $f(x) = 1 - (1/2)(x-1) + (3/8)(x-1)^2 - (5/16)(x-1)^3 + \cdots$

13. $f(x) = x^{-2}; f(1) = 1$

$f'(x) = -2x^{-3}; f'(1) = -2$

$f''(x) = 6x^{-4}; f''(1) = 6$

$f'''(x) = -24x^{-5}; f'''(1) = -24$

Thus $f(x) = 1 - 2(x-1) + \dfrac{6(x-1)^2}{2!} - \dfrac{24(x-1)^3}{3!} + \cdots$

$= 1 - 2(x-1) + 3(x-1)^2 - 4(x-1)^3 + \cdots$

15. $f(x) = \cos x; f(\pi) = -1$

$f'(x) = -\sin x; f'(\pi) = 0$

$f''(x) = -\cos x; f''(\pi) = 1$

$f'''(x) = \sin x; f'''(\pi) = 0$

$f^{(4)}(x) = \cos x; f^{(4)}(\pi) = -1$

Thus $f(x) = -1 + \dfrac{(x-\pi)^2}{2!} - \dfrac{(x-\pi)^4}{4!} + \cdots$

Note: Due to space restrictions in this manual, answers only are provided in Sections 10.8 and 10.9.

Section 10.8

1. 1.10517   3. 0.99985   5. −0.68229   7. 1.0488   9. 3.66832

11. 0.48481   13. 0.029996

Sections 10.7 – 10.8

Section 10.9

1. $f(x) = -\pi + 2 \sin x + \sin 2x + (2/3) \sin 2x/3 + \cdots$

    $f(x) = -x,\ 0 \le x < 2\pi$

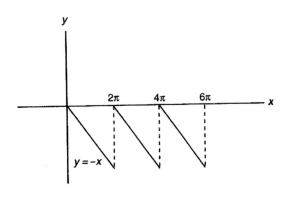

3. $f(x) = \pi/3 - (2/3) \sin x - (1/3) \sin 2x - (2/9) \sin 3x - \cdots$

    $f(x) = \dfrac{1}{3}x,\ 0 \le x < 2\pi$

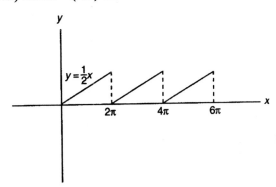

5. $f(x) = 1/2 - (2/\pi) \sin x - (2/3\pi) \sin 3x - (2/5\pi) \sin 5x - \cdots$

    $f(x) = \begin{cases} & 0 \le x < \pi \\ 1 & \pi \le x < 2\pi \end{cases}$

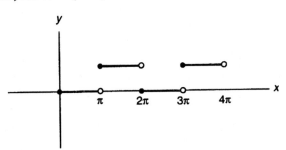

7. $f(x) = (4/\pi) \sin x + (4/3\pi) \sin 3x + (4/5\pi) \sin 5x + \cdots$

    $f(x) = \begin{cases} 1 & 0 \le x < \pi \\ -1 & \pi \le x < 2\pi \end{cases}$

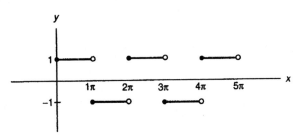

Section 10.9

9. $f(x) = 3 + \dfrac{12}{\pi} \sin \dfrac{\pi x}{5} + \dfrac{4}{\pi} \sin \dfrac{3\pi x}{5} + \dfrac{12}{5\pi} \sin \pi x + \dfrac{12}{7\pi} \sin \dfrac{7\pi x}{5} + \cdots$

$f(x) = \begin{cases} 0 & -5 \le x < 0 \\ 6 & 0 \le x < 5 \end{cases}$

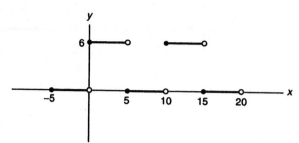

11. $f(x) = \pi/2 - (4/\pi) \cos x - (4/9\pi) \cos 3x - (4/25\pi) \cos 5x - \cdots$

$f(x) = \begin{cases} x & 0 \le x < \pi \\ 2\pi - x, & \pi \le x < 2\pi \end{cases}$

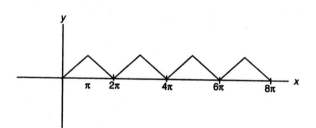

13. $f(x) = \dfrac{e^{2\pi} - 1}{2\pi} + \dfrac{e^{2\pi} - 1}{2\pi} \cos x + \dfrac{e^{2\pi} - 1}{5\pi} \cos 2x + \dfrac{e^{2\pi} - 1}{10\pi}$

$\cos 3x + \cdots + \dfrac{1}{\pi} \cdot \dfrac{-e^{2\pi} + 1}{2} \sin x + \dfrac{1}{\pi} \cdot \dfrac{-e^{2\pi} + 2}{5} \sin 2x + \cdots$

$f(x) = e^x \quad 0 \le x < 2\pi$

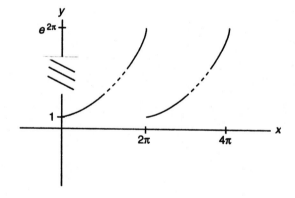

Section 10.9

15. $\dfrac{1}{\pi} + \dfrac{1}{2}\sin x - \dfrac{2}{\pi}\left\{\dfrac{1}{3}\cos 2x + \dfrac{1}{15}\cos 4x + \cdots\right\}$

$f(x) = \begin{cases} \sin x & 0 \le x < \pi \\ 0 & \pi \le x < 2\pi \end{cases}$

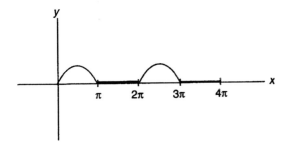

Chapter 10 Review

1. $-2 - 5 - 8 - 11 - 14 - 17$ 2. $2 + 3/2 + 4/3 + 5/4 + \cdots + (n+1)/n$ 3. $\displaystyle\sum_{n=1}^{7} 1/3^n$

4. $\displaystyle\sum_{n=1}^{10} \dfrac{n}{n+3}$ 5. Converges; p-series with $p = 3$

6. Diverges; p-series with $p = 1/4$

7. Converges; $\displaystyle\lim_{n\to\infty} \dfrac{1/(6n^2 + 2)}{1/n^2} = \lim_{n\to\infty} \dfrac{n^2}{6n^2 + 2} = \dfrac{1}{6}$ Since the given series and $\displaystyle\sum 1/n^2$ have the same order of magnitude and $\displaystyle\sum 1/n^2$ converges, the given series also converges by the limit comparison test.

8. Converges; Compare with $\displaystyle\sum \sqrt{n}/n^2 = \sum 1/n^{3/2}$, which is a convergent p-series.

$\displaystyle\lim_{n\to\infty} \dfrac{\sqrt{n}/(n^2 - 1)}{\sqrt{n}/n^2} = \lim_{n\to\infty} \dfrac{n^2}{n^2 - 1} = 1$ Since both series have the same order of magnitude, both series converge by the limit comparison test.

9. Diverges; using the integral test, $\displaystyle\lim_{b\to\infty} \int_{2}^{b}\dfrac{\ln x\, dx}{x} = \lim_{b\to\infty} \dfrac{\ln^2 x}{2}\Big|_{2}^{b} = \lim_{b\to\infty}\left\{\dfrac{\ln^2 b}{2} - \dfrac{\ln^2 2}{2}\right\} = \infty$

10. Converges; $r = \displaystyle\lim_{n\to\infty} \dfrac{5n + 7}{(3n + 4)4^{n+1}} \cdot \dfrac{(3n+1)4^n}{5n + 2} = \lim_{n\to\infty} \dfrac{(5n + 7)(3n + 1)}{4(3n + 4)(5n + 2)}$

$= \displaystyle\lim_{n\to\infty} \dfrac{(5 + 7/n)(3 + 1/n)}{4(3 + 4/n)(5 + 2/n)} = \dfrac{1}{4} < 1$

The series converges by the ratio test.

Section 10.9 – Chapter 10 Review

11. Converges; $r = \lim\limits_{n \to \infty} \dfrac{(n+1)^3}{2^{n+1}} \cdot \dfrac{2}{n^3} = \lim\limits_{n \to \infty} \dfrac{(n+1)^3}{2n^3} = \dfrac{1}{2} < 1$

This series converges by the ratio test.

12. Diverges; $\lim\limits_{n \to \infty} \dfrac{3+1/n}{4-5/n} = \dfrac{3}{4}$

13. Diverges; compare with $\sum 1/n$, which diverges. $\lim\limits_{n \to \infty} \dfrac{(n+1)/(n^2+4n)}{1/n} = \lim\limits_{n \to \infty} \dfrac{n+1}{n+4} = 1$

Since both series have the same order of magnitude, both series diverge by the limit comparison test.

14. Converges; compare with $\sum 1/n^2$. $\dfrac{|\sin n|}{n^2} \leq \dfrac{1}{n^2}$ Thus the given series converges by the comparison test.

15. Converges absolutely; $r = \lim\limits_{n \to \infty} \dfrac{3^{n+1}}{(n+1)!} \cdot \dfrac{n!}{3^n} = \lim\limits_{n \to \infty} \dfrac{3}{n+1} = 0 < 1$

Thus the series $\sum 3^n/n!$ converges by the ratio test.

16. Converges absolutely; $r = \lim\limits_{n \to \infty} \dfrac{2^{n+1}}{5^{n+1}(n+2)} \cdot \dfrac{5^n(n+1)}{2^n} = \lim\limits_{n \to \infty} \dfrac{2(n+1)}{5(n+2)} = \dfrac{2}{5} < 1$ Thus the given

series $\dfrac{2^n}{5^n(n+1)}$ converges by the ratio test.

17. Diverges; $\lim\limits_{n \to \infty} \dfrac{n+1}{n-1} = 1$

18. Conditionally convergent; the alternating series converges by the alternating series test. The corresponding series of positive terms diverges by the limit comparison test with $\sum 1/n$.

19. $\lim\limits_{n \to \infty} \left| \dfrac{[(n+1)/2](x-2)^{n+1}}{(n/2)(x-2)^n} \right| = |x-2| \lim\limits_{n \to \infty} \dfrac{n+1}{n} = |x-2| < 1$ or $1 < x < 3$. For $x = 3$, $\sum n/2$

diverges. For $x = 1$, $\sum(-1)^n(n/2)$ diverges. Thus the interval of convergence is $1 < x < 3$.

20. $\lim\limits_{n \to \infty} \left| \dfrac{(n+1)^2(x-3)^{n+1}}{n^2(x-3)^n} \right| = |x-3| \lim\limits_{n \to \infty} \dfrac{(n+1)^2}{n^2} = |x-3| < 1$ or $2 < x < 4$. For $x = 4$, the series

$\sum n^2$ diverges. For $x = 2$, the series $\sum(-1)^n n^2$ diverges. The interval of convergence is $2 < x < 4$.

Chapter 10 Review

21. $\lim\limits_{n\to\infty}\left|\dfrac{(x-1)^{n+1}}{(n+1)!}\cdot\dfrac{n!}{(x-1)^n}\right|=|x-1|\lim\limits_{n\to\infty}\dfrac{1}{n+1}=0$

Thus the interval of convergence is $-\infty<x<\infty$.

22. $\lim\limits_{n\to\infty}\left|\dfrac{3^{n+1}(x-4)^{n+1}}{(n+1)^2}\cdot\dfrac{n^2}{3^n(x-4)^n}\right|=3|x-4|\lim\limits_{n\to\infty}\dfrac{n^2}{(n+1)^2}=3|x-4|<1$ or $11/3<x<13/3$.

For $x=13/3$, the series $\sum 1/n^2$ converges. For $x=11/3$, the series $\sum(-1)^n/n^2$ also converges. Thus the interval of convergence is $11/3\le x\le 13/3$.

23. $f(x)=1/(1-x);\,f(0)=1$
$f'(x)=1/(1-x)^2;f'(0)=1$
$f''(x)=2/(1-x)^3;f''(0)=2$
$f'''(x)=6/(1-x)^4;f'''(0)=6$
Thus $\dfrac{1}{1-x}=1+x+x^2+x^3+\cdots$

24. $f(x)=(x+1)^{1/2};\,f(0)=1$
$f'(x)=(1/2)(x+1)^{-1/2};f'(0)=1/2$
$f''(x)=(-1/4)(x+1)^{-3/2};f''(0)=-1/4$
$f'''(x)=(3/8)(x+1)^{-5/2};f'''(0)=3/8$
$f^{(4)}(x)=(-15/16)(x+1)^{-7/2};f^{(4)}(0)=-15/16$
Thus $\sqrt{x+1}=1+x/2-x^2/8+x^3/16-5x^4/128+\cdots$

25. $f(x)=\sin x+\cos x;\,f(0)=1$
$f'(x)=\cos x-\sin x;f'(0)=1$
$f''(x)=-\sin x-\cos x;f''(0)=-1$
$f'''(x)=-\cos x+\sin x;f'''(0)=-1$
$f^{(4)}(x)=\sin x+\cos x;f^{(4)}(0)=1$
Thus $\sin x+\cos x=1+x-x^2/2!-x^3/3!+x^4/4!+\cdots$

26. $f(x)=e^x\sin x;\,f(0)=0$
$f'(x)=e^x(\cos x+\sin x);f'(0)=1$
$f''(x)=2e^x\cos x;f''(0)=2$
$f'''(x)=2e^x(\cos x-\sin x);f'''(0)=2$
Thus $e^x\sin x=0+x+2x^2/2!+2x^3/3!+\cdots=x+x^2+x^3/3+\cdots$

27. $\dfrac{1-e^x}{x}=\dfrac{1}{x}-\dfrac{e^x}{x}=\dfrac{1}{x}-\dfrac{1}{x}\left[\dfrac{1+x+x^2/2!+x^3/3!+\cdots}{x}\right]=\dfrac{1}{x}-\left[\dfrac{1}{x}+1+\dfrac{x}{2!}+\dfrac{x^2}{3!}+\cdots\right]=-1-\dfrac{x}{2!}-\dfrac{x^2}{3!}-\cdots$

28. Substitute x^2 for x in equation (3)in Section 10.6.
$\cos x^2=1-x^4/2!+x^8/4!-x^{12}/6!+\cdots$

29. Substitute $3x$ for x in equation (2) in Section 10.6.
$\sin 3x=3x-9x^3/2+81x^5/40-\cdots$

Chapter 10 Review

30. Substitute $\sin x$ for x in equation (1) in Section 10.6.

$$e^{\sin x} = 1 + \sin x + \frac{\sin^2 x}{2!} + \frac{\sin^3 x}{3!} + \cdots$$

31. Use equation (4) in Section 10.6.

$$\int_0^{0.1} \frac{\ln(x+1)}{x} dx = \int_0^{0.1} (1 - x/2 + x^2/3 - x^3/4) dx = (x - x^2/4 + x^3/9 - x^4/16) \Big|_0^{0.1} = 0.09772$$

32. Use equation (2) in Section 10.6.

$$\int_0^{0.1} \frac{\sin t}{t} dt = \int_0^{0.1} (1 - t^2/3! + t^4/5!) dt = \left[t - \frac{t^3}{3!(3)} + \frac{t^5}{5!(5)} \right] \Big|_0^{0.1} = 0.09994$$

33. $f(x) = \cos 2x; f(\pi/6) = 1/2$
$f'(x) = -2 \sin 2x; f'(\pi/6) = -\sqrt{3}$
$f''(x) = -4 \cos 2x; f''(\pi/6) = -2$
$f'''(x) = 8 \sin 2x; f'''(\pi/6) = 4\sqrt{3}$
$f^{(4)}(x) = 16 \cos 2x; f^{(4)}(\pi/6) = 8$

Thus $f(x) = \frac{1}{2} - \sqrt{3}\left(x - \frac{\pi}{6}\right) - 2\left(x - \frac{\pi}{6}\right)^2 + 4\sqrt{3}\left(x - \frac{\pi}{6}\right)^3 + 8\left(x - \frac{\pi}{6}\right)^4 + \cdots$

34. $f(x) = \ln x; f(4) = \ln 4$
$f'(x) = 1/x; f'(4) = 1/4$
$f''(x) = -1/x^2; f''(4) = -1/16$
$f'''(x) = 2/x^3; f'''(4) = 1/32$

Thus $f(x) = \ln 4 + \dfrac{x-4}{4} - \dfrac{(x-4)^2}{32} + \dfrac{(x-4)^3}{192} - \cdots$

35. $f(x) = e^{x^2}; f(1) = e$
$f'(x) = 2x \, e^{x^2}; f'(1) = 2e$
$f''(x) = e^{x^2}(4x^2 + 2); f''(1) = 6e$
$f'''(x) = e^{x^2}(8x^3 + 12x); f'''(1) = 20e$

Thus $f(x) = e\left[1 + 2(x-1) + \dfrac{6(x-1)^2}{2!} + \dfrac{20(x-1)^3}{3!} + \cdots \right]$

36. $f(x) = \sin x; f(3\pi/2) = -1$
$f'(x) = \cos x; f'(3\pi/2) = 0$
$f''(x) = -\sin x; f''(3\pi/2) = 1$
$f'''(x) = -\cos x; f'''(3\pi/2) = 0$
$f^{(4)}(x) = \sin x; f^{(4)}(3\pi/2) = -1$

Thus $f(x) = -1 + \dfrac{(x - 3\pi/2)^2}{2!} - \dfrac{(x - 3\pi/2)^3}{4!} + \cdots$

Chapter 10 Review

Solutions to Odd-Numbered Exercises

37. $(31° = 31\pi/180)$ Use equation (2) in Section 10.6.

$$\sin 31° = 31\ \pi/180 - \frac{(31\pi/180)^3}{3!} + \frac{(31\pi/180)^5}{5!} = 0.5150$$

38. $e^{x-1} = e\left[1 + (x-1) + \frac{(x-1)^2}{2!} + \frac{(x-1)^3}{3!}\right]$ (used Taylor)

$= e[1 + 0.2 + (0.2)^2/2! + (0.2)^3/3!] = 3.3199$

39. Use equation (4) in Section 10.6 and let $x = 0.2$.

$\ln 1.2 = \ln(1 + 0.2) = 0.2 - (0.2)^2/2 + (0.2)^3/3 - (0.2)^4/4 = 0.18227$

40. Use Exercise 4 in Section 10.7 with $x - a = 4.1 - 4 = 0.1$

$$\sqrt{4.1} = 2 + \frac{0.1}{4} - \frac{(0.1)^2}{2!(32)} + \frac{3(0.1)^3}{3!(256)} = 2.024846$$

41. $f(x) = -1/2 + (2/\pi)\sin x + (2/3\pi)\sin 3x + (2/5\pi)\sin 5x + \cdots$

$$f(x) = \begin{cases} 0 & 0 \le x < \pi \\ -1 & \pi \le x < 2\pi \end{cases}$$

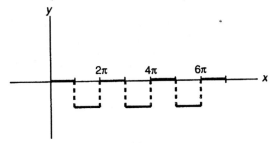

42. $f(x) = \pi^2/6 - 2\cos x + (1/2)\cos 2x - (2/9)\cos 3x + (1/8)\cos 4x -$

$$\cdots + \frac{\pi^2 - 4}{\pi}\sin x - \frac{\pi}{2}\sin 2x + \frac{9\pi^2 - 4}{27\pi}\sin 3x - \cdots$$

$$f(x) = \begin{cases} x^2 & 0 \le x < \pi \\ 0 & \pi \le x < 2\pi \end{cases}$$

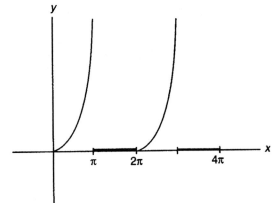

Chapter 10 Review

CHAPTER 11

Section 11.1

1. order 1, degree 1

3. order 2, degree 1

5. order 3, degree 1

7. order 2, degree 3

9. $y' = 3 = dy/dx$

11. $y' = 2x - 4$; $xy' - 2y = 4x$; $x(2x - 4) - 2(x^2 - 4) = 4x$; $4x = 4x$

13. $y' = -(x + 2)e^{-x} + e^{-x}$; $y' + y = e^{-x}$;
$[-(x + 2)e^{-x} + e^{-x}] + (x + 2)e^{-x} = e^{-x}$
$$e^{-x} = e^{-x}$$

15. $y' = 4C_1 \cos 4x - 4C_2 \sin 4x$; $y'' = -16C_1 \sin 4x - 16C_2 \cos 4x$;
$y'' + 16y = 0$; $(-16C_1 \sin 4x - 16C_2 \cos 4x) + 16(C_1 \sin 4x + C_2 \cos 4x) = 0$
$$0 = 0$$

17. $y' = \cos x - \sin x + e^{-x}$; $y' + y - 2 \cos x = 0$;
$(\cos x - \sin x + e^{-x}) + (\sin x + \cos x - e^{-x}) - 2 \cos x = 0$;
$$0 = 0$$

19. $y' = \cos^2 x - \sin^2 x$; $y'' = -4 \sin x \cos x$; $(y'')^2 + 4(y')^2 = 4$;
$(-4 \sin x \cos x)^2 + 4(\cos^2 x - \sin^2 x)^2 = 4$;
$16 \sin^2 x \cos^2 x + 4 \cos^4 x - 8 \cos^2 x \sin^2 x + 4 \sin^4 x = 4$;
$4 \cos^4 x + 8 \sin^2 x \cos^2 x + 4 \sin^4 x = 4$;
$4(\cos^4 x + 2 \sin^2 x \cos^2 x + 4 \sin^4 x) = 4$;
$4(\cos^2 x + \sin^2 x)^2 = 4$; $4 = 4$

21. $y' = 4e^{4x}$; $y'' = 16e^{4x}$; $y'' - 5y' + 4y = 0$;
$16e^{4x} - 5(4e^{4x}) + 4e^{4x} = 0$; $0 = 0$

23. $y' = e^{-x}(1 - x)$; $y'' = -e^{-x}(2 - x)$; $y'' + 2y' + y = 0$;
$[-e^{-x}(2 - x)] + 2[e^{-x}(1 - x)] + xe^{-x} = 0$;
$-2e^{-x} + xe^{-x} + 2e^{-x} - 2xe^{-x} + xe^{-x} = 0$; $0 = 0$

Section 11.2

1. $x \, dy - y^2 \, dx = 0$; $\int dy/y^2 = \int dx/x$; $-1/y = \ln x - C$;
$y \ln x + 1 = Cy$

3. $x \, dy + y \, dx = 0$; $\int dy/y + \int dx/x = 0$; $\ln y + \ln x = \ln C$;
$\ln yx = \ln C$; $xy = C$

5. $dy/dx = y^{3/2}$; $= \int y^{-3/2} dy = \int dx$; $-2y^{-1/2} = x + C$

7. $dy/dx = x^2(1 + y^2)$; $\int dy/(1+y^2) = \int x^2 dx$; Arctan $y = x^3/3 + C$

9. $x\, dy/dx + y = 3$; $x\, dy = (3 - y)dx$; $\int dy/(3 - y) = \int dx/x$;
   $- \ln (3 - y) = \ln x - \ln C$; $\ln C = \ln x(3-y)$; $x(3 - y) = C$

11. $dy/dx = x^2/y$; $\int y\, dy = \int x^2 dx$; $y^2/2 = x^3/3 + C$; $3y^2 = 2x^3 + C$

13. $dy/dx + y^3 \cos x = 0$; $\int dy/y^3 + \int \cos x\, dx = 0$;
    $- 1/(2y^2) + \sin x + C = 0$; $1 = 2y^2(\sin x + C)$

15. $e^{3x}\, dy/dx - e^x = 0$; $\int dy = \int e^{-2x} dx$; $y = (-1/2)\, e^{-2x} + C$; $2y = -e^{-2x} + C$

17. $(1 + x^2)\, dy - dx = 0$; $\int dy = \int dx/(1 + x^2)$; $y =$ Arctan $x + C$

19. $dy/dx = 1 + x^2 + y^2(1 + x^2)$; $dy/dx = (1 + x^2)(1 + y^2)$
    $\int dy/(1 + y^2) = \int (1 + x^2)dx$; Arctan $y = x + x^3/3 + C$

21. $dy/dx = e^x \cdot e^{-y}$; $\int e^y dy = \int e^x dx$; $e^y = e^x + C$; $e^y - e^x = C$

23. $(x + 1)\, dy/dx = y^2 + 4$; $\int dy/(y^2 + 4) = \int dy/(x + 1)$;
    $(1/2)$ Arctan $(y/2) = \ln (x + 1) + C$; Arctan $(y/2) = 2 \ln (x + 1) + C$

25. $(4xy + 12x)\, dx = (5x^2 + 5)\, dy$; $4x(y + 3)\, dx = 5(x^2 + 1)\, dy$;
    $\int \dfrac{4x\, dx}{x^2 + 1} = \int \dfrac{5\, dy}{y + 3}$; $2 \ln (x^2 + 1) = 5 \ln (y + 3) + \ln C$;
    $\ln (x^2 + 1)^2 = \ln (y + 3)^5 + \ln C$; $\ln \dfrac{(x^2 + 1)^2}{(y + 3)^5} = \ln C$; $(x^2 + 1)^2 = C(y + 3)^5$

27. $dy/dx = x^2 y^4$; $\int dy/y^4 = \int x^2 dx$; $(-1/3)y^{-3} = (1/3)x^3 + C$;
    for $x = 1$, $y = 1$, $C = -2/3$; Thus $0 = x^3 y^3 + 1 - 2y^3$; $y^3(2 - x^3) = 1$

29. $\dfrac{dy}{dx} = \dfrac{2x}{y + x^2 y}$; $\int y\, dy = \int \dfrac{2x\, dx}{x^2 + 1}$; $y^2/2 = \ln(x^2 + 1) + C$;
    for $x = 0$, $y = 4$, $C = 8$; Thus $y^2 = 2 \ln (x^2 + 1) + 16$

31. $y\, dy/dx = e^x$; $\int y\, dy = \int e^x dx$; $y^2/2 = e^x + C$; for $x = 0$, $y = 6$, $C = 17$ Thus $y^2 = 2e^x + 34$

Section 11.2

33. $\sqrt{x} + \sqrt{y}\, dy/dx = 0$; $\int x^{1/2}\, dx + \int y^{1/2}\, dy = 0$;

$(2/3)x^{3/2} + (2/3)y^{3/2} = C$; for $x = 1$, $y = 4$, $C = 6$

Thus $x^{3/2} + y^{3/2} = 9$

35. $xy\, dy/dx = \ln x$; $\int y\, dy = \int \dfrac{\ln x}{x}\, dx$; $y^2/2 = (1/2)\ln^2 x + C$;

$y^2 = \ln^2 x + C$; for $x = 1$, $y = 0$, $C = 0$; Thus $y^2 = \ln^2 x$

Section 11.3

1. $x\, dy + y\, dx = y^2\, dy$; $\int d(xy) = \int y^2\, dy$; $xy = y^3/3 + C$

$3xy = y^3 + C$

3. $x\, dy - y\, dx = 5x^2\, dy$; $\dfrac{x\, dy - y\, dx}{x^2} = 5\, dy$; $\int d(y/x) = 5\int dy$;

$y/x = 5y + C$; $y = 5xy + Cx$

5. $y\, dx - x\, dy + y^2\, dx = 3\, dy$; $\dfrac{y\, dx - x\, dy}{y^2} + dx = \dfrac{3}{y^2}\, dy$;

$\int d(x/y) + \int dx = 3\int y^{-2}\, dy$; $x/y + x = -3/y - C$;

$x + xy = -3 - Cy$; $x + xy + Cy + 3 = 0$

7. $x\sqrt{x^2 + y^2}\, dx - 2x\, dx = 2y\, dy$; $x\sqrt{x^2 + y^2}\, dx = 2(x\, dx + y\, dy)$;

$x\, dx = 2\dfrac{x\, dx + y\, dy}{\sqrt{x^2 + y^2}}$; $\int x\, dx = \int \dfrac{d(x^2 + y^2)}{\sqrt{x^2 + y^2}}$;

$x^2/2 = 2(x^2 + y^2)^{1/2} + C$; $x^2 = 4\sqrt{x^2 + y^2} + C$

9. $x\, dx + y\, dy = x(x^2 + y^2)\, dy + y(x^2 + y^2)\, dx$; $\dfrac{x\, dx + y\, dy}{x^2 + y^2} = x\, dy + y\, dx$;

$\int d(\ln\sqrt{x^2 + y^2}) = \int d(xy)$; $\ln\sqrt{x^2 + y^2} = xy + C$

11. $x\, dy + y\, dx = 2(x\, dx + y\, dy)$; $\int d(xy) = \int d(x^2 + y^2)$;

$xy = x^2 + y^2 + C$; for $x = 0$, $y = 1$, $C = -1$;

Thus $xy = x^2 + y^2 - 1$

13. $x\, dy - y\, dx = (x^3 + y^2 x)\, dy + (x^2 y + y^3)\, dx$;

$x\, dy - y\, dx = x(x^2 + y^2)dy + y(x^2 + y^2)\, dx$; $\dfrac{x\, dy - y\, dx}{x^2 + y^2} = x\, dy + y\, dx$;

$\int d(\text{Arctan } y/x) = \int d(xy)$; Arctan $y/x = xy + C$; for $x = 2$, $y = 2$, $C = \dfrac{\pi - 16}{4}$;

Arctan $\dfrac{y}{x} = xy + \dfrac{\pi - 16}{4}$

Sections 11.2 – 11.3

Section 11.4

1. $dy/dx - 5y = e^{3x}$; $dy - 5y\,dx = e^{3x}\,dx$; $P(x) = -5$, $Q(x) = e^{3x}$,

$$\int P(x)\,dx = -5x, \quad ye^{-5x} = \int e^{3x}e^{-5x}\,dx\ ;$$

$$ye^{-5x} = (-1/2)e^{-2x} + C; \quad 2y + e^{3x} = Ce^{5x}$$

3. $dy/dx + 3y/x = x^3 - 2$; $P(x) = 3/x$, $Q(x) = x^3 - 2$,

$$\int P(x)\,dx = \int (3/x)dx = 3\ln x = \ln x^3;$$

$$ye^{\ln x^3} = \int (x^3 - 2)x^3\,dx\ ; \quad yx^3 = y^7/7 - x^4/2 + C;$$

$$14yx^3 = 2x^7 - 7x^4 + C$$

5. $dy/dx + 2xy = e^{3x}(3 + 2x)$; $P(x) = 2x$, $Q(x) = e^{3x}(3 + 2x)$,

$$\int P(x)\,dx = x^2;$$

$$ye^{x^2} = \int e^{3x}(3 + 2x)e^{x^2}\,dx = \int e^{x^2+3x}(3 + 2x)\,dx = e^{x^2+3x} + C; \quad y = e^{3x} + Ce^{-x^2}$$

7. $dy - 4y\,dx = x^2\,e^{4x}\,dx$; $P(x) = -4$, $Q(x) = x^2\,e^{4x}$,

$$\int P(x)\,dx = \int -4x\ , \quad ye^{-4x} = \int x^2 e^{4x}e^{-4x}\,dx = \int x^2\,dx = x^3/3 + C\ ; \quad 3y = x^3\,e^{4x} + Ce^{4x}$$

9. $x\,dy - 5y\,dx = (x^6 + 4x)\,dx$; $dy - (5/x)\,y\,dx = (x^5 + 4)\,dx$;

$$P(x) = -5/x, \quad Q(x) = x^5 + 4, \quad \int P(x)\,dx = \int (-5/x)\,dx = -5\ \ln/x = \ln x^{-5},\ e^{P(x)\,dx} = e^{\ln x^{-5}} = x^{-5};$$

$$yx^{-5} = \int (x^5 + 4)x^{-5}\,dx = \int (1 + 4x^{-5})\,dx = x - 1/x^4 + C; \quad y = x^6 - x + Cx^5$$

11. $(1 + x^2)\,dy + 2xy\,dx = 3x^2\,dx$; $P(x) = \dfrac{2x}{x^2+1}$;

$$Q(x) = \frac{3x^2}{x^2+1}; \quad \int P(x)\,dx = \int \frac{2x\,dx}{x^2+1} = \ln(x^2 + 1);$$

$$ye^{\ln(x^2+1)} = \int \frac{3x^2}{x^2+1}(x^2 + 1)\,dx = x^3 + C;$$

$$y(x^2 + 1) = x^3 + C$$

13. $dy/dx + (2/x)\,y = (x^2 - 7/x^2)$; $P(x) = 2/x$, $Q(x) = x^2 - 7x^{-2}$,

$$\int P(x)\,dx = \int (2/x)\,dx = 2\ln x = \ln x^2;\ e^{\int P(x)(dx)} = e^{\ln x^2} = x^2\ ;$$

$$yx^2 = \int (x^4 - 7)dx = x^5/5 - 7x + C; \quad 5yx^2 = x^5 - 35x + C$$

15. $dy/dx + 2y = e^{-x}$; $P(x) = 2$, $Q(x) = e^{-x}$, $\int P(x)\,dx = 2x$;

$$ye^{2x} = \int e^{-x}e^{2x}\,dx = \int e^x\,dx = e^x + C; \quad y = e^{-x} + Ce^{-2x}$$

Section 11.4

17. $dy/dx - (1/x) y = 3x$; $P(x) = -1/x$, $Q(x) = 3x$,

$\int P(x)\,dx = \int (-1/x)\,dx = -\ln x = \ln x^{-1}; e^{\ln x^{-1}} = x^{-1}$;

$yx^{-1} = \int 3x\, x^{-1}dx = \int 3\,dx = 3x + C$; $y = 3x^2 + Cx$

19. $dy/dx + y \cos x = \cos x$; $P(x) = \cos x$; $Q(x) = \cos x$;

$\int P(x)\,dx = \int \cos x\,dx = \sin x; ye^{\sin x} = \int \cos x\, e^{\sin x}dx = e^{\sin x} + C; (y-1)e^{\sin x} = C$

21. $dy/dx - 3y = e^{2x}$; $P(x) = -3$, $Q(x) = e^{2x}$, $\int P(x)\,dx = -3x$;

$ye^{-3x} = \int e^{2x}e^{-3x}\,dx = \int e^{-x}\,dx = -e^{-x} + C$;

$y = -e^{2x} + Ce^{3x}$; for $x = 0$, $y = 2$, $C = 3$;

$y = -e^{2x} + 3e^{3x} = e^{2x}(3e^x - 1)$

23. $dy/dx + y \cot x = \csc x$; $P(x) = \cot x$, $Q(x) = \csc x$, $\int P(x)\,dx = \int \cot x\,dx = \ln (\sin x)$;

$ye^{\ln(\sin x)} = \int \csc x\, e^{\ln(\sin x)}\,dx$;

$y \sin x = \int \csc x \sin x\,dx = \int dx = x + C$; $y \sin x = x + C$;

for $x = \pi/2$, $y = 3\pi/2$, $C = \pi$, $y \sin x = x + \pi$

25. $dy/dx + y = e^x$; $P(x) = 1$, $Q(x) = e^x$, $\int P(x)\,dx = x$;

$ye^x = \int e^x e^x\,dx = \int e^{2x}\,dx = (1/2)e^{2x} + C$; for $x = 0$, $y = 3/2$, $C = 1$;

$ye^x = (1/2)e^{2x} + 1$; $y = e^x/2 + e^{-x}$

27. $dy/dx + (1/x) y = 3/x$; $P(x) = 1/x$, $Q(x) = 3/x$,

$\int P(x)\,dx = \int (1/x)dx = \ln x$; $ye^{\ln x} = \int (3/x)x\,dx = 3x + C$;

$yx = 3x + C$; for $x = 1$, $y = -2$, $C = -5$, $yx = 3x - 5$;

$y = 3 - 5/x$

29. $dy/dx + (1/x)y = 4x^2$; $P(x) = 1/x$, $Q(x) = 4x^2$, $\int P(x)\,dx = \int (1/x)\,dx = \ln x$

$ye^{\ln x} = \int 4x^2 x\,dx$; $yx = x^4 + C$; for $x = 2$, $y = 3$, $C = -10$;

$xy = x^4 - 10$

Section 11.5

1. $v = \int dv = \int a\,dt = \int 5\,dt = 4t + C$; $v = 10$ m/s at $t = 0$ so $C = 10$;

$v = 5t + 10\Big|_{t=3} = 25$ m/s

3. $dy/dx = (x^2 - y)/x$; $x\,dy = x^2\,dx - y\,dx$; $x\,dy + y\,dx = x^2\,dx$;

$\int d(xy) = \int x^2\,dx$; $xy = x^3/3 + C$; at $(1, 1)$, $C = 2/3$, $xy = x^3/3 + 2/3$; $3xy = x^3 + 2$

Sections 11.4 – 11.5

5. $di/dt + (R/L)i = V/L$; $di/dt + 800i = 1200$ since $L = 0.1H$,
   $R = 80\ \Omega$, $V = 120\ V$. $P(t) = 800$, $Q(t) = 1200$, $\int P(t)\,dt = 800t$;

   $ie^{800t} = \int 1200e^{800t}\,dt = (3/2)\,e^{800t} + C$; at $t = 0$; $i = 2A$

   Thus $C = 1/2$; $ie^{800t} = (3/2)\,e^{800t} + 1/2$; $i = (1/2)(3 + e^{-800t})$

7. $\dfrac{di}{dt} + \left(\dfrac{R}{L}\right)i = V/L$; $L = 1H$, $R = 4\Omega$, $V = 10\sin 2t$ volts

   $\dfrac{di}{dt} + \dfrac{4}{1}i = 10\sin 2t$

   $ie^{\int 4dt} = \int 10\sin 2t\, e^{\int 4dt}\,dt$

   $ie^{4t} = 10\int \sin 2t\left(e^{4t}\right)dt$

   First find $\int e^{4t}\,\sin 2t\,dt$

   Let $u = \sin 2t$ $dv = e^{4t}\,dt$

   $\qquad du = 2\cos 2t\,dt$ $v = \dfrac{1}{4}e^{4t}$

   $\int e^{4t}\,\sin 2t\,dt = \dfrac{1}{4}e^{4t}\sin 2t - \int \dfrac{1}{2}e^{4t}\cos 2t\,dt$

   Let $u = \cos 2t$ $dv = \dfrac{1}{2}e^{4t}\,dt$

   $\qquad du = -2\sin 2t$ $v = \dfrac{1}{8}e^{4t}$

   $\int e^{4t}\sin 2t\,dt = \dfrac{1}{4}e^{4t}\sin 2t - \left[\dfrac{1}{8}e^{4t}\cos 2t - \int \dfrac{-1}{4}e^{4t}\sin 2t\,dt\right]$

   $\int e^{4t}\sin 2t\,dt = \dfrac{1}{4}e^{4t}\sin 2t - \dfrac{1}{8}e^{4t}\cos 2t - \dfrac{1}{4}\int e^{4t}\sin 2t\,dt$

   $\dfrac{5}{4}\int e^{4t}\sin 2t\,dt = \dfrac{1}{4}e^{4t}\sin 2t - \dfrac{1}{8}e^{4t}\cos 2t$

   thus: $\int e^{4t}\sin 2t\,dt = \dfrac{1}{5}e^{4t}\sin 2t - \dfrac{1}{10}e^{4t}\cos 2t$

   So $ie^{4t} = 10\int e^{4t}\sin 2t\,dt$ becomes

   $ie^{4t} = 10\left[\dfrac{1}{5}e^{4t}\sin 2t - \dfrac{1}{10}e^{4t}\cos 2t\right]$

   $ie^{4t} = 2e^{4t}\sin 2t - e^{4t}\cos 2t + C$
   $i = 2\sin 2t - \cos 2t + Ce^{-4t}$
   At $t = 0$, $i_o = 0$, so
   $0 = 2\sin 0 - \cos 0 + Ce^0$
   $0 = -1 + C$ thus $C = 1$
   Solution is: $i = 2\sin 2t - \cos 2t + e^{-4t}$

Section 11.5

9. $dQ/dt = kQ$; $Q = Q_o e^{kt}$; $Q_o = 1$ g, $Q = 0.5$ g when $t = 4.5 \times 10^9$ yr

$0.5 = 1 e^{(4.5 \times 10^9) k}$; $k = \dfrac{\ln 0.5}{4.5 \times 10^9} = -1.54 \times 10^{-10}$;

$Q = 1 e^{-1.54 \times 10^{-10} t}$

11. $dQ/dt = kQ$; $Q = Q_o e^{kt}$; at $t = 0$, $Q_o = 5$ g, $Q = 5 e^{kt}$; at $t = 36$.

$Q = 4.5$ so $4.5 = 5 e^{36k}$; $k = \dfrac{\ln 0.9}{36} = -0.00293$;

Half-life: $t = \dfrac{-\ln 2}{k} = \dfrac{-\ln 2}{-0.00293} = 237$ yr

13. Let Q = amt of salt at time t. dQ/dt = rate of gain − rate of loss.
    (2 lb/gal)(2 gal/min) − (Q lb/50 gal)(2 gal/min) = $4 - Q/25$;

$\dfrac{dQ}{dt} = \dfrac{100 - Q}{25}$; $\displaystyle\int \dfrac{dQ}{100 - Q} = \int \dfrac{dt}{25}$; $Q = C e^{-0.04t} + 100$; $t = 0$,

$Q = 10$, $C = -90$; $Q = -90^{-0.04t} + 100$; find Q at $t = 30$ min;
$Q = -90 e^{(-0.04)(30)} + 100 = 72.9$ lb

15. $dT/dt = k(T - T_o) = k(T - 10)$; $\displaystyle\int dT/(T - 10) = \int k \, dt$;

$\ln (T - 10) = kt + \ln C$; $\ln \dfrac{T - 10}{C} = kt$; $\dfrac{T - 10}{C} = e^{kt}$;

$T = C e^{kt} + 10$; At $t = 0$, $T = 90$ so $90 = C e^{k \cdot 0} + 10$; $C = 80$;
$T = 80 e^{kt} + 10$; find k when $t = 5$ and $T = 70$: $70 = 80 e^{5k} + 10$;
$k = -0.0575$; then at $t = 30$: $T = 80 e^{(-0.0575)(30)} + 10 = 24.3°$ C

17. $y' = ky$; $\displaystyle\int dy/y = \int k \, dt$; $\ln y = kt + \ln C$; $y = C e^{kt}$; at $t = 0$, $y = 2{,}000{,}000 = C$

Since the population doubles in 20 yr, $2 = e^{20k}$; $k = 0.0347$;
in 80 yr, $y = 2{,}000{,}000 \, e^{(0.0347)(80)} = 3.2 \times 10^7$

19. $dp/dV = kp/V$; $dp/p = k \, dV/V$; $\displaystyle\int dp/p = k \int dV/V$;
    $\ln p = k \ln V + \ln C$; $\ln p = \ln CV^k$; Thus $p = CV^k$

Chapter 11 Review

1. order 2, degree 1

2. order 1, degree 2

3. order 2, degree 3

4. order 1, degree 2

5. $y = x^3 - 2x$; $y' = 3x^2 - 2$; $y'' = 6x$; $y'' + 3y = 3x^3$;
    $6x + 3(x^3 - 2x) = 3x^3$; $3x^3 = 3x^3$

6. $y = 2e^x - 3xe^x + e^{-x}$; $y' = -e^x - 3xe^x - e^{-x}$;
    $y'' = -4e^x - 3xe^x + e^{-x}$;
    $y'' - 2y' + y = 4e^{-x}$;
    $(-4e^x - 3xe^x + e^{-x}) - 2(-e^x - 3xe^x - e^{-x}) + (2e^x - 3xe^x + e^{-x}) = 4e^{-x}$; $4e^{-x} = 4e^{-x}$

Section 11.5 – Chapter 11 Review

7. $y = \sin x + x^2$; $y' = \cos x + 2x$; $y'' = -\sin x + 2$;
   $y'' + y = x^2 + 2$; $(-\sin x + 2) + (\sin x + x^2) = x^2 + 2$;
   $x^2 + 2 = x^2 + 2$

8. $y = e^{2x}(x^5 - 1)$; $y' = 2e^{2x}(x^5 - 1) + 5x^4 e^{2x}$; $y' - 2y = 5x^4 e^{2x}$;
   $2e^{2x}(x^5 - 1) + 5x^4 e^{2x} - 2e^{2x}(x^5 - 1) = 5x^4 e^{2x}$;
   $5x^4 e^{2x} = 5x^4 e^{2x}$

9. $\int dy/y = \int dx/x^3$; $\ln y = -1/(2x^2) + C$; $2x^2 \ln y = -1 + Cx^2$

10. $\int dy/(3y) = \int e^{-2x} dx$; $(1/3) \ln y = (-1/2) e^{-2x} + C$; $2 \ln y + 3e^{-2x} = C$

11. $\int y^2 dy + \int \dfrac{x\, dx}{x^2 + 9} = 0$; $y^3/3 + (1/2) \ln (x^2 + 9) = C$;
    $2y^3 + 3 \ln(x^2 + 9) = C$

12. $\int e^{-y} dy = \int \sec^2 x \, dx$; $-e^{-y} = \tan x + C$; $\tan x + e^{-y} = C$

13. $\dfrac{x\, dy - y\, dx}{x^2} = 3x^2 dx$; $\int d(y/x) = 3 \int x^2 dx$; $y/x = x^3 + C$; $y = x^4 + Cx$

14. $dy/dx + 6x^2 y = 12x^2$; $P(x) = 6x^2$, $Q(x) = 12x^2$, $\int P(x) \, dx = 2x^3$;
    $ye^{2x^3} = \int 12x^2 \, e^{2x^3} dx = 2e^{2x^3} + C$; $(y - 2)e^{2x^3} = C$

15. $dy/dx - (1/x^2) \, y = 5/x^2$; $P(x) = -1/x^2$, $Q(x) = 5/x^2$,
    $\int P(x) \, dx = 1/ x$; $ye^{1/x} = \int (5/x^2) e^{1/x} dx = -5e^{1/x} + C$; $(y + 5)e^{1/x} = C$

16. $\dfrac{x\, dx + y\, dy}{x^2 + y^2} = y \, dy$; $\int d(\ln \sqrt{x^2 + y^2}) = \int y \, dy$; $\ln \sqrt{x^2 + y^2} = y^2/2 + C$;
    $(1/2) \ln (x^2 + y^2) = (1/2)y^2 + C$; $\ln (x^2 + y^2) = y^2 + C$

17. $\dfrac{x\, dy - y\, dx}{x^2 + y^2} = x^2 dx$; $\int d (\text{Arctan } y/x) = \int x^2 dx$; $3 \text{ Arctan } y/x = x^3 + C$

18. $x \, dy + y \, dx = 14x^5 dx$; $\int d(xy) = \int 14x^5 dx$; $xy = (7/3)x^6 + C$; $3xy = 7x^6 + C$

19. $dy/dx + 3y = e^{-2x}$; $P(x) = 3$; $Q(x) = e^{-2x}$, $\int P(x) \, dx = 3x$;
    $ye^{3x} = \int e^{-2x} \, e^{3x} dx = e^x + C$; $y = e^{-2x} + Ce^{-3x}$

Chapter 11 Review

20. $dy/dx - (5/x)y = x^3 + 7$; $P(x) = -5/x$, $Q(x) = x^3 + 7$,

$\int P(x)\,dx \int(-5/x)\,dx = -5 \ln x = \ln x^{-5}$;

$ye^{\ln x^{-5}} = \int(x^3 + 7)e^{\ln x^{-5}}\,dx$; $yx^{-5} = \int(x^{-2} + 7x^{-5})dx =$

$-x^{-1} - (7/4)x^{-4} + C$; $4y = -4x^4 - 7x + Cx^5$

21. $x\,dy - 3y\,dx = (4x^3 - x^2)\,dx$; $dy/dx - (3/x)\,y = 4x^2 - x$;

$P(x) = -3/x$, $Q(x) = 4x^2 - x$, $\int P(x)\,dx \int(-3/x)\,dx = \ln x^{-3}$; $ye^{\ln x^{-3}} = \int(4x^2 - x)e^{\ln x^{-3}}\,dx$;

$yx^{-3} = \int(4x^2 - x)x^{-3}dx = \int(4/x - 1/x^2)dx = 4 \ln x + 1/x + C$;

$y = 4x^3 \ln x - x^2 + Cx^3$

22. $dy/dx + y \cot x = \cos x$; $P(x) = \cot x$, $Q(x) = \cos x$, $\int P(x)\,dx = \ln (\sin x)$;

$ye^{\ln (\sin x)} = \int \cos x \cdot e^{\ln (\sin x)}\,dx$; $y \sin x = \int \cos x \sin x\,dx = \dfrac{1}{2}\sin^2 x + C$

$2y \sin x = \sin^2 x + C$

23. $\int 3x\,dx = \int dy/y^2$; $3x^2/2 = 1/y + C$; for $x = -2$,

$y = -1$, $C = 5$; $3x^2/2 = -1/y + 5$; $3x^2y - 10y + 2 = 0$

24. $\int y\,dy = -3\int e^{-2x}dx$; $y^2/2 = (3/2)\,e^{-2x} + C$; for $x = 0$,

$y = -2$, $C = 1/2$; $y^2/2 = (3/2)e^{-2x} + 1/2$; $y^2 = 3e^{-2x} + 1$

25. $\dfrac{x\,dy - y\,dx}{x^2} = x^3dx$; $\int d(y/x) = \int x^3dx$;

$y/x + x^4/4 + C$; for $f(1) = 1$, $C = 3/4$; $y/x = x^4/4 + 3/4$;
$4y = x^5 + 3x$

26. $x\,dy + y\,dx = x \ln x\,dx$; $\int d(xy) = \int x \ln x\,dx$;

$xy = (x^2/2) \ln x - \int(x^2/2)(1/x)\,dx = (x^2/2) \ln x - (1/4)x^2 + C$;
For $f(1) = 3/4$, $C = 1$; $xy = (x^2/2) \ln x - x^2/4 + 1$;
$xy = (x^2/4)(2 \ln x - 1) + 1$

27. $dy/dx - y = e^{5x}$; $P(x) = -1$, $Q(x) = e^{5x}$, $\int P(x)\,dx = -x$;

$ye^{-x} = \int e^{5x}\,e^{-x}\,dx = (1/4)e^{4x} + C$; $4y = e^{5x} + Ce^x$;
for $f(0) = -3$, $C = -13$, $4y = e^{5x} - 13e^x$

Chapter 11 Review

28. $dy/dx - (2/x) y = x^3 - 5x$; $P(x) = -2/x$, $Q(x) = x^3 - 5x$,

$\int F'(x)\, dx = \int (-2/x)\, dx = \ln x^{-2}$; $ye^{\ln x^{-2}} = \int (x^3 - 5x)e^{\ln x^{-2}}\, dx$;

$yx^{-2} = \int (x - 5/x)\, dx = x^2/2 - 5\ln x + C$;

$y = x^4/2 - 5x^2 \ln x + Cx^2$; for $f(1) = 3$, $C = 5/2$

$2y = x^4 - 10x^2 \ln x + 5x^2$

29. $di/dt + (R/L)i = V/L$; $L = 0.2\ H$, $R = 60\ \Omega$, $V = 120\ V$;

$di/dt + 300i = 600$; $P(t) = 300$, $Q(t) = 600$, $\int P(t)\, dt = 300\ t$;

$ie^{300t} = \int 600e^{300t}\, dt = 2e^{300t} + C$; at $t = 0$; $i = 1$, $C = -1$;

$ie^{300t} = 2e^{300t} - 1$; $i = 2 - e^{-300t}$

30. $dQ/dt = kQ$; $Q = Q_o e^{kt}$; at $t = 0$, $Q = Q_o = 1$ and $Q = e^{kt}$;

$k = \dfrac{-\ln 2}{8.8 \times 10^8} = -7.88 \times 10^{-10}$; $Q = e^{-7.88 \times 10^{-10}t}$

31. $dQ/dt =$ rate of gain $-$ rate of loss

$(0\ lb/L)(1\ L/min) - (Q\ lb/200\ L)(1\ L/min) = -Q/200$;

$dQ/dt = -Q/200$; $200 \int dQ/Q = -\int dt$; $200 \ln Q = -t + \ln C$;

$Q = Ce^{-t/200}$; at $t = 0$, $Q = 5$, $C = 5$, $Q = 5e^{-t/200}$;

at $t = 20$ min, $Q = 5e^{(-20/200)} = 4.524$ kg

32. $dT/dt = k(T - 12)$; $\int dT/(T - 12) = k \int dt$;

$\ln(T - 12) = kt + \ln C$; $\ln \dfrac{T - 12}{C} = kt$; $T = Ce^{kt} + 12$;

at $t = 0$, $T = 30$, $C = 18$, $T = 18e^{kt} + 12$; find k if $T = 25$

when $t = 20$; $25 = 18e^{20k} + 12$; $k = \dfrac{\ln(13/18)}{20} = -0.0163$

$T = 18e^{-0.0163t} + 12$; at $t = 45$ min, $T = 18e^{(-0.0163)(45)} + 12 = 20.6°$ C

33. $m = dy/dx = x^2 y$; $\int dy/y = \int x^2 dx$; $\ln y = x^3/3 + C$; for $x = 0$,

$y = 2$, and $C = \ln 2$; $\ln y = x^3/3 + \ln 2$; $\ln (y/2) = x^3/3$;

$y = 2e^{x^3/3}$ or $y^3 = 8e^{x^3}$

34. $a = \int dv = \int a\, dt = \int 4t^3 dt = t^4 + C$; at $t = 0$, $v = 0$, $C = 0$

$v = t^4 \Big|_{t = 3\ s} = 81$ m/s^2

Chapter 11 Review

CHAPTER 12

Section 12.1

1. Homogeneous; 4

3. Homogeneous; 3

5. Homogeneous; 2

7. Nonhomogeneous; 3

9. $m^2 - 5m - 14 = 0$; $(m - 7)(m + 2) = 0$; $m = -2, 7$;
$y = k_1 e^{7x} + k_2 e^{-2x}$

11. $m^2 - 2m - 8 = 0$; $(m - 4)(m + 2) = 0$; $m = -2, 4$
$y = k_1 e^{4x} + k_2 e^{-2x}$

13. $m^2 - 1 = 0$; $(m - 1)(m + 1) = 0$; $m = -1, 1$; $y = k_1 e^x + k_2 e^{-x}$

15. $m^2 - 3m = 0$; $m(m - 3) = 0$; $m = 0, 3$; $y = k_1 + k_2 e^{3x}$

17. $2m^2 - 13m + 15 = 0$; $(2m - 3)(m - 5) = 0$; $m = 3/2, 5$;
$y = k_1 e^{3x/2} + k_2 e^{5x}$

19. $3m^2 - 7m + 2 = 0$; $(3m - 1)(m - 2) = 0$; $m = 2, 1/3$;
$y = k_1 e^{2x} + k_2 e^{x/3}$

21. $m^2 - 4m = 0$; $m(m - 4) = 0$; $m = 0, 4$; $y = k_1 + k_2 e^{4x}$;
$y' = 4k_2 e^{4x}$ substitute $y = 3$, $x = 0$, and $y' = 4$ into y and y'
equations and solve the resulting system of equations:
$3 = k_1 + k_2 e^{4(0)}$; $4 = 4k_2 e^{4(0)}$;
$k_1 = 2$, $k_2 = 1$; Thus $y = 2 + e^{4x}$

23. $m^2 - m - 2 = 0$; $(m - 2)(m + 1) = 0$; $m = -1, 2$;
$y = k_1 e^{2x} + k_2 e^{-x}$; $y' = 2k_1 e^{2x} - k_2 e^{-x}$; substitute $y = 2$,
$x = 0, y' = 1$; $2 = k_1 + k_2$; $1 = 2k_1 - k_2$; $k_1 = 1$; $k_2 = 1$
Thus $y = e^{2x} + e^{-x}$

25. $m^2 - 8m + 15 = 0$; $(m - 5)(m - 3) = 0$; $m = 3, 5$;
$y = k_1 e^{3x} + k_2 e^{5x}$; $y' = 3k_1 e^{3x} + 5k_2 e^{5x}$; substitute $y = 4$, $x = 0$,
$y' = 2$; $4 = k_1 + k_2$; $2 = 3k_1 + 5k_2$; $k_1 = 9$; $k_2 = -5$,
Thus $y = 9e^{3x} - 5e^{5x}$

Section 12.2

1. $m^2 - 4m + 4 = 0$; $(m - 2)^2 = 0$; $m = 2, 2$; $y = k_1 e^{2x} + k_2 x e^{2x}$;
$y = e^{2x}(k_1 + k_2 x)$

3. $m^2 - 4m + 5 = 0$; $m = 2 \pm j$; $y = e^{2x}(k_1 \sin x + k_2 \cos x)$

5. $4m^2 - 4m + 1 = 0$; $(2m - 1)^2 = 0$; $m = 1/2, 1/2$;
$y = e^{x/2}(k_1 + k_2 x)$

Sections 12.1 – 12.2

7. $m^2 - 4m + 13 = 0$; $m = 2 \pm 3j$; $y = e^{2x}(k_1 \sin 3x + k_2 \cos 3x)$

9. $m^2 - 10m + 25 = 0$; $(m - 5)^2 = 0$; $m = 5, 5$; $y = e^{5x}(k_1 + k_2 x)$

11. $m^2 + 9 = 0$; $m = \pm 3j$; $y = k_1 \sin 3x + k_2 \cos 3x$ 13. $m^2 = 0$; $m = 0, 0$; $y = k_1 + k_2 x$

15. $m^2 - 6m + 9 = 0$; $(m - 3)^2 = 0$; $m = 3, 3$; $y = k_1 e^{3x} + k_2 x e^{3x}$;
    $y' = 3k_1 e^{3x} + k_2 e^{3x} + 3k_2 x e^{3x}$; substitute $y = 2$, $x = 0$, $y' = 4$;
    $2 = k_1$, $4 = 3k_1 + k_2$; $k_1 = 2$, $k_2 = -2$; $y = 2e^{3x} - 2x e^{3x} = 2e^{3x}(1 - x)$

17. $m^2 + 25 = 0$; $m = \pm 5j$; $y = k_1 \sin 5x + k_2 \cos 5x$;
    $y' = 5k_1 \cos 5x - 5k_2 \sin 5x$; substitute $y = 2$, $x = 0$, $y' = 0$;
    $2 = k_2$, $0 = k_1$, $y = 2 \cos 5x$

19. $m^2 - 12m + 36 = 0$; $(m - 6)^2 = 0$; $m = 6, 6$; $y = k_1 e^{6x} + k_2 x e^{6x}$;
    $y' = 6k_1 e^{6x} + k_2 e^{6x} + 6k_2 x e^{6x}$; substitute $y = 1$, $x = 0$, $y' = 0$;
    $1 = k_1$; $0 = 6k_1 + k_2$; $k_1 = 1$; $k_2 = -6$;
    $y = e^{6x} - 6x e^{6x} = e^{6x}(1 - 6x)$

Section 12.3

1. yc: $m^2 + m = 0$; $m(m + 1) = 0$; $m = 0, -1$; $y_c = k_1 + k_2 e^{-x}$;
    y_p: $y_p = A \sin x + B \cos x$; $y'_p = A \cos x - B \sin x$;
    $y''_p = -A \sin x - B \cos x$; substitute into the given differential equation:
    $(-A \sin x - B \cos x) + (A \cos x - B \sin x) = \sin x$
    $(-A - B) \sin x + (A - B) \cos x = \sin x$; equate coefficients:
    $-A - B = 1$, $A - B = 0$; $A = -1/2$, $B = -1/2$;
    $y = k_1 + k_2 e^{-x} - (1/2) \sin x - (1/2) \cos x$

3. yc: $m^2 - m - 2 = 0$; $(m - 2)(m + 1) = 0$; $m = -1, 2$;
    $y_c = k_1 e^{2x} + k_2 e^{-x}$
    y_p: $y_p = A + Bx$; $y'_p = B$; $y_p'' = 0$; substitute into the given differential equation:
    $0 - B - 2(A + Bx) = 4x$;
    $(-2A - B) - 2Bx = 4x$; equate coefficients: $-2A - B = 0$,
    $-2B = 4$; $B = -2$; $A = 1$, $y = k_1 e^{2x} + k_2 e^{-x} - 2x + 1$

5. y_c: $m^2 - 10m + 25 = 0$; $(m - 5)^2 = 0$; $m = 5, 5$; $y_c = e^{5x}(k_1 + k_2 x)$
    y_p: $y_p = Ax + B$; $y'_p = A$; $y_p'' = 0$; substitute into given D.E.
    $0 - 10A + 25(Ax + B) = x$; $25Ax + (-10A + 25B) = x$;
    equate coefficients: $25A = 1$, $-10A + 25B = 0$; $A = 1/25$, $B = 2/125$;
    $y = e^{5x}(k_1 + k_2 x) + x/25 + 2/125$

7. y_c: $m^2 - 1 = 0$; $m = \pm 1$; $y_c = k_1 e^x + k_2 e^{-x}$
    y_p: $y_p = Ax^2 + Bx + C$; $y'_p = 2Ax + B$; $y''_p = 2A$; substitute into D.E.
    $2A - (Ax^2 + Bx + C) = x^2$; $-Ax^2 - Bx + (2A - C) = x^2$;
    equate coefficients; $-A = 1$, $-B = 0$, $2A - C = 0$; $A = -1$, $B = 0$, $C = -2$;
    $y = k_1 e^x + k_2 e^{-x} - x^2 - 2$

Sections 12.2 – 12.3

9. y_c: $m^2 + 4 = 0$; $m = \pm 2j$; $y_c = k_1 \sin 2x + k_2 \cos 2x$
y_p: $y_p = Ae^x + B$; $y'_p = Ae^x$; y''_p; substitute into D.E.
$Ae^x + 4(Ae^x + B) = e^x - 2$; $5Ae^x + 4B = e^x - 2$; $A = 1/5$,
$B = -1/2$; $y = k_1 \sin 2x + k_2 \cos 2x + (1/5) e^x - 1/2$

11. y_c: $m^2 - 3m - 4 = 0$; $(m - 4)(m + 1) = 0$; $m = -1, 4$;
$y_c = k_1 e^{4x} + k_2 e^{-x}$; y_p: $y_p = Ae^x = y'_p = y''_p$; substitute into D.E.
$Ae^x - 3Ae^x - 4Ae^x = 6e^x$; $-6Ae^x = 6e^x$; $A = -1$;
$y = k_1 e^{4x} + k_2 e^{-x} - ex$

13. y_c: $m^2 + 1 = 0$; $m = \pm j$; $y_c = k_1 \sin x + k_2 \cos x$
y_p: $y_p = A + B \cos 3x + C \sin 3x$; $y'_p = -3B \sin 3x + 3C \cos 3x$;
$y''_p = -9B \cos 3x - 9C \sin 3x$; substitute into D.E.
$(-9B \cos 3x - 9C \sin 3x) + (A + B \cos 3x + C \sin 3x) = 5 + \sin 3x$;
$A - 8B \cos 3x - 8c \sin 3x = 5 + \sin 3x$;
$A = 5, B = 0, C = -1/8$
$y = k_1 \sin x + k_2 \cos x + 5 - (1/8) \sin 3x$

15. y_c: $m^2 - 1 = 0$; $m = \pm 1$; $y_c = k_1 e^x + k_2 e^{-x}$; note $g(x) = e^x$ is a solution of the homogeneous equation
$y'' - y = 0$; use note 2 from table: y_p: $y_p = Axe^x$; $y'_p = Axe^x + Ae^x$; $y''_p = Axe^x + 2Ae^x$;
substitute into D.E. $(Axe^x + 2Ae^x) - Axe^x = e^x$; $2Ae^x = e^x$;
$2A = 1$; $A = 1/2$; $y_p = (1/2)xe^x$; thus $y = k_1 e^x + k_2 e^{-x} + (1/2)xe^x$

17. y_c: $m^2 + 4 = 0$; $m = \pm 2j$; $y_c = k_1 \sin 2x + k_2 \cos 2x$;
note $g(x) = \cos 2x$ is a solution of the homogeneous equation
$y'' + 4y = 0$; use note 2 from table: y_p: $y_p = Ax \cos 2x + Bx \sin 2x$;
$y'_p = A \cos 2x - 2Ax \sin 2x + B \sin 2x + 2Bx \cos 2x$;
$y''_p = -4A \sin 2x - 4Bx \sin 2x - 4Ax \cos 2x + 4B \cos 2x$; substitute into D.E:
$(-4A \sin 2x - 4Bx \sin 2x - 4Ax \cos 2x + 4B \cos 2x)$
$-4(Ax \cos 2x + Bx \sin 2x) = \cos 2x$;
$-4A \sin 2x + 4B \cos 2x = \cos 2x$; $-4A = 0, 4B = 1$; $A = 0, B = 1/4$;
$y = k_1 \sin 2x + k_2 \cos 2x + (1/4)x \sin 2x$

19. y_c: $m^2 + 1 = 0$; $m = \pm j$; $y_c = k_1 \sin x + k_2 \cos x$
y_p: $y_p = Ae^{2x}$; $y'_p = 2Ae^{2x}$; $y''_p = 4Ae^{2x}$; substitute into D.E.
$4Ae^{2x} + Ae^{2x} = 10e^{2x}$; $A = 2$; $y = k_1 \sin x + k_2 \cos x + 2e^{2x}$;
$y' = k_1 \cos x - k_2 \sin x + 4e^{2x}$; substitute $y = 0, x = 0, y' = 0$;
$0 = k_2 + 2$; $0 = k_1 + 4$; $k_1 = -4, k_2 = -2$
$y = -4 \sin x - 2 \cos x + 2e^{2x}$

21. y_c: $m^2 + 1 = 0$; $m = \pm j$; $y_c = k_1 \sin x + k_2 \cos x$
y_p: $y_p = Ae^x = y'_p = y''_p$; substitute into D.E.: $Ae^x + Ae^x = e^x$;
$A = 1/2$; $y = k_1 \sin x + k_2 \cos x + (1/2) e^x$;
$y' = k_1 \cos x - k_2 \sin x + (1/2) e^x$; substitute $y = 0, x = 0$
$y' = 3$; $k_2 + 1/2 = 0$; $k_1 + 1/2 = 3$; $k_1 = 5/2, k_2 = -1/2$;
$y = (5/2) \sin x - (1/2) \cos x + (1/2) e^x$ or
$y = (1/2)(5 \sin x - \cos x + e^x)$

Section 12.3

1. $m = \dfrac{W}{g} = \dfrac{4\,\text{lb}}{32\,\text{ft/s}^2} = \dfrac{1}{8}\,\text{lb-s}^2/\text{ft};\ \ k = \dfrac{F}{s} = \dfrac{4\,\text{lb}}{(1/4)\,\text{ft}} = 16\,\text{lb/ft};$

   $x = C_1 \sin 8\sqrt{2}\,t + C_2 \cos 8\sqrt{2}\,t\,;$

   $dx/dt = 8\sqrt{2}\,C_1 \cos 8\sqrt{2}\,t - 8\sqrt{2}\,C_2 \sin 8\sqrt{2}\,t$

   substitute $x = 2$ in. $= 1/6$ ft, $dx/dt = 0$, $t = 0$;

   $1/6 = C_1 \sin 0 + C_2 \cos 0;\ \ 0 = 8\sqrt{2}\,C_1 \cos 0 - \sqrt{8}\,C_2 \sin 0\,;$

   $C_1 = 0,\ C_2 = 1/6;\ x = (1/6)\cos 8\sqrt{2}\,t$

3. $p = \dfrac{F_{resistant}}{dx/dt} = \dfrac{5\,\text{lb}}{1/3\,\text{ft/s}} = 15\,\text{lb-s}^2/\text{ft};\ \ m = \dfrac{W}{g} = \dfrac{20\,\text{lb}}{32\,\text{ft/s}^2} = 0.625\,\text{lb-s}^2/\text{ft}$

   $k = \dfrac{F}{s} = \dfrac{20\,\text{lb}}{1/2\,\text{ft}} = 40\,\text{lb/ft};$

   $d = \left\{\dfrac{p}{2m}\right\}^2 - \dfrac{k}{m} = \left\{\dfrac{15}{2(0.625)}\right\}^2 - \dfrac{40}{0.625} = 80;$ since $d > 0$, we have two real roots of the

   auxiliary equation

   $\overline{m}_1 = (-p/2m) + \sqrt{d} = -12 + \sqrt{80} = -3.06$

   $\overline{m}_2 = -12 - \sqrt{80} = -20.9$

   $x = C_1 e^{-3.06t} + C_2 e^{-20.9t}$ (Case 1--overdamped)

5. Let $x =$ downward displacement; the change in submerged volume is $\pi(1.5)^2 x;$ the buoyant force

   is $62.4\pi(1.5)^2 x;$ and $W =$ weight of buoy in lbs; $-62.4\pi(1.5)^2 x = \dfrac{W}{g}\dfrac{d^2 x}{dt^2}$ or $\dfrac{d^2 x}{dt^2} + \dfrac{4521\pi}{W}x = 0;$

   $x = C_1 \sin\sqrt{4521\pi/W}\,t;\ C_2 \cos\sqrt{4521\pi/W}\,t;$ since $p = 6 = \dfrac{2\pi}{\sqrt{4521\pi/W}}$

   $\pi/3 = \sqrt{4521\pi/W}\,;\ W = 9(4521)/\pi = 13{,}000\,\text{lb}$

7. Let $x =$ downward displacement, $\pi r^2 x =$ change in submerged volume,

   $62.4\pi r^2 x =$ the buoyant force, $W = 1000$ lb, find r.

   $-62.4\pi r^2 x = \dfrac{1000}{g}\dfrac{d^2 x}{dt^2};\ \dfrac{d^2 x}{dt^2} - \dfrac{2009\pi r^2}{1000}x = 0$

   $x = C_1 \sin\sqrt{2009\pi r^2/1000}\,t + C_2 \cos\sqrt{2009\pi r^2/1000}\,t$

   $p = 2 = \dfrac{2\pi}{\sqrt{2009\pi r^2/1000}};\ \pi^2 = 2009\pi r^2/1000$

   $r = \sqrt{1000\pi/2009} = 1.25\,\text{ft}$

Section 12.4

9. $L = 0.1$ H, $R = 50$ Ω, $C = 2 \times 10^{-4}F$; $d = \dfrac{R^2}{4L^2} - \dfrac{1}{CL} = \dfrac{50^2}{4(0.1)^2} - \dfrac{1}{(2 \times 10^{-4})(0.1)} = 12{,}500;$

$-\dfrac{R}{2L} = -250;$ since $d > 0$, $m_1 = -\dfrac{R}{2L} + \sqrt{d} = -250 + \sqrt{12500} = -138;$ $m_2 = -250 - \sqrt{12500}$

$= -362$ Thus $i = k_1 e^{-138t} + k_2 e^{-362t}$

11. $L = 0.4$ H, $R = 200$ Ω, $C = 5 \times 10^{-5}F$;

$d = \dfrac{R^2}{4L^2} - \dfrac{1}{CL} = \dfrac{(200)^2}{4(0.4)^2} - \dfrac{1}{(5 \times 10^{-5})(0.4)} = 12500;$ $\quad -\dfrac{R}{2L} = \dfrac{-200}{2(0.4)} = -250;$

since $d > 0$, $m_1 = -250 + \sqrt{12500} = -140;$

$m_2 = -250 - \sqrt{12500} = -360;$ Thus $i = k_1 e^{-140t} + k_2 e^{-360t};$

$di/dt = -140 k_1 e^{-140t} - 360 k_2 e^{-360t};$

Since $i = 0$ at $t = 0$, $0 = k_1 + k_2;$

using $L\, di/dt + Ri + (1/C) \displaystyle\int_0^t i\, dt = V;$

$(0.4)\, di/dt + R(0) + (1/C) \displaystyle\int_0^t 0\, dt = 12;$ $\quad 0.4\, di/dt = 12;$ $\quad di/dt = 30;$

$30 = -140 k_1 e^{-140t} - 360 k_2 e^{-360t};$ at $t = 0$, $i = 0;$

$30 = -140 k_1 - 360 k_2;$ solve system with $0 = k_1 + k_2$ from above;

$k_1 = 0.136$, $k_2 = -0.136;$ Thus $i = 0.136 e^{-140t} - 0.136 e^{-360t}$

Section 12.5

$\mathscr{L}$ is our notation for the Laplace transform.

1. $\mathscr{L}[f(t)] = \displaystyle\int_0^\infty e^{-st} \cdot t\, dt = \lim_{t \to \infty} \int_0^t e^{-st} \cdot t\, dt$

$= \lim_{t \to \infty} \dfrac{[e^{-st}(-st - 1)]}{s^2} \bigg|_0^t = \lim_{t \to \infty} \left\{ \left[\dfrac{e^{-st}}{s^2}(-st - 1) \right] - \left[\dfrac{e^0}{s^2}(-0 - 1) \right] \right\}$

$= \lim_{t \to \infty} \left\{ \left[\dfrac{-1}{e^{st} s^2}(st + 1) \right] + \dfrac{1}{s^2} \right\} = \dfrac{1}{s^2}$

3. #8; $\dfrac{3}{s^2 + 9}$ $\qquad$ 5. #13; $\dfrac{-4}{s(s - 4)}$ $\qquad$ 7. #6; $\mathscr{L}(t)^2 = 2\,\mathscr{L}(t^2/2) = 2/s^3$

9. #20; $\dfrac{2 \cdot 2^3}{(s^2 + 4)^2} = \dfrac{16}{(s^2 + 4)^2}$

11. #5, 7; $\mathscr{L}(t - e^{2t}) = \mathscr{L}(t) - \mathscr{L}(e^{2t}) = \dfrac{1}{s^2} - \dfrac{1}{s - 2} = \dfrac{-s^2 + s - 2}{s^2(s - 2)}$

Sections 12.4 – 12.5

13. #18; $\dfrac{8s}{(s^2+16)^2}$

15. #11; $\dfrac{s+3}{(s+3)^2+25} = (s+3)/(s^2+6s+34)$

17. #5, 6; $\mathscr{L}(8t+4t^3) = 8\,\mathscr{L}(t) + 24\,\mathscr{L}(t^3/3!) = 8/s^2 + 24/s^4 = \dfrac{8s^2+24}{s^4}$

19. $\mathscr{L}(y''-3y') = \mathscr{L}(y'') - 3\,\mathscr{L}(y') = [s^2\,\mathscr{L}(y) - s\,y(0) - y'(0)] - 3[s\,\mathscr{L}(y) - y(0)] = (s^2-3s)\,\mathscr{L}(y)$
   Since $y(0) = y'(0) = 0$

21. $\mathscr{L}(y''+y'+y) = [s^2\,\mathscr{L}(y) - s\,y(0) - y'(0)] + [s\,\mathscr{L}(y) - y(0)] + \mathscr{L}(y)$ $[y(0) = 0, y'(0) = 1]$
   $= s^2\,\mathscr{L}(y) - 1 + s\,\mathscr{L}(y) + \mathscr{L}(y) = (s^2+s+1)\,\mathscr{L}(y) - 1$

23. $\mathscr{L}(y''-3y'+y) = \mathscr{L}(y'') - 3\,\mathscr{L}(y') + \mathscr{L}(y)$ $[y(0) = 1, y'(0) = 0]$
   $= [s^2\,\mathscr{L}(y) - s\,y(0) - y'(0)] - 3[s\,\mathscr{L}(y) - y(0)] + \mathscr{L}(y) = (s^2-3s+1)\,\mathscr{L}(y) - s + 3$

25. $\mathscr{L}(y''+8y'+2y) = \mathscr{L}(y'') + 8\,\mathscr{L}(y') + 2\,\mathscr{L}(y)$ $[y(0) = 4, y'(0) = 6]$
   $= [s^2\,\mathscr{L}(y) - s\,y(0) - y'(0)] + 8[s\,\mathscr{L}(y) - y(0)] + 2\,\mathscr{L}(y)$
   $= s^2\,\mathscr{L}(y) - 4s - 6 + 8s\,\mathscr{L}(y) - 32 + 2\,\mathscr{L}(y)$
   $= (s^2+8s+2)\,\mathscr{L}(y) - 4s - 38$

27. $\mathscr{L}(y''-6y') = \mathscr{L}(y'') - 6\,\mathscr{L}(y')$ $[y(0) = 3, y'(0) = 7]$
   $= [s^2\,\mathscr{L}(y) - s\,y(0) - y'(0)] - 6[s\,\mathscr{L}(y) - y(0)]$
   $= s^2\,\mathscr{L}(y) - 3s - 7 - 6s\,\mathscr{L}(y) + 18 = (s^2-6s)\,\mathscr{L}(y) - 3s + 11$

29. $\mathscr{L}(y''+8y'-3y) = \mathscr{L}(y'') + 8\,\mathscr{L}(y') - 3\,\mathscr{L}(y)$ $[y(0) = -6, y'(0) = 2]$
   $= [s^2\,\mathscr{L}(y) - s\,y(0) - y'(0)] + 8[s\,\mathscr{L}(y) - y(0)] - 3\,\mathscr{L}(y)$
   $= s^2\,\mathscr{L}(y) + 6s - 2 + 8s\,\mathscr{L}(y) + 48 - 3\,\mathscr{L}(y)$
   $= (s^2+8s-3)\,\mathscr{L}(y) + 6s + 46$

31. #4; 1   33. #15; te^{5t} 35. #10; $\cos 8t$ 37. #12; $e^{6t} - e^{2t}$

39. #9; $\mathscr{L}^{-1}\left\{\dfrac{2}{(s-3)^2+4}\right\} = e^{3t}\sin 2t$ 41. #20; $\sin t - t\cos t$

43. use partial fractions: $\mathscr{L}^{-1}\left\{\dfrac{1}{s-1} - \dfrac{s+1}{s^2+1}\right\}$ #7, 8, 10

   $= \mathscr{L}^{-1}\left\{\dfrac{1}{s-1}\right\} - \mathscr{L}^{-1}\left\{\dfrac{s}{s^2+1}\right\} - \mathscr{L}^{-1}\left\{\dfrac{s}{s^2+1}\right\} = e^t - \cos t - \sin t$

Section 12.5

45. use partial fractions: $6 \, \mathscr{L}^{-1}\left\{\dfrac{1/2}{s} + \dfrac{1/6}{s+6} + \dfrac{1/3}{s+3}\right\}$ #4, 7

$$= 3 \, \mathscr{L}^{-1}\left\{\dfrac{1}{s}\right\} + \mathscr{L}^{-1}\left\{\dfrac{1}{s+6}\right\} + 2 \, \mathscr{L}^{-1}\left\{\dfrac{1}{s+3}\right\} = 3 + e^{-6t} + 2e^{-3t}$$

47. #9, 11; $\mathscr{L}^{-1}\left\{\dfrac{s+11}{(s+5)^2 + 2^2}\right\} = \mathscr{L}^{-1}\left\{\dfrac{(s+5)+6}{[s-(-5)]^2 + 2^2}\right\}$

$$= \mathscr{L}^{-1}\left\{\dfrac{s+5}{(s+5)^2 + 2^2}\right\} + 3 \, \mathscr{L}^{-1}\left\{\dfrac{2}{(s+5)^2 + 2^2}\right\} = e^{-5t}\cos 2t + 3e^{-5t}\sin 2t = e^{-5t}(\cos 2t + 3\sin 2t)$$

49. $\mathscr{L}[f''(t)] = \displaystyle\int_0^\infty e^{-st} f''(t)dt$. Integrate by parts.

$$u = e^{-st} \qquad\qquad dv = f''(t)dt$$
$$du = -se^{-st}dt \qquad\qquad v = f'(t)$$

$$\int_0^\infty e^{-st} f''(t)dt = e^{-st}f'(t)\,\Big|_0^\infty - \int_0^\infty f'(t)(-se^{-st}dt)$$

$$= -f'(0) + s \int_0^\infty e^{-st}f'(t)dt$$

$$= -f'(0) + s \, \mathscr{L}[f'(t)]$$

$$= -f'(0) + s\{s \, \mathscr{L}[f(t)] - f(0)\}$$

$$= s^2 \, \mathscr{L}[f(t)] - s \, f(0) - f'(0)$$

Section 12.6

1. $\mathscr{L}(y') - \mathscr{L}(y) = 0$; $s \, \mathscr{L}(y) - y(0) - \mathscr{L}(y) = 0$;
    $y(0) = 2$; $s \, \mathscr{L}(y) - 2 - \mathscr{L}(y) = 0$; $(s-1) \, \mathscr{L}(y) - 2 = 0$;
    $\mathscr{L}(y) = 2/(s-1)$; #7, $y = 2e^t$

3. $4 \, \mathscr{L}(y') + 3 \, \mathscr{L}(y) = 0$; $4[s \, \mathscr{L}(y) - y(0)] + 3 \, \mathscr{L}(y) = 0$; $y(0) = 1$;
    $4s \, \mathscr{L}(y) - 4 + 3 \, \mathscr{L}(y) = 0$; $(4s+3) \, \mathscr{L}(y) = 4$;
    $\mathscr{L}(y) = \dfrac{4}{4s+3} = \dfrac{1}{s+3/4}$ #7; $y = e^{(-3/4)t}$

5. $\mathscr{L}(y') - 7\,\mathscr{L}(y) = \mathscr{L}(e^t);\ s\,\mathscr{L}(y) - y(0) - 7\,\mathscr{L}(y) = \dfrac{1}{s-1}:$

   $y(0) = 5;\ s\,\mathscr{L}(y) - 5 - 7\,\mathscr{L}(y) = \dfrac{1}{s-1};\ (s-7)\,\mathscr{L}(y) = \dfrac{1}{s-1} + 5;$

   $\mathscr{L}(y) = \dfrac{1}{(s-7)(s-1)} + \dfrac{5}{s-7};$

   $y = \dfrac{1}{6}\,\mathscr{L}^{-1}\left\{\dfrac{6}{(s-7)(s-1)}\right\} + 5\,\mathscr{L}^{-1}\left\{\dfrac{1}{s-7}\right\}$

   $y = (1/6)[e^{7t} - e^t] + 5e^{7t}$ or $6y = 31e^{7t} - e^t$ (#7, #12)

7. $\mathscr{L}(y'') + \mathscr{L}(y) = 0;\ s^2\,\mathscr{L}(y) - s\,y(0) - y'(0) + \mathscr{L}(y) = 0;$

   $y(0) = 1, y'(0) = 0;\ (s^2 + 1)\,\mathscr{L}(y) = s;\ \mathscr{L}(y) = \dfrac{s}{s^2 + 1};$

   $y = \cos t$ (#10)

9. $\mathscr{L}(y'') - 2\,\mathscr{L}(y') = 0;\ [s^2\,\mathscr{L}(y) - s\,y(0) - y'(0)] - 2[s\,\mathscr{L}(y) - y(0)] = 0;$

   $y(0) = 1, y'(0) = -1;$

   $[s^2\,\mathscr{L}(y) - s + 1] - 2[s\,\mathscr{L}(y) - 1] = 0;\ \mathscr{L}(y) = \dfrac{s-3}{s^2 - 2s};$

   $y = \dfrac{1}{2}\,\mathscr{L}^{-1}\left\{\dfrac{2s}{(s-0)((s-2)}\right\} + \dfrac{3}{2}\,\mathscr{L}^{-1}\left\{\dfrac{(-3)(2/3)}{(s-0)(s-2)}\right\}$

   $= e^{2t} + \dfrac{3}{2} - \dfrac{3}{2}e^{2t}$ or $2y = 3 - e^{2t}$ (#13, #14)

11. $\mathscr{L}(y'') + 2\,\mathscr{L}(y') + \mathscr{L}(y) = 0;\ [s^2\,\mathscr{L}(y) - s\,y(0) - y'(0)] - 2[s\,\mathscr{L}(y) - y(0)] + \mathscr{L}(y) = 0;$

   $y(0) = 1, y'(0) = 0;$

   $(s^2 + 2s + 1)\,\mathscr{L}(y) - s - 2 = 0;\ \mathscr{L}(y) = \dfrac{s+2}{s^2 + 2s + 1}$

   $y = \mathscr{L}^{-1}\left\{\dfrac{s+2}{(s+1)^2}\right\} = \mathscr{L}^{-1}\left\{\dfrac{s}{(s+1)^2}\right\} + 2\,\mathscr{L}^{-1}\left\{\dfrac{1}{(s+1)^2}\right\}$

   $= e^{-t}(1 - t) + 2te^{-t};\ y = e^{-t} + te^{-t} = e^{-t}(1 + t)$

13. $\mathscr{L}(y'') - 4\,\mathscr{L}(y') + 4\,\mathscr{L}(y) = \mathscr{L}(te^{2t});\ y(0) = 0, y'(0) = 0$

   $(s^2\,\mathscr{L}(y) - s\,y(0) - y'(0)) - 4[s\,\mathscr{L}(y) - y(0)] + 4\,\mathscr{L}(y) = \dfrac{1}{(s-2)^2};$

   $(s^2 - 4s + 4)\,\mathscr{L}(y) = \dfrac{1}{(s-2)^2};\ \mathscr{L}(y) = \dfrac{1}{(s-2)^4};$

   $y = \mathscr{L}^{-1}\left\{\dfrac{1}{(s-2)^4}\right\} = \dfrac{t^3 e^{2t}}{3!} = (1/6)t^3 e^{2t}$ (#16)

Section 12.6

15. $\mathcal{L}(y'') + 2 \mathcal{L}(y') + \mathcal{L}(y) = \mathcal{L}(3te^{-t}); \; y(0) = 4, y'(0) = 2;$

$[s^2 \mathcal{L}(y) - s\,y(0) - y'(0)] + 2[s\,\mathcal{L}(y) - y(0)] + \mathcal{L}(y) = \dfrac{3}{(s+1)^3}$

$(s^2 + 2s + 1)\,\mathcal{L}(y) - 4s - 10 = \dfrac{3}{(s+1)^2}; \; \mathcal{L}(y) = \dfrac{4s+10}{(s+1)^2} + \dfrac{3}{(s+1)^4};$

$y = \mathcal{L}^{-1}\left\{\dfrac{4s+10}{(s+1)^2}\right\} + \mathcal{L}^{-1}\left\{\dfrac{3}{(s+1)^4}\right\}$

$= \mathcal{L}^{-1}\left\{\dfrac{4s}{(s+1)^2}\right\} + \mathcal{L}^{-1}\left\{\dfrac{10}{(s+1)^2}\right\} + \mathcal{L}^{-1}\left\{\dfrac{3}{(s+1)^4}\right\}$ (#17, 15, 16)

$= 4e^t(1-t) + 10te^{-t} + \dfrac{3t^3 e^{-t}}{6} = 4e^{-t} + 6te^{-t} + (1/2)t^3 e^{-t}$ or $y = e^{-t}(4 + 6t + (1/2)t^3)$

17. $\mathcal{L}(y'') + 3\,\mathcal{L}(y') - 4\,\mathcal{L}(y) = 0; \; [y(0) = 1, y'(0) = -2]$

$[s^2 \mathcal{L}(y) - s\,y(0) - y'(0)] + 3[s\,\mathcal{L}(y) - y(0)] - 4\,\mathcal{L}(y) = 0;$

$(s^2 + 3s - 4)\,\mathcal{L}(y) - s - 1 = 0; \; \mathcal{L}(y) = \dfrac{s+1}{s^2+3s-4} = \dfrac{s+1}{(s+4)(s-1)}$

$y = \mathcal{L}^{-1}\left\{\dfrac{s+1}{(s^2+4)(s-1)}\right\} = \mathcal{L}^{-1}\left\{\dfrac{3/5}{s+4} + \dfrac{2/5}{s-1}\right\}$ using partial fractions

$y = (3/5)e^{-4t} + (2/5)e^t$ (#7) or $5y = 3e^{-4t} + 2e^t$

19. $L\dfrac{di}{dt} + Ri = V$

$0.2\dfrac{di}{dt} + 20i = 12$

$\dfrac{di}{dt} + 100i = 60$

$\mathcal{L}\left(\dfrac{di}{dt} + 100i\right) = \mathcal{L}(60)$

$s\,\mathcal{L}(i) - 3 + 100\,\mathcal{L}(i) = \dfrac{60}{s}$ \qquad $(i = 3$ at $t = 0)$

$(s + 100)\,\mathcal{L}(i) = \dfrac{60}{s} + 3$

$\mathcal{L}(i) = \left(\dfrac{60+3s}{s}\right)\left(\dfrac{1}{s+100}\right)$

$\mathcal{L}(i) = \dfrac{60+3s}{s(s+100)}$

$\mathcal{L}(i) = \dfrac{0.6}{s} + \dfrac{2.4}{s+100}$

$I = 0.6 + 2.4e^{-100t}$

Section 12.6

21. $L\dfrac{di}{dt} + Ri + \dfrac{1}{C}q = V$

    * $0.1\dfrac{di}{dt} + 45i + \dfrac{1}{2(10^{-4})}q = 12$

        $0.1\dfrac{di}{dt} + 45(0) + \dfrac{1}{2(10^{-4})}(0) = 12$ note: $\dfrac{di}{dt} = 120$ when $t = 0$

Differentate both sides of * with respect to t.

    $0.1\dfrac{d^2i}{d^2t} + 45\dfrac{di}{dt} + \dfrac{1}{2(10^{-4})}i = 0$

    $\dfrac{d^2i}{dt^2} + 450\dfrac{di}{dt} + \dfrac{10}{2(10^{-4})}i = 0$

    $\mathscr{L}\left(\dfrac{d^2i}{dt^2} + 450\dfrac{di}{dt} + 50000i\right) = \mathscr{L}(0)$

    $s^2\,\mathscr{L}(i) - s(0) - 120 + 450s\,\mathscr{L}(i) - 0 + 50000\,\mathscr{L}(i) = 0$

    $(s^2 + 450s + 50000)\,\mathscr{L}(i) = 120$

    $\mathscr{L}(i) = \dfrac{120}{(s + 200)(s + 250)} = \dfrac{2.4}{s + 200} + \dfrac{-2.4}{s + 250}$

    $I = 2.4e^{-200t} - 2.4e^{-250t}$

Chapter 12 Review

1. homogeneous, 2   2. nonhomogeneous, 2   3. nonhomogeneous, 1

4. homogeneous, 3

5. $m^2 + 4m - 5 = 0$; $(m + 5)(m - 1) = 0$; $m = -5, 1$;
    $y = k_1e^{-5x} + k_2e^x$

6. $m^2 - 5m + 6 = 0$; $(m - 3)(m - 2) = 0$; $m = 2, 3$;
    $y = k_1e^{2x} + k_2e^{3x}$

7. $m^2 - 6m = 0$; $m(m - 6) = 0$; $m = 0, 6$; $y = k_1 + k_2e^{6x}$

8. $2m^2 - m - 3 = 0$; $(2m - 3)(m + 1) = 0$; $m = 3/2, -1$;
    $y = k_1e^{3x/2} + k_2e^{-x}$

9. $m^2 - 6m + 9 = 0$; $(m - 3)^2 = 0$; $m = 3, 3$; $y = k_1e^{3x} + k_2xe^{3x}$

10. $m^2 + 10m + 25 = 0$; $(m + 5)^2 = 0$; $m = -5, -5$; $y = k_1e^{-5x} + k_2xe^{-5x}$

11. $m^2 - 2m + 1 = 0$; $(m - 1)^2 = 0$; $m = 1, 1$; $y = k_1e^x + k_2xe^x$

Section 12.6 – Chapter 12 Review

12. $9m^2 - 6m + 1 = 0$; $(3m - 1)^2 = 0$; $m = \dfrac{1}{3}, \dfrac{1}{3}$; $y = k_1 e^{x/3} = k_2 x e^{x/3}$

13. $m^2 + 16 = 0$; $m = \pm 4j$; $y = k_1 \sin 4x + k_2 \cos 4x$

14. $4m^2 + 25 = 0$; $m = \pm(5/2)j$; $y = k_1 \sin 5x/2 + k_2 \cos 5x/2$

15. $m^2 - 2m + 3 = 0$; $m = 1 \pm \sqrt{2}\,j$; $y = e^x(k_1 \sin \sqrt{2}\,x + k_2 \cos \sqrt{2}\,x)$

16. $m^2 - 3m + 8 = 0$; $m = 3/2 \pm (\sqrt{23}/2)j$;
    $y = e^{3x/2}[k_1 \sin (\sqrt{23}/2)x + k_2 \cos (\sqrt{23}/2)x]$

17. $m^2 + m - 2 = 0$; $(m + 2)(m - 1) = 0$; $m = -2, 1$;
    $y_c = k_1 e^x = k_2 e^{-2x}$
    y_p: $y_p = Ax + B$; $y'_p = A$; $y''_p = 0$; $A - 2(Ax + B) = x$;
    $-2Ax + A - 2B = x$; $-2A = 1$; $A - 2B = 0$; $A = -1/2$, $B = -1/4$;
    $y = k_1 e^x + k_2 e^{-2x} - x/2 - 1/4$

18. $m^2 - 6m + 9 = 0$; $(m - 3)^2 = 0$; $m = 3, 3$; $y_c = k_1 e^{3x} + k_2 x e^{3x}$;
    y_p: $y_p = Ae^x = y'_p = y''_p$; $Ae^x - 6Ae^x + 9Ae^x = e^x$;
    Thus $4Ae^x = e^x$; $A = 1/4$; $y = k_1 e^{3x} + k_2 x e^{3x} + (1/4)e^x$

19. $m^2 + 4 = 0$; $m = \pm 2j$; $y_c = k_1 \sin 2x + k_2 \cos 2x$;
    y_p: $y_p = A \cos x + B \sin x$; $y'_p = -A \sin x + B \cos x$;
    $y''_p = -A \cos x - B \sin x$
    $-A \cos x - B \sin x + 4(A \cos x + B \sin x) = \cos x$;
    $3A \cos x + 3B \sin x = \cos x$; $A = 1/3$, $B = 0$;
    $y = k_1 \sin 2x + k_2 \cos 2x + (1/3) \cos x$

20. $m^2 - 2m + 3 = 0$; $m = 1 \pm \sqrt{2}\,j$; $y_c = e^x(k_1 \sin \sqrt{2}x + k_2 \cos \sqrt{2}\,x)$
    y_p: $y_p = Ae^{2x}$; $y'_p = 2Ae^{2x}$; $y'' = 4Ae^{2x}$; $4Ae^{2x} - 2(2Ae^{2x}) + 3A^{2x} = 6e^{2x}$;
    $3A = 6$; $A = 2$; $y = e^x(k_1 \sin \sqrt{2}\,x + k_2 \cos \sqrt{2}\,x) + 2e^{2x}$

21. $m^2 + 2m - 8 = 0$; $(m + 4)(m - 2) = 0$; $m = -4, 2$;
    $y = k_1 e^{-4x} + k_2 e^{2x}$; $y' = -4k_1 e^{-4x} + 2k_2 e^{2x}$; substitute $y = 6$,
    $x = 0$, $y' = 0$; $k_1 + k_2 = 6$; $-4k_1 + 2k_2 = 0$; $k_1 = 2$, $k_2 = 4$;
    $y = 2e^{-4x} + 4e^{2x}$

22. $m^2 - 3m = 0$; $m(m - 3) = 0$; $m = 0, 3$; $y = k_1 + k_2 e^{3x}$;
    $y' = 3k_2 e^{3x}$; substitute $y = 3$, $x = 0$, $y' = 6$; $k_1 + k_2 = 3$; $3k_2 = 6$; $k_1 = 1$, $k_2 = 2$; $y = 1 + 2e^{3x}$

23. $m^2 - 4m + 4 = 0$; $(m - 2)^2 = 0$; $m = 2, 2$; $y = k_1 e^{2x} + k_2 x e^{2x}$;
    $y' = 2k_1 e^{2x} + k_2 e^{2x} + 2k_2 x e^{2x}$; substitute $y = 0$, $x = 0$, $y' = 3$;
    $k_1 = 0$; $2k_1 + k_2 = 3$; $k_1 = 0$, $k_2 = 3$; $y = 3xe^{2x}$

Chapter 12 Review

24. $m^2 + 6m + 9 = 0$; $(m + 3)^2 = 0$; $m = -3, -3$;
 $y = k_1e^{-3x} + k_2xe^{-3x}$; $y' = -3k_1e^{-3x} + 3k_2xe^{-3x} + k_2e^{-3x}$;
 substitute $y = 8$, $x = 0$, $y' = 0$; $k_1 = 8$; $-3k_1 + k_2 = 0$; $k_1 = 8$, $k_2 = 24$;
 $y = 8e^{-3x} + 24xe^{-3x}$; $y = 8e^{-3x}(1 + 3x)$

25. $m^2 + 4 = 0$; $m = \pm 2j$; $y = k_1 \sin 2x + k_2 \cos 2x$;
 $y' = 2k_1 \cos 2x - 2k_2 \sin 2x$; substitute $y = 1$, $x = 0$, $y' = -4$;
 $k_2 = 1$; $2k_1 = -4$; $k_1 = -2$, $k_2 = 1$; $y = -2 \sin 2x + \cos 2x$

26. $m^2 - 8m + 25 = 0$; $m = 4 \pm 3j$; $y = e^{4x}(k_1 \sin 3x + k_2 \cos 3x)$
 $y' = (3e^{4x}k_1 + 4e^{4x}k_2) \cos 3x + (4e^{4x}k_1 - 3e^{4x}k_2) \sin 3x$;
 substitute $y = 2$, $x = 0$, $y' = 11$; $k_2 = 2$; $3k_1 + 4k_2 = 11$;
 $k_1 = 1$, $k_2 = 2$; $y = e^{4x}(\sin 3x + 2 \cos 3x)$

27. $m^2 - m = 0$; $m(m - 1) = 0$; $m = 0, 1$; $y_c = k_1 + k_2e^x$;
 y_p: $y_p = Ae^{2x}$; $y'_p = 2Ae^{2x}$; $y''_p = 4Ae^{2x}$; $4Ae^{2x} - 2Ae^{2x} = e^{2x}$; $2Ae^{2x} = e^{2x}$;
 $A = 1/2$; $y = k_1 + k_2e^x + (1/2)e^{2x}$; $y' = k_2e^x + e^{2x}$;
 substitute $y = 1$, $x = 0$, $y' = 0$; $1 = k_1 + k_2 + 1/2$;
 $0 = k_2 + 1$; $k_2 = 3/2$, $k_2 = -1$; $y = 3/2 - e^x + (1/2)e^{2x}$ or $2y = e^{2x} - 2e^x + 3$

28. $m^2 + 4 = 0$; $m = \pm 2j$; $y_c = k_1 \sin 2x + k_2 \cos 2x$
 y_p: $y_p = A \sin x + B \cos x$; $y'_p = A \cos x - B \sin x$;
 $y''_p = -A \sin x - B \cos x$;
 $-A \sin x - B \cos x + 4(A \sin x + B \cos x) = \sin x$;
 $3A \sin x + 3B \cos x = \sin x$; $3A = 1$; $3B = 0$; $A = 1/3$, $B = 0$;
 $y = k_1 \sin 2x + k_2 \cos 2x + (1/3) \sin x$;
 $y' = 2k_1 \cos 2x - 2k_2 \sin 2x + (1/3) \cos x$; substitute $y = 2$,
 $x = 0$, $y' = 7/3$; $2 = k_2$; $7/3 = 2k_1 + 1/3$; $k_1 = 1$; $k_2 = 2$;
 $y = \sin 2x + 2 \cos 2x + (1/3) \sin x$

29. $m = \dfrac{W}{g} = \dfrac{16\,\text{lb}}{32\,\text{ft/s}^2} = \dfrac{1}{2}$ lb-s^2/ft; $k = \dfrac{F}{s} = \dfrac{16\,\text{lb}}{1/3\,\text{ft}} = 48$ lb/ft;
 $x = C_1 \sin 4\sqrt{6}\,t + C_2 \cos 4\sqrt{6}\,t$;
 $dx/dt = 4\sqrt{6}\,C_1 \cos 4\sqrt{6}\,t - 4\sqrt{6}\,C_2 \sin 4\sqrt{6}\,t$ substitute $x = 6$ in. $= 1/2$ ft, $dx/dt = 0$, $t = 0$;
 $1/2 = C_2$; $0 = 4\sqrt{6}\,C_1$; $C_1 = 0$; $x = (1/2) \cos 4\sqrt{6}\,t$

30. $m = \dfrac{W}{g} = \dfrac{12\,\text{lb}}{32\,\text{ft/s}^2} = 3/8$ lb-s^2/ft; $k = \dfrac{F}{s} = \dfrac{12\,\text{lb}}{1/6\,\text{ft}} = 72$ lb/ft;
 $P = \dfrac{8\,\text{lb}}{1/3\,\text{ft/s}} = 24$ lb-s/ft; $d = \left\{\dfrac{P}{2m}\right\}^2 - \dfrac{k}{m} = \left\{\dfrac{24}{2(3/8)}\right\}^2 - \dfrac{72}{3/8} = 832 > 0$;
 since $d > 0$, we have two real roots of the auxiliary equation
 $\overline{m}_1 = -p/2m + \sqrt{d} = -32 + \sqrt{832} = -3.16$; $\overline{m}_2 = -32 - \sqrt{832} = -60.8$;
 $y = k_1e^{-3.16t} + k_2e^{-60.8t}$

Chapter 12 Review

31. $L = 2$ H, $R = 400$ Ω, $C = 10^{-5}$F; from Section 12.4:

$$d = \frac{R^2}{4L^2} - \frac{1}{CL} = \frac{(400)^2}{4(2)^2} - \frac{1}{(10^{-5})(2)} = -4 \times 10^4 < 0;$$

$$-\frac{R}{2L} = -\frac{400}{2(2)} = -100; \ \omega = \sqrt{\left|-4 \times 10^4\right|} = 200;$$

$$i = e^{-100t}(k_1 \sin 200t + k_2 \cos 200t)$$

32. $L = 1$ H, $R = 2000$ Ω, $C = 4 \times 10^{-6}$F; from Section 29.4:

$$d = \frac{R^2}{4L^2} - \frac{1}{CL} = \frac{(2000)^2}{4(1)^2} - \frac{1}{(1)(4 \times 10^{-6})} = 7.5 \times 10^5 > 0;$$

since $d > 0$, $-\dfrac{R}{2L} = \dfrac{-2000}{2(1)} = -1000;$ $\overline{m}_1 = -1000 + \sqrt{7.5 \times 10^5} = -134;$

$\overline{m}_2 = -1000 + \sqrt{7.5 \times 10^5} = -1866;$

$i = k_1 e^{-134t} + k_2 e^{-1866t}$

33. $\dfrac{1}{s-6};$ #7   **34.** $\dfrac{s+2}{(s+2)^2 + 9};$ #11

35. $\mathscr{L}(t^3) + \mathscr{L}(\cos t) = \dfrac{6}{s^4} + \dfrac{s}{s^2 + 1};$ #6, 10

36. $3\,\mathscr{L}(t) - \mathscr{L}(e^{5t}) = \dfrac{3}{s^2} - \dfrac{1}{s-5};$ #5, 7   **37.** $\mathscr{L}^{-1}(1/s^2) = t$

38. $-\dfrac{3}{2}\,\mathscr{L}^{-1}\left\{\dfrac{-2}{s(s-2)}\right\} = -\dfrac{3}{2}(1 - e^{2t}) = \dfrac{3}{2}(e^{2t} - 1)$ #13

39. $-2\,\mathscr{L}^{-1}\left\{\dfrac{-1}{(s-3)(s-4)}\right\} = -2(e^{3t} - e^{4t}) = 2(e^{4t} - e^{3t})$ #12

40. $\sin 3t$ #8

41. $4\,\mathscr{L}(y') - 5\,\mathscr{L}(y) = 0;$ $4[s\,\mathscr{L}(y) - y(0)] - 5\,\mathscr{L}(y) = 0;$ $y(0) = 2;$

$(4s - 5)\,\mathscr{L}(y) - 8 = 0;$ $\mathscr{L}(y) = \dfrac{8}{4s-5};$ $y = \mathscr{L}^{-1}\left\{\dfrac{8}{4s-5}\right\} = \mathscr{L}^{-1}\left\{\dfrac{2}{s-5/4}\right\} = 2e^{5t/4}$

Chapter 12 Review

42. $\mathcal{L}(y'') + 9\ \mathcal{L}(y) = 0;\ \ y(0) = 3;\ \ y'(0) = 0;$

$[s^2\ \mathcal{L}(y) - s\ y(0) - y'(0)] + 9\ \mathcal{L}(y) = 0;\ s^2\ \mathcal{L}(y) - 3s + 9\ \mathcal{L}(y) = 0;$

$(s^2 + 9)\ \mathcal{L}(y) = 3s;\ \mathcal{L}(y) = \dfrac{3s}{s^2 + 9};\ \ y = \mathcal{L}^{-1}\left\{\dfrac{3s}{s^2 + 9}\right\} = 3\cos 3t$

43. $\mathcal{L}(y'') + 5\ \mathcal{L}(y') = 0;\ \ y(0) = 0;\ \ y'(0) = 2;$

$[s^2\ \mathcal{L}(y) - s\ y(0) - y'(0)] + 5\ [s\ \mathcal{L}(y) - y(0)] = 0;\ s^2\ \mathcal{L}(y) - 2 + 5s\ \mathcal{L}(y) = 0;$

$(s^2 + 5s)\ \mathcal{L}(y) = 2;\ \mathcal{L}(y) = \dfrac{2}{s(s+5)};\ \ y = \mathcal{L}^{-1}\left\{\dfrac{2}{s(s+5)}\right\} = \dfrac{2}{5}\ \mathcal{L}^{-1}\left\{\dfrac{5}{s(s+5)}\right\} = \dfrac{2}{5}(1 - e^{-5t})$

44. $\mathcal{L}(y'') + 4\ \mathcal{L}(y') + 4\ \mathcal{L}(y) = \mathcal{L}(e^{-2t});\ \ y(0) = 0;\ \ y'(0) = 0;$

$[s^2\ \mathcal{L}(y) - s\ y(0) - y'(0)] + 4[s\ \mathcal{L}(y) - y(0)] + 4\ \mathcal{L}(y) = \dfrac{1}{s + 2};$

$(s^2 + 4s + 4)\ \mathcal{L}(y) = \dfrac{1}{s + 2};\ \mathcal{L}(y) = \dfrac{1}{(s + 2)^3};$

$y = \mathcal{L}^{-1}\left\{\dfrac{1}{(s + 2)^3}\right\} = \dfrac{t^2 e^{-2t}}{2}$

Chapter 12 Review